AF658260

Encyclopaedia of Environment and Development

Encyclopaedia of Environment and Development

VOLUME - I
Water Resource Management

FALENDRA K. SUDAN

SERIALS PUBLICATIONS
NEW DELHI

Published - 2009

ISBN: 978-81-8387-195-2

Published by

SERIALS PUBLICATIONS
4830/24, Prahlad Street, Ansari Road
Darya Ganj, New Delhi-110002 (India)
Phone : 23245225. Fax : 91-11-23272135
E-mail: serials@satyam.net.in

CONTENTS

PREFACE

In the paper titled "Institutional and socio-economic aspects of participatory watershed management", Falendra K. Sudan reviews the participatory processes adopted under Integrated Watershed Development Project (IWDP), Hills-II, Jammu and Kashmir, and examines the institutional and socio-economic dimensions of participatory watershed management. The participatory processes adopted under IWDP, Hills-II, J&K includes issues such as reviewing existing information; initial participatory rural appraisal; making implementation agreements; participatory monitoring and evaluation; progressive withdrawal of support; and promoting forums for collaborative watershed management. The institutional aspects of watershed management identified methodological steps in participation, which includes: formation of village development committee (VDC), decision making in watershed management; system of financial management; and interaction of VDC with project functionaries and other agencies. In order to examine the socio-economic dimension of participation, factor analysis has been carried out. The dominating variables in social factor are all related to VDC members' attitude towards the meeting. The economic factor represents VDC members' perceived economic benefits and contribution and their acceptance of the institutional arrangement. The level of resources is always positively linked to participation, which shows that participation is enhanced when people perceive their resource as being of a good quality. This suggests that improving levels of common property resources would strengthen people's participation. A higher level of dependence on common resources means that people have a higher stake in the common resources, which is reflected in their higher level of participation.

The paper titled "Fostering research on wetland management in Bangladesh: A pod of ecological resource development" by M Zulfiquar Ali Islam is designed to explore the ways of fostering research on wetland management in Bangladesh. It tries to show that wetland is extremely productive and eco-friendly for a variety of animals and plant species.

But unfortunately it's hidden value and usefulness as a unique ecological resource is not even known to our people and thus it is often filled in, or dredged for housing, recreation, industry, and garbage damps. The paper is wedded to assert that the research works by both natural and social scientists can explore the hidden values of wetland and this effort may help the development of wetland as an ecological resource. Also it attempts to find out the fact that the community people and government and non-governmental organization collaboratively have to play a viable and effective role in this regard to develop sustainable indigenous and technologically sound strategies for restoring and conserving the resourceful ecosystem of wetland in Bangladesh. The paper endeavors to find out the possible areas and benefits of interdisciplinary analyses that might pave the ways for sustainable management and development of wetland. Finally, it concludes with some policy recommendations, which may be included by the policy planners and implementers in the contents of their development programs in Bangladesh.

In the paper titled "Nature, extent and determinants of collective action over water and grazing management in Botswana" by Patricia M. Makepe, collective action is defined as any action taken jointly by syndicate members to achieve a common objective, whose outcome depends on the interdependence and reciprocity of syndicate members. Collective action occurs whenever members take steps actively to work together in managing water and grazing resources. An attempt has been made to examine the determinants of collective action over the management of water and grazing resources in Kgatleng district. After presenting some descriptive statistics of some indicators of collective action, an index of collective action has been developed at the syndicate level. The econometric methods have also been used to examine the determinants of collective action over borehole water and grazing management. The main concern is to establish whether cattle owners in Kgatleng District actively and jointly take steps together to manage their water and grazing resources. Recently, policy makers and researchers have acknowledged that a critical dimension to environmental sustainability is the success with which local communities husband collectively managed natural resources. By looking at collective action as a possible alternative to privatization, the study contributes to the search for appropriate communal resource management strategies.

The paper titled "Sustainable livestock development in semi-arid watersheds of Shivaliks, Jammu and Kashmir" by Falendra K. Sudan reveals that watershed development programmes while exhibiting significant impacts in terms of increased crop production and productivity, appear to have accentuated the inter-household inequities, by ignoring or sometimes negatively affecting landless and livestock, especially small-ruminant, owning households. Enhancing the production potential of a watershed for sustainable livestock production, would contribute to livelihood- and equity-enhancing impacts of the programmes. Appropriate policy and investment decisions are crucial to making this plausible. Sudan's another paper titled "Biophysical gains of participatory watershed development: Evidence from Jammu and Kashmir" reveals that the aim of participatory integrated watershed development is to promote proven locally adopted technologies and mechanical structures through active beneficiary involvement to conserve water and reduce soil erosion. Broadly, the treatments undertaken under IWDP Hills-II are activities for arable land and private non-arable, common and forestland and livestock improvement and animal husbandry practices, and rural infrastructure development. The various project interventions have improved significantly the biophysical resources in the shape of: (i) significant improvement in vegetal cover and bio-mass thereof; (ii) multifold increase in the production of local as well as HYV of grasses; (iii) decrease in biotic pressure on the natural forests; (iv) revised the trend of deforestation; (v) slowing down of water run-off and increase in water regime of the catchments; and (vi) increase in yield of different crops, milk, wool and meat. The soil loss has decreased from 23.70 tons to 9.65 tons and on average grass biomass has recorded a growth of 6 quintals per hectares. The treated areas under afforestation, rejuvenation and silvi-pastures interventions have also produced more biomass in the shape of large-scale production of grasses and bushes. With more use of these alternative sources of energy, the pressure on CPRs has been reduced significantly. Besides above, some of the project interventions such as rehabilitation of gravity based irrigation system, water harvesting structures, enclosures on common property resources and village common lands, soil and water conservation measures required collective action through group formation, as the benefits from such interventions cannot be reaped singly but collectively, for which user groups have formulated suitable resource use rules to ensure equity in watershed development.

The paper titled "Improving irrigation efficiency in Kerala through participatory irrigation management: An analysis" by George Chackacherry and K. Madhava Chandran reveals that though many States in India advanced in the implementation of participatory irrigation management (PIM), serious initiatives had not been taken up in Kerala State for irrigation management transfer to farmers through PIM. Neither the farmer organization set up, nor the government agency concerned is physically or socially equipped fro taking up PIM and joint management of irrigation systems. It was in this background that a study was undertaken by the Centre for Water Resources Development and Management (CWRDM) and Command Area Development Authority (CADA) of the State, with the main objectives of suggesting measures for reorganizing/restructuring the existing farmer associations for implementing PIM in the State and to suggest improvements in the irrigation agency set up for supporting the process. The study was carried out in three irrigation projects in Kerala, namely, Neyyar, Chalakkudy and Malampuzha. More than 650 farmers, 300 officials, members of Legislative Assembly, leaders of political parties, etc. were consulted under the study. The study has identified the following prerequisites for implementation of PIM in Kerala: (i) favourable orientation of politicians and administrators; (ii) changes in the organizational set up; (iii) legislative backing; (iv) clear water rights; (v) rehabilitation of irrigation system; (vi) meaningful transfer agreement; (vii) coordination between stakeholders; (viii) consultation and involvement of farmers; (ix) motivation and incentive system; (x) training and awareness programmes; (xi) involvement of women; and (xii) mechanism to settle disputes. Based on the study, strategy was designed for the implementation of PIM, and accordingly, two pilot projects are being presently implemented at Neyyar and Malampuzha irrigation commands.

Falendra K. Sudan paper titled "Participatory irrigation water management in Jammu and Kashmir" reveals that with operationalizing the gravity based irrigation channels and renovating existing water harvesting structures, the water regimes and conditions of farms have improved and crop production has increased in Shivaliks. The run-off soil loss on barren hills was 23.70 tons in baseline period, which fell to 9.65 tons per hectare after treatment. The area under irrigated cropping as well as crop intensity and crop productivity has also increased significantly in project area with WUAs than project area without WUAs and non-project area. If only crop water access is secured, investment in soil fertility, crop and timing of operation will pay off in terms of substantially increased soil productivity. Due to uncertain rainfall in semi-arid and fragile Shivalik hills, a shift is needed from 'soil conservation' approaches to 'productivity' approaches, where surface run-off generated in one part of a watershed is collected and used as a resource for both agriculture and domestic purposes. Such planning is complex among small farmers since even a small run-off collecting system will involve multiple landowners. At present, there is little or no attention given to ownership of locally produced surface run-off, but one may expect this to become an issue of importance if run-off is to be successfully managed on a larger scale for local production purposes through user groups.

The paper titled "benefit-cost analysis of semi-intensive prawn farms in West Bengal: An empirical study" by Debabrata Lahiri reveals that environmental impact of prawn farming in South-East Asia has been mostly related to mangrove destruction and estimation of loss to the society due to aggressive progression of prawn farming. Prawn farming also induces soil salinity to land. Further among the three forms of technology available for prawn farming e.g. extensive, semi-intensive and intensive, the least harmful to environment has been the extensive form. In 24-

Paraganas South of West Bengal semi-intensive prawn farms have increased soil salinity. Due to soil salinity a major part of the net sown area has gone fallow and production of crops has also reduced to a great extent. The financial benefit-cost analysis indicates that there has been benefit to prawn farmers over years. But social cost indicates that it has been more than that of financial one thereby reducing social benefits to a great extent.

The paper titled "Agricultural land to shrimp pond in Orissa: Problems and prospects" by Tushar Kanti Das and Lopamudra Mishra reveals that the rapid expansion of shrimp aquaculture in the coastal areas of Orissa has been an important cause for the conversion of agriculture land to shrimp ponds in the past decade, generating environmental as well as social concern. This study describes the costs and benefits of shrimp culture and the implications of this land conversion. The study draws the conclusion by arguing that it is not the lack of environmental legislations rather lack of enforcement of these legislations, which led to the environmental and social problems and recommends that the people who are adversely affected by shrimp culture should influence the state to ban shrimp culture by forming different collectives. The paper titled "Modelling coastal degradation in a small island developing state: The case of fisheries resources in Mauritius" by Sanjeev K. Sobhee is about the implications of human interaction with nature and explores the socio-economic characteristics of coastal Mauritius as being, amongst others, the major elements that cause environmental degradation. Our emphasis is the fisheries resources of the island on which the coastal community at large depends for its living. What is interesting in this Small Island Developing State (SIDS) case study is how family size, poverty and educational attainment interact to explain depletion of the island's fisheries resources. We postulate and prove four theorems that would particularly track the channels through which the interaction between human beings and nature would occur in the case of Mauritius.

The paper titled "Cumulative impact assessment in the Sharavathi river basin" by Ramachandra et. al. advocates that sustainable development of a region requires a synoptic ecosystem approach that relates to the dynamics of natural variability and the effects of human interventions on key indicators of biodiversity and productivity. The concept of cumulative effects of the incremental reduction and erosion of the integrity of natural systems from the interactions of developmental activities provides a perspective to redirect impact analysis to deal with the driving causes of unsustainable development. An explicit attempt to monitor cumulative changes at catchment level due to hydroelectric projects is expected to lend itself to the establishment of notional ecological thresholds. This paper focuses on the cumulative effects that have taken place in the river ecosystem due to developmental projects over the last four decades.

Rakesh Tiwary's paper titled "Energy cooperation in the Ganga Brahmaputra basin: Opportunities and challenges" reveals that the Ganga-Brahmaputra Basin states are going to face huge energy crunch in coming decades. There are multiple energy sources available, which show high interdependence and complimentarity among the nations for their development and utilization. Though there are numerous constraints: the lack of conducive geopolitical environment, mistrust, absence of regulatory mechanisms, financial and technological gaps etc, but there are huge prospects and opportunities of power trading among riparian states. The interconnections of energy systems of the riparian countries of Ganga-Brahmaputra basin can become crucial agent of economic and social development of the entire region. There is need to take political initiatives at highest levels then create institutional structures for sustained mutual interactions. There is a need to build up interconnections step by step. India, being largest nation in the region has larger role to play without hurting the sensitivities of smaller nations. Gradual opening up of the region and sectoral reforms can help in filling up the institutional, technical and investment gaps. We need to have the vision to seize the opportunity.

M. Mishra and N. C Sahu's study titled "Environmental cost of drinking water in Angul-Talcher in Orissa: A case study" aims at assessing the environmental damage cost of water pollution through valuing drinking water that involves extra labour time involved in fetching water from a distance place in Angul-Talcher industrial agglomeration in Orissa, India. The economic value of the hardship in fetching drinking water is assessed through labour time spent or wage lost in the activity. The cost of fetching drinking water involves time, which has been valued at the existing wage rate. As per the findings of the study damage cost of water comes to Rs.5 lakhs per day for 50,000 rural females, implying Rs.10/- per household per day that is a higher amount for a rural agricultural and daily labourer community. In the conclusion it is suggested that since industries injure the aquatic ecosystem they need to pay for alternative supply (through piped supply or hand pumps) of drinking water.

Maheen Sabreena Karim's study titled "socio-economic impact of arsenicosis on rural communities of Bangladesh" reveals that arsenic contamination of ground water is a serious health hazard in Bangladesh. Its pollution is not only causing serious health problem to the people, but also it has been affecting the environment by creating enormous social problems in the community. It generates multifarious socio-cultural problems for the arsenic victims and their kin members in the rural areas of the country. The victims and their families virtually become fully isolated from the community, and nobody wants to keep any good contact with them. The villagers do not want to make any matrimonial alliances with these families. Their children are the victims of direct avoidance of their classmates. They are often ignored by their teachers who indirectly discourage them to come in public. It has also been reported that the rural people refuse to accept food prepared by an arsenic patient. Because of illiteracy, many people confuse skin lesions with leprosy, which they believe is totally incurable and also contagious. Under this situation, this paper attempts to explain how arsenicosis creates impediments in terms of their economic activities and how their children and family members are struggling with this adverse situation as they debarred form social we have evidences where it creates a kind of disintegration in the village life, which requires immediate intervention from social reformists point of view. And accordingly, the paper give evidences from an anthropologists perspective and thus provide suggestions in context to remedial procedure by rendering them proper assistance and knowledge about it.

The paper titled "The southeastern Anatolia project and the question of large dams: The dilemma of developmet in southeastern Turkey" by Eddie J. Girdner reveals that the Southeastern Anatolia Project (Gunay Adadolu Projesi) includes 22 large dams and 19 hydro-electrical plants. The project is expected to promote industrial development in the region, creating some three million jobs and producing socio-economic change. Benefits from the production of electricity and agriculture are estimated at $1.8 billion a year between 2001 and 2010 and at $3.3 billion per year after completion of the project. Without land reform in the area, however, there is a danger that the project may strengthen feudalism and strengthen existing inequality. The project is likely to significantly erode the fragile ecology of the region. In 2000, the World Commission on Dams reported that the world would be significantly poorer without the 45,000 large dams which have been built. However, in many cases, there might have been cheaper and better ways to produce the energy, food and fiber. Since 1970, there has been a decline in the building of large dams and some dams in North America have been decommissioned. That trend may continue with the increase in social and political conflicts over dams and the decrease in available state funding.

F. A. Shaheen's study on "Asia's largest freshwater lake vanishing: Problems and remedies" reveals that Wullar lake is known as Asia's largest and probably most picturesque lake. Situated in the northern part of Jammu and Kashmir state of India, it has a distinct position and status among all other lakes. It sustains the livelihoods of a large section of population by providing numerous valuable products besides acting as an important flood absorption basin and natural reservoir for downstream hydro-power plants. However, lake is plagued by a number of problems such as encroachment, anthropogenic pressure, sewage and sewerage disposal, weed infestation, deteriorating water quality, catchment problems, sedimentation, excessive willow plantations, shrinking area etc. The lake has experienced considerable deterioration and decline in the last century, especially, in the last few decades. The lake was given RAMSAR status in 1990, but still no visible measures to restore its lost legacy are seen on ground. Keeping in view the great magnitude of problems involved in the catchment and main lake body, there is an immediate need of a comprehensive multi-dimensional action plan with both short term and long term objectives. The primary objective of management is to strike a healthy balance between the conservation of the lake and its economic use on sustainable basis. Lakes and wetlands representing dynamic and living aquatic eco-systems can in no case have isolated engineering or administrative solutions. There must be a coordinated effort by all the principal actors to achieve the objective. Till now, the approach has remained only an isolated one without any impressive impact. Besides this, political will and coordinated legal, administrative, and scientific measures coupled with earnest management and conservation strategies on a comprehensive scale are needed to restore the lake to its pristine glory

Akongbowa Bramwell Amadasun's paper titled "What role should the European Union play in the development of the Niger delta?" focuses on the challenges of promoting the desired sustainable development for the Niger Delta through forward-looking partnership between Nigeria and the European Union (EU). The paper posits that the region parades low-income levels or negative economic growth, fratricidal inter- and intra-ethnic nationalities armed conflicts, catastrophic environmental degradation and high level of poverty. Which makes it impossible to overlook deep-seated links between long-term development and spiral of violence in the region. The paper examined the development element in the Niger Delta with a view to determine what are the constitutive elements in poverty reduction and development architecture deficits. Included in these are environmental degradation and their economic and political consequences, knack for faulty implementation of development programmes, poor incentive framework which hamper inflow of foreign capital, institutionalization of corruption, lack of credible governance, human resources underdevelopment, lack of critical infrastructures amongst others. In examining these issues, the paper posits that the EU is a major recipient of crude oil from the region and a major global economic power remains a critical partner for growing sustainable development and poverty reduction in the Niger Delta. Consequently the paper opined that the most fundamental challenge is how to overcome the marked development deficit through a productive partnership with the European Union.

F.A. Shaheen and R.L. Shiyani's paper titled "Chasing the elixir: Case of groundwater resource in north Gujarat region" reveals that in India, there is a severe crisis in groundwater management. In many areas water levels are going down rapidly, the quality of water is deteriorating. The North Gujarat region of the country has made a prominent place in groundwater over-exploited zones. The region is heavily dependant on the groundwater resources to meet the domestic, industrial and irrigation demand. The chronic overdraft of the resource with high rates of water table decline had many implications with regard to sustainability, equity, energy consumption and rising costs of groundwater extraction. The present study examines the investment costs incurred by the farmers in borewell as well as the pace with which the resource is chased at higher depths.

Falendra K. Sudan and Jennifer McKay's paper titled "Institutional aspects of groundwater governance: Experiences from south Australia and lessons for India" reveals that a national agreement on water reform was reached in 1994 and the COAG agreed to a comprehensive water reform agenda that explicitly linked, for the first time, economic and environmental issues with a package of reform measures. The agreement focused on establishing water allocations and entitlements separate from land tenure, backed by secure access to water. It also provided for trading in water entitlements, making water available for ecosystems, as well as institutional reform, public consultation and education, and research. Based on the 1994 COAG agenda, there has been some progress across all jurisdictions, but much remains to be done if the goal is to meet the reasonable expectations of future generations. A strong commitment and ongoing cooperation of governments and stakeholders will be a key factor in sustaining and driving the momentum of groundwater governance reform. The continuing efforts on integrated water management, cross-border cooperation, improving irrigation practices and water use efficiency will be necessary to ensure improved productivity and environmental sustainability of water resources and secure access to water for all. A number of important lessons have been learnt by analysing the institutional reform process in South Australia, which can be applicable in Indian context too and these includes integrated water resource management, recognising the connectivity between surface and ground water systems, effectively balancing management of the environment and the need for certainty of access for water users, system of water markets and water trading, continual improvement of data and scientific information, and extensive public consultation for effective decision making.

The paper titled "Institutional and legal groundwater management framework: lessons learnt from south Australia for Pakistan" by Bakhshal Lashari, Jennifer McKay and Karen Villholth reveals that the socio economic situation and awareness in South Australia is much stronger than in Pakistan and hence the groundwater management is more sustainable. The market system of selling the products is very sustainable and reliable as compared to Pakistan. The farmers of South Australia are using groundwater with pressurized irrigation system such as sprinkler and drip irrigation system which have irrigation efficiency more than 90 per cent, whereas in Pakistan most of the groundwater is being used in furrow and basin irrigation system having irrigation efficiency not more than 60 per cent.

LIST OF CONTRIBUTORS

Falendra K. Sudan, Reader, Department of Economics, University of Jammu, Jammu, Jammu and Kashmir – 180 006, India.

M Zulfiquar Ali Islam, Associate Professor, Department of Sociology, University of Rajshahi, Rajshahi 6205, Bangladesh

Patricia M. Makepe, Lecturer, Department of Economics, University of Botswana, Private Bag UB 00705, Gaborone, Botswana.

George Chackacherry, Scientist & Officer in Charge, Centre for Water Resources Development & Management (CWRDM), Sub Centre, Neyyattinkara, Thiruvananthapuram, Kerala – 695121, India.

K. Madhava Chandran, Scientist, Centre for Water Resources Development & Management (CWRDM), CWRDM Headquarters, Kozhikode, Kerala, India

Debabrata Lahiri, Associate Professor, Rural Development Centre, Indian Institute of Technology, Kharagpur: 721302, West Bengal, India.

Tushar Kanti Das, Lecturer, P.G. Department of Business Administration, Sambalpur University, Jyoti Vihar, Sambalpur, Orissa-768019, Indis.

Lopamudra Mishra, Research Associate (UGC), P.G. Department of Economics, Sambalpur University, Jyoti Vihar, Sambalpur, Orissa-768019, India.

Sanjeev K. Sobhee, Department of Economics and Statistics, University of Mauritius, Reduit, Mauritius.

Ramachandra T.V., Centre for Ecological Sciences, Indian Institute of Science, Bangalore 560 018, India.

Subhash Chandran M.D., Centre for Ecological Sciences, Indian Institute of Science, Bangalore 560 018, India.

Sreekantha, Diwakar Mesta, Centre for Ecological Sciences, Indian Institute of Science, Bangalore 560 018, India.

G.R. Rao, Centre for Ecological Sciences, Indian Institute of Science, Bangalore 560 018, India.

Sameer Ali, Centre for Ecological Sciences, Indian Institute of Science, Bangalore 560 018, India.

Rakesh Tiwary, International Water Management Institute (IWMI), South Asia Program - Anand Office, Elecon, Anand-Sojitra Road, Vallabh Vidyanagar, Anand, Gujarat-388120, India.

M. Mishra, Reader, Department of Economics, Benaras Hindu University, Vanarasi, Uttar Pradesh, India.

N. C. Sahu, Professor, Department of Economics, Berhampur University, Berhampur-760007, Ganjam, Orissa, India.

Maheen Sabreena Karim, Lecturer in Anthropology, School of Liberal Arts, Independent University, Dhaka, Bangladesh.

Eddie J. Girdner, Professor, Department of Political Science and International Relations, Baskent University, Eskisehir Yolu, Ankara, Turkey.

F. A. Shaheen, Assistant Professor, Sher-e-Kashmir University of Agriculture Sciences and Technology (SKUAST), Kashmir, Shalimar, Srinagar, Jammu and Kashmir, India.

Akongbowa Bramwell Amadasun, Executive Director, Center for Development Integration, Africa.

R.L. Shiyani, Professor, Department of Agricultural Economics, Junagadh Agricultural University, Gujarat, India.

Jennifer McKay, Professor, Centre for Comparative Water Policies and Laws, University of South Australia, Adelaide, South Australia.

Bakhshal Lashari, Professor, Mehran University of Engineering and Technology, Jamshoro, Sindh, Pakistan.

Karen Villholth, Director, Groundwater Program, International Water Management Institute, Sri Lanka.

1

INSTITUTIONAL AND SOCIO-ECONOMIC ASPECTS OF PARTICIPATORY WATERSHED MANAGEMENT#

Falendra K. Sudan

INTRODUCTION

During the recent past, there has been fundamental change in the beliefs about the role of the state in developing countries, from regulator to facilitator in the process of development in which a coalition of different actors and institutions are involved (Carney and Farrington, 1998). The emphasis has been laid on "partnership" approach, in which state has to work together with NGOs, private sector, and people's organizations, so that comparative advantage of each party can be exploited (World Bank, 1997). The need for change is justified on numerous grounds such as low growth rate, lower living standard of the majority of population, poverty, environmental degradation, financial mismanagement, waste and corruption, etc. The natural resource management is an area where the institutional reforms have been implemented vigorously, and as a part of the change, participatory approaches has been implemented in numerous projects related to agricultural development, horticultural development, forest development, watershed development, etc. With the institutional change, some of the activities performed by the government functionaries are supposed to be taken over by the participants in the reform process. Under IWDP (Hills-II), Jammu and Kashmir institutional reforms have been implemented for watershed development and protection in various sub-sectors such as provisioning of agricultural services, forestry, water resources, etc. The local level stakeholders have been entrusted with the task of actual management of natural resources to make more productive

This paper is a part of the research project titled "Participatory Approaches and Environmental and Economic Impact: With Special Reference to IWDP, Hills-II, Jammu and Kashmir, India", completed in May 2003 and sponsored by Ministry of Environment and Forests, New Delhi and World Bank funded "Environmental Management Capacity Building Technical Assistance (EMCaB) Project".

use of those resources. It has been emphasized that the institutional change will facilitate reduction in transaction costs involved in operationalizing the programme of participatory watershed development and protection. The users' participation in resource management would ensure proper maintenance and sustainability of the assets created over the project period and would facilitate smooth withdrawal and improve the performance of the project. However, poorly planned withdrawal clearly does not bring such advantages and may pose threats particularly to the livelihoods of the poorest and to the environment.

The recent reform in user participation in natural resource management has highlighted the importance of user groups. The existing natural resources in the selected sub-watersheds of Jammu and Kashmir share three essential features, which support participatory approach in natural resource management. The ownership of forests, water resources and pastures and grazing lands is vested with the state, which does not have the capacity to regulate their access and use. The individual resource management is difficult due to lack of defined boundaries in case of water resources and pastures and grazing lands, and at the same time the cost of supervision is supposed to be high. The most natural resources have multiple users and uses; therefore private management ignores the equity aspects of natural resource management. These issues are now widely accepted and user participation in natural resource management has been widely encouraged under the reforms. It has also been argued that common property resources (CPRs) are more effectively and efficiently managed through users' participation than state's regulation and control. The user groups have a direct stake in the future of the resource, which they are managing, better and easy access to information about local needs, so that they can take prompt action as per changed situation. The stability and robustness are essential qualities for user groups for successful management of natural resources (Wade, 1988 and Ostrom, 1990). However, the successful management of CPRs must be integrated into users' livelihood strategies. For instance, in case of water user groups the success can be defined in terms of increased cropping area or intensity, rising real per capita incomes, increased water use efficiency, improved productivity per unit of water, or improved equity and access to irrigation (Turral, 1995). Thus, the determinants of success should be identified to judge the performance of the user groups and overall success of reforms. Besides, the scaling up and replication of the success of the reforms is also important. Keeping above in view, an attempt has been made to review the participatory processes adopted under IWDP, Hills-II, J&K, and to examine the institutional and socio-economic dimensions of participatory watershed management.

STUDY AREA AND METHODOLOGY

Numerous schemes for development of land resources are in operation in the state of Jammu and Kashmir. Some are watershed based while others are based on administrative boundaries. The objective of all the schemes is to put emphasis on management of soil and land resources for sustainable crop production and protection of environment (Sadhu and Sudan, 1999). In order to prevent and reverse the degradation process of the *Shivaliks and Karewas,* Integrated Watershed Development Project (IWDP), Hills-I was undertaken in 1990. Seeing the achievement of physical targets, the World Bank decided to further extend the programme in the form of IWDP (Hills-II) in 1999 for five years to other rainfed areas of the state viz. Akhnoor and Ramnagar in *Shivaliks* (Jammu region) and Rajwar and Rambiyara in *Karewas* (Kashmir Valley). The main objective of the IWDP (Hills-II) is to restore the productive potential of the *Shivaliks* and *Karewas* by evolving participatory approaches. The focus is on improving *in-situ* moisture conservation, water harvesting and land management practices. An associated objective of the project is to assist with the institutional development and consolidate the progress already made under IWDP (Hills-I) (World Bank, 1999).

The present study is confined to two sub-watersheds of Akhnoor (Jammu district) and Ramnagar (Udhampur district). While Ramnagar is in the inner *Shivaliks,* Akhnoor is on its outer part. Ramnagar sub-watershed is the catchment area of Ramnagarwali *Khad* in the middle catchment of river Tawi. It has an area of 32,630 Ha. and is subdivided into 39 micro-watersheds. Akhnoor sub-watershed with an area of 42350 Ha. is sub divided into 37 micro-watersheds. It covers about a dozen rivulets (*nallahs*) and a large number of small nallahs originating from *Kalidhar* ridge and draining into Chinab river on the western portion of *Kalidhar* ridge, drains join Manawar Tawi which in turn also meets the Chinab river. The *Shivaliks,* locally known, as *'Kandi'* comprise of piedmont deposits made of boulder and pebbles, gravel and sand with minor clays mixed in varying proportions. One of the characteristic features of the area is the stream called *choes* or *khads,* which remain dry for most of the year. Though total rainfall is high (average 1000-1500 mm.), but its distribution is very erratic resulting in frequent droughts. This area is subject to soil erosion due to undulating topography, steep slopes, poor vegetative cover (scrub forest) and coarse to medium texture of the sedimentary material. *Shivalik* hills consist of highly erodible sandstones, conglomerates, and siltstones.

The uncontrolled deforestation and overgrazing in the past has resulted in the reduction of vegetative cover and accelerated erosion in the *Shivaliks.* The area experiences paucity of water for plant growth due to excessive run-off, high evaporation especially during summer, unpredictable and erratic rainfall. The eroded material from *Shivalik* hills, brought down by the seasonal rivulets *(choes)* is deposited in the sloping piedmont plain and also in the area around *choes.* The repeated deposition of coarse sediments renders these areas comparatively low in agricultural productivity. Due to lack of irrigation, subsistence rainfed agriculture is the prevalent production system is selected sub-watersheds. In terms of physiographic features, there may not be 100 per cent similarity between the two sub-watersheds but, certainly, in term of accessibility, natural resource endowment, infrastructure development and general index of socio-economic awareness, two sub-watersheds are comparable.

During the Khon Kaen Conference, 1985, different types of Rapid Rural Appraisal (RRA) techniques were identified, one of which was labeled "participatory". The orientation of a participatory rural appraisal was to facilitate or stimulate community awareness and capability regarding a problem or issue. Particular attention was given to enabling local people to conduct their own analysis of problems, and to share their findings. The role of the outsider became one of catalyst, rather than one of expert. Stimulation of community awareness and capability was also intended to reduce the extractive nature of RRA, and to help local people to empower themselves. In that regard, what became known, as Participatory Rural Appraisal (PRA) is consistent with some of the basic aspects of sustainable development (local empowerment, equity, social justice) (Conway and McCracken, 1990). The primary data has been collected using well-structured and pre-tested questionnaires, participatory rural appraisal (PRA) techniques and group meetings.

Through participatory observations and focus group discussions, it is possible to describe what goes on, who and what is involved, when and where things happen, how they occur, and why-at least from the stand point of participants-things happens as they do in particular situations. The methodology of participatory observation and focus group discussions has been used to ascertain important differences between the views of insiders as opposed to outsiders. Keeping in view, the objectives of the proposed study in mind, the participant observation inquiry has been helpful in an open-ended logic and process of enquiry as well as qualitative description of the phenomenon under study. The sample size has been restricted to the members of the village development committees (VDCs) selected under IWDP (Hills-II) in the selected sub-watersheds. Three VDCs each has randomly been selected from the sub-watersheds of Akhnoor and Ramnagar. Therefore, a total of six VDCs have been selected with the consultation of project functionaries at sub-watershed level. A total of 15, 12 and 19 VDC's members have been selected from village Sunetar, Johnu and Dehari respectively in Ramnagar sub-watershed and 21, 12 and 11 VDC's members have been selected from village Chohara, Maira and Ambarain respectively in Akhnoor sub-watershed.

The team strategy has been used to collect data and information and to gain access to multiple perspectives. Both the unfocused and focused observation techniques have been used. The unfocused initial has been used to become increasingly familiar with the insider's world to refine and focus subsequent observation and data collection. All the observations have been recorded on site and misunderstanding, if any has been corrected thereof. In order to collect primary data and information on participatory processes, PRA exercises and group meetings have been held separately for male and female members of VDCs in a smaller and homogeneous group. Highly formal interviews have been conducted using structured interview schedules. In-depth interviews have also been conducted to elicits opinion regarding strengthening of participatory processes, of stakeholders with extensive knowledge of the phenomenon under study. Besides, study has been supplemented by utilizing potentially rich sources of secondary data and research materials. The content analysis technique has been used to analyze the data and information qualitatively and quantitatively (using descriptive statistics). The content analysis technique has been supplemented by use of code and labels, field notes, sorting, shifting, constructing and reconstructing these materials, in order to prepare the paper.

BACKGROUND OF IWDP (HILLS-II), JAMMU AND KASHMIR

In the last decade, in order to prevent and reverse the degradation process of the *Shivaliks* and *Karewas*, Integrated Watershed Development Project (IWDP), Hills-I was undertaken. Originally, the project was for a period of 7 years (1990-97), but was extended for another year up to 1998. The project was supported by the World Bank and was implemented in the four states: Jammu and Kashmir, Himachal Pradesh, Punjab and Haryana (World Bank, 1999). In the state of Jammu and Kashmir, IWDP (Hills-I) was initiated in three sub-watersheds of Devak and Ramkote in *Shivalik* hills (Jammu region) and Dudh Ganga in *Karewas* (Kashmir Valley). The overwhelming evidence from natural resource management projects is that without people's involvement the benefits are not sustainable in the long run. After the funds from the government or donors dry up, conservation structures disappear, committees are disbanded or abandoned and the livelihood base of the people remains only marginally improved at all (Saxena, 1999). Fortunately, these shortcomings were taken care of, to some extent during the implementation of IWDP (Hills-I) (Sadhu and Sudan, 1999). IWDP (Hills-I) was implemented for soil and water conservation in the watershed areas aiming at proper land use according to land potential, protection of land against all kinds of deterioration, building and maintaining soil fertility, conserving water, proper management of water for drainage, flood protection, sediment reduction and increasing productivity from all kinds of land uses. Besides, improved vegetative coverage of treated areas, increased community participation and formulation of engaged village users' group, increased crop yields and increased household incomes of marginal and small farmers, the landless and the women were included as major objectives of IWDP (Hills-I) (World Bank, 1998).

IWDP (Hills-I) has made only a modest impact and it can't be denied that as it has been the first phase of this experiment, some shortcomings were bound to remain, but, at the same time, it must have enriched the experience of the project functionaries which should equip them to carry out their role more efficiently in second phase (Sadhu and Sudan, 1999). The principal lessons learnt from IWDP (Hills-I) were:

(i) the need to develop an integrated approach to the *Shivalik* Watersheds,

(ii) the need for a coordinated approach to the use of available funds from all sources,

(iii) the need to involve stakeholders in planning, implementation and maintenance,

(iv) the need to increase stakeholder awareness of environmental and socio-economic considerations in articulating communities' needs,

(v) the need to focus on measures to arrest soil erosion and promote *in-situ* moisture conservation from ridge to valley,

(vi) the need to assess marketing prospects, especially of horticultural products,

(vii) the need to improve infrastructure within watersheds, particularly rural links to markets, water harvesting and drinking water,

(viii) the need to develop monitorable project implementation and development objectives' indicators, and

(ix) the need to involve local NGOs in the project as facilitators and trainers in the planning and implementation of the project (World Bank, 1998).

Seeing the physical targets being achieved by the participating states, the World Bank decided to further extend the programme in the form of IWDP (Hills-II) not only in other areas of states selected in phase-I, but also including the state of Uttar Pradesh, in 1999, for five years. This was aimed at providing a uniform integrated rural development platform to address the social and natural resources problems of the entire *Shivaliks* in India (World Bank, 1999). The experience gained through implementation of IWDP (Hills-I), J&K encouraged the State Government to extend the watershed development project to other rainfed areas of the State viz. Akhnoor and Ramnagar in *Shivaliks* (Jammu region) and Rajwar and Rambiyara in *Karewas* (Kashmir Valley). The main objective of the IWDP (Hills-II) is to restore the productive potential of the *Shivaliks* and *Karewas*. This would be attained by using evolving watershed development and community participation approaches. The focus is on improving *in-situ* moisture conservation, water harvesting and land management practices. An associated objective of the project is to assist with the institutional development and consolidate the progress already made under IWDP (Hills-I). Although the main objective remains to be the restoration of the productive potential of the hills in five participating states, the priority plans of different states differ. Following are the objectives of IWDP (Hills-II), Jammu and Kashmir:

(i) to restore on a sustainable basis the productive potential of the region and improving the quality of life,

(ii) to reduce soil erosion and improve availability of water,

(iii) to help increase production and income, grain crops, horticulture, fodder, fibre, fuelwood, livestock and household based products,

(iv) to promote holistic and sustainable agro-ecological development involving people's participation,

(v) to strengthen community participation, and

(vi) to develop local level institutions to enhance the sustainability of the model (World Bank, 1999).

PARTICIPATORY PROCESSES UNDER IWDP (HILLS-II), J&K

In the context of IWDP (Hills-II), J&K, institutional reforms for watershed development have been carried out by identifying the ways through which the poor people have been supported in identifying their emergent needs and priorities, with the goal of creating robust institutions to manage the CPRs in sustainable manner. In order to operationalise the institutional arrangement, the following participatory processes have been followed in the selected sub-watersheds with minor variations across the forested and agricultural watersheds.

Initial Preparation

Though some of the local professionals and field workers already had some knowledge of, and experience with, the local setting, to avoid preconceptions IWDP (Hills-II) has made efforts to maintain a receptive and open frame of mind. Indeed, especially at the beginning of the project, there were many unknowns and uncertainties about the environmental and social situation of the selected sub-watersheds. Therefore, information gathering was considered necessary before launching the participatory process. This preparatory activity included:

- Review of existing information, complemented, when necessary, by rapid appraisals of specific environmental or social issues;
- Tentative identification of the communities and sites in the project area that, in light of both technical and political considerations, were the most suitable for implementing the process of participatory and integrated watershed management; and
- Preliminary visit to the short-listed local communities, aimed at validating the initial choice.

At the beginning of the project, a period of three to six months was devoted to accomplishing these tasks.

Reviewing Existing Information

Gaining an understanding of the environmental, social and institutional situation of the project area was obviously essential for identifying the key issues, opportunities and constraints to be considered in designing and organizing subsequent phases of the participatory and integrated watershed management process.

To this end, the IWDP (Hills-II) considered the following basic information:

- General physical and environmental features of the area: size, altitude, rivers, climate, rainfall, slopes, soil, vegetation, natural resource degradation, etc.;
- Population trends: size, natural growth rates, migration, density, type of settlements, etc.;
- Social and economic situation: social stratification, indigenous social organization, cultural background, literacy, the status of women, food security, health conditions, land tenure, farming system, off-farm activities, etc.; infrastructure and social services: roads, communications systems, credit, health, education, etc.; and
- Institutional setting: line agencies, NGOs, grassroots organizations, rural banks, ongoing development initiatives in the area, etc.

Conducting Complementary Studies

At the beginning of the project, the project-implementing agency found that it was not necessary to carry out more structured preliminary inquiries. However, at a later stage, the validity and accuracy of the project's background information was often criticized as being too superficial to support informed decision-making. Project staff working in the field sometimes questioned the biophysical and demographic data gathered from existing sources. Furthermore, major gaps were often found in the information regarding community social organization, farming systems and local knowledge. The project has hired the services of local consulting firms to carry out complementary studies such as 'Shivalik Development Strategy", Retrospective Study, IWDP (Hills-I), "Prospective Study on Social Assessment', Tribal and Transhumance Study', etc. To obtain sounder information, both conventional and Rapid Rural Appraisal (RRA) methods were used to carry out complementary studies according to specific needs. The focus of these studies differed, however, there were three common areas of research: farming systems research, environmental analysis, and social surveys.

Short-listing of Communities

At the beginning of the project, IWDP (Hills-II) attempted to select participating communities according to strictly technical and managerial criteria, such as high environmental risk (degraded upland zones generally received a higher priority), and relatively strong community organization (the existence of grassroots organizations that could potentially become project partners was considered an important asset). However, it quickly became clear that the project could not adhere strictly to these criteria. In particular, kinship structure, ethnic or caste interests, and affiliation with political parties and interest groups proved to be elements strongly influencing the final decision about where to start the participatory process and who to involve.

Initial Visits to Short-listed Communities

An especially important element in site selection was the initial visit to the sort listed communities, with the threefold aim of:

- Exploring the community's interest in collaborating with the project;
- Identifying potential actors in starting the participatory process (grassroots organizations, interest groups, concerned individuals); and
- Making organizational arrangements for subsequent phases of the participatory process.

In some cases, extension workers made introductory visits of one or two days to the communities. On other occasions, these visits continued over a longer period of time, often through informal contacts with community members. Despite these differences, most visits included the following three core activities:

- Formal visits to community leaders and influential people: Staff members presented the project to local authorities and expressed their intention of establishing a partnership with the community. The community's problems were also discussed and the interest of the community leadership in collaborating with the project was explored.
- Informal interviewing: Conversations with individuals and spontaneous discussions with groups of people often provided important insights into the local situation, the problems at stake, the basic features of social organization and potential sources of conflict. They were also essential in identifying key informants to be consulted during subsequent phases of the process.
- Community meetings: At the end of the visit, community-wide meetings were held with the support of local leaders. In these

meetings, project staff presented the rationale and purpose of participation appraisal and planning exercises. Gender was a major consideration during the initial visits, allowing the project to promote the involvement of women in the subsequent phases of the participatory process.

Initial Participatory Rural Appraisal

The main goal of the initial participatory rural appraisal (PRA) exercise was to support community members in better assessing their situation and identifying the most important and urgent goals to be pursued through participatory action. Another objective was to collect additional and more detailed information on the local environmental, social and economic setting, according to the local perception. To carry out the PRA exercise representatives of other local institutions were also involved. These teams typically included the project's social development functionaries, mid-level technical staff and extension officers. A hired consultant with solid experience in participatory action-research provided technical assistance to the team, especially at the beginning of the project. The consultant acted both as a trainer and as a team coordinator. This was highly instrumental in later allowing local team members to acquire the necessary skills for conducting similar exercises on their own. Initial participatory exercises tended to be comprehensive, covering the community's environmental, social and economic conditions.

Based on preliminary studies, field visits and the consultant's inputs, team members identified specific needs to be addressed, which included:

- Community's population and social organization;
- Income generation and the distribution of wealth;
- Education and literacy;
- Gender issues;
- Accessibility and use of social services;
- Community's infrastructure;
- Functioning and productivity of farming systems; and
- Management of common natural resources.

A number of participatory research and action-learning techniques were used in the framework of these exercises, including:

- Thematic group discussions based on a list of topics or on an open ended questionnaire;
- Time-line analysis to collect information on events occurring in the far past, project staff often asked community elders to take part in the exercise or interviewed them separately;

- Transect walks to explore significant environmental and agricultural features; and
- Participatory mapping to prepared maps of the community's territory, the village or a sample of forms, which highlighted important environmental, social or agricultural features.

Participatory Planning Meetings

The participatory rural appraisal ended with a series of shorter meetings. The main activities carried out during these meetings were as follows:

- Feedback of participatory rural appraisal information to communities;
- Identification, analysis and prioritization of problems by participants;
- Identification of solutions; and
- Drafting of a tentative action plan.

Participatory Feasibility Analysis

Through participatory feasibility analysis, project management and field staff reviewed participating community members' ideas for action that were developed during the participatory planning meetings, with the aim of assessing the feasibility of the proposed action in the light of the existing conditions. In most cases, this required an intensive discussion with community members on technical aspects of the proposed activities, available incentives and subsidies, and requisites the community should meet for engaging in a partnership with the project. This interaction ensured that the process continued to be participatory. The purpose of the participatory feasibility analysis was to carry out a rapid and practical analysis that would allow for a decision to be made on whether or not to implement the activity in question. Procedures commonly adopted for conducting the rapid analysis included:

- Short-listed activities: Very of ten, plans prepared during the participatory planning meetings were redundant and over ambitious, and certain activities seemed to be technically unfeasible. Furthermore, staff often felt the need to focus on those initiatives that best responded to the project's mandate and operational capabilities, such as activities related to the management of CPRs, improvements in the efficiency and sustainability of farming systems, the diversification of income generation, the empowerment of rural women and the building of small-scale infrastructure.
- On-site verifications: Project prepared a list of minimal conditions for implementing short-listed activities and entrusted field staff

with the responsibility of the verifying whether these conditions had been met on site. Field staff generally carried out this verification in a participatory fashion, with the concerned community or interest group actively involved in decision-making

Making Implementation Agreements

Following the feasibility study, operational agreements for implementing the activity were made. Making implementation agreements basically entails making decisions about practical issues, such as use of resources. The process of making an implementation agreement required the following:

- Preparing the design of the activity: Based on the results of the feasibility analysis, a technical design of the activity was prepared that included the definition of services (training, extension services), materials and incentives to be provided by the project, the contributions to be made by the interest group and a tentative schedule.
- Group strengthening; For the sake of transparency, to promote sustainability of the interest groups and to facilitate conflict management, project helped formalize the structure and the operations of the user groups or village development committees (VDCs) collaborating on the activities. This usually involved establishing clear rules for membership, watershed revolving fund, record-keeping systems and election of a formal leadership body. Fulfillment of these requirements was often stated as a condition for working with the project.
- Negotiation of task-sharing and cost-sharing arrangements: Project staff held a few meetings with the concerned VDCs to review in detail the village development plans (VDPs) and to amend it according to their comments and suggestions. At this time, negotiations determined the inputs to be provided by each party and established simple participatory monitoring procedures.
- Finalization of agreements: Based on these negotiations, the participants defined the terms of reference for reciprocal collaboration. When money was involved, this usually took the form of a written 'implementation contract' between the project and the VDCs or user groups.

Making implementation agreements was at times a long and tiring process. The main problem in this process was the 'non-participatory' attitude of local institutions, communities and sometimes project staff.

Top-down decision-making and bureaucracy often hindered the collaboration with governmental agencies. Strengthening VDCs and user groups and meeting basic needs figured prominently at the beginning of the participatory and integrated watershed management process whereas later, attention progressively shifted towards environmental issues.

- Strengthening VDCs and user groups: Project made significant efforts towards facilitating the formation and development of user groups and community organizations. The role of VDCs in project implementation was in general less direct than that of user groups. However, VDCs were often highly instrumental in coordinating the activities of user groups and in providing a local forum for the exchange of experiences among groups and individuals.

The participatory development functionaries of the project played a major role in the strengthening of user groups. They attended all relevant training activities and were part of the participatory monitoring, evaluation and replanning team, and provided constant support to all user groups formed in the project area. They assisted user groups in such areas as organizing and conducting meetings, defining group constitutions and keeping records. Activities carried out by the project to strengthen the structure and operational capacities of VDCs and user groups included:

- Support for internal operations: When necessary, the project supported user groups and VDCs in defining their procedures for carrying out work and sharing tasks. The staff in charge facilitated the group's operations, especially at the beginning. However, the project tended to withdraw this type of support as the groups progressed towards self-reliance. In exceptional cases, the project acted as a mediator to negotiate internal conflicts.
- Managerial capacity-building: Project staff held training courses and workshops on such topics as short-term planning, monitoring, financial management, record-keeping, leadership and communication, the management of meetings and conflict resolution.
- Micro-capitalization: Establishing a small watershed fund was considered an essential element of VDCs' self-reliance. Moreover, these common funds were found to have positive effects on the group's cohesion, commitment and sense of responsibility. Micro-capital was usually generated through membership fees. A treasurer was appointed and trained to manage these funds, and, when feasible, a savings account was opened in a local bank to familiarize people with basic banking procedures. Once a significant amount of savings was available, re-investment was encouraged, often in the form of loans to members wanting to engage in individual income generating activities.

- Facilitation of linkages among user groups and VDCs: Meetings, social events and reciprocal visits among user groups and VDCs in the project area provided opportunities for exchanging ideas and experiences.
- Communication activities: Information on the activities and achievements of user groups and VDCs was disseminated. All of these activities were instrumental in motivating more people to join the existing groups and VDCs or to form new associations.
- Meeting basic needs: Project was committed to supporting activities aimed at meeting basic needs not directly related to natural resource management. This included income-generating activities and improvements in local infrastructure. The project paid special attention to initiatives promoting the economic independence of women, decreasing their workload and improving their living conditions.
- Income generating activities: Community members' needs for increasing their cash revenue were a strong incentive for the formation of user groups. To fulfill this need, the project supported and promoted a variety of short-term income generating activities. These included both on-farm activities (such as courtyard animal raising, commercial vegetable gardening and agro- forestry production) and off-farm activities (such as cottage industries, handicraft production). All of these income-generating activities were small scale, low budget initiatives targeting the local market (i.e. the exchanges taking place in the local community or, at best, in the immediately surrounding area). Because of their small scale, the project and participants were able to start these activities with very limited initial investment and risk. However, their small scale also meant that they only generated a small income. These activities were useful for satisfying the immediate personal needs of participants, especially women, and increasing the groups' self-reliance. Only in a few cases was a significant capitalization process started through these initiative. Usually, the project and/or line departments provided external support, including technical and administrative training, extension services, selected inputs and credit.
- Community infrastructure: The enhancement of the community's infrastructure was often identified as an additional need in participatory planning exercises. A number of user groups spontaneously formed to build water supply systems, to improve roads and bridges, or to construct small public buildings. Responding to the demands for support in this area often provided the project with a good entry point for establishing a

partnership with the community. In some cases, these activities also played major role in creating environmental awareness.

- Strengthening communities competence and awareness in natural resource management: Two main areas of activity were part of this core component of the project's implementation strategy: improvement of farming systems; and Management of common property resources.
- Improving farming systems: Most of the project- supported initiatives for improving farming systems developed out of negotiations between participants who wanted to have better yields, earn a higher income and save time, and the project's core mandate for promoting the conservations of water, soil and vegetation cover. Therefore, these initiatives were 'conservation by use' activities that sought a healthy balance between environmental and economic needs.
- Managing common property resources: In the project area, most initiatives for managing common property resources (CRPs) have been more or less directly associated with farming system improvement. Thus, with regard to water and soil conservation and agro-forestry initiatives, it is often difficult to make a sharp distinction between CPR interventions and those for farming system improvement. However, three main types of activities focusing specifically on CPRs can be identified:
    - Regeneration of public forests and rangelands, which resulted from entrusting local communities with the responsibility for management. This included, 'social fencing' (i.e. restrictions on land use imposed through regulations, not physical barriers), fire control and the introduction of household technologies to decrease fuelwood consumption;
    - Control of the effects of erosion, such as landslides and gullies, which cause major damage to agriculture and property; and
    - Management of streams, including measures for controlling water power and down stream erosion or facilitating the recharging of the local water table.

During participatory planning, user groups and VDCs often identified and then implemented activities in these areas in response to a specific need related to their livelihoods. On several occasions during participatory planning exercises, participants expressed the need for soil and water retention measures for preventing landslides, gullies and river floods from affecting arable land, houses and roads. To address this need, the project supported forestry plantations and small-scale environmental engineering works by providing technical assistance and incentives. The project also promoted the sustainable use of forestland by testing agro-forestry techniques for fodder production. Major efforts were also made to sensitize people to the environmental risks related to deforestation and overgrazing. As part of the integrated system of agriculture, livestock rearing and agroforestry the project prompted additional reforestation and erosion control measures of communal interest, such as roadside tree planting and the establishment of erosion control hedges along the borders of agricultural fields. Moreover, participatory planning exercises identified the need to control the consequences of erosion, such as landslides and gullies that were affecting communal grazing areas and private agricultural plots. To this end, technical assistance, training and incentives were provided to the user groups concerned. Actions facilitating better water management were undertaken in the framework of farming system improvement initiatives.

Participatory Monitoring

Participatory monitoring consisted of the continued follow-up of the organizational and technical aspects of an activity's implementation. The local participants conducted it, with some support from project staff. Participatory monitoring paralleled the implementation process and was intended to enable participants and staff to conduct the following tasks: assess the progress made in implementation; identify and address difficulties and constraints in implementation; and revise the implementation plans accordingly. Participatory monitoring practice progressively focused on three selected elements: self-monitoring by VDCs, the monitoring of implementation agreements and technical monitoring of natural resource management activities.

Self-monitoring consisted of the following activities:

- Verifying whether ordinary duties mandated by internal regulations were honoured by all group members;
- Assessing whether commitments made in previous meetings were accomplished appropriately and on time;
- Solving problems progressively met during the implementation of the activity; and
- Organizing the continuation of ongoing work

The periodic monitoring of implementation agreements was highly instrumental in:

- Providing participants with external feedback on their organizational and technical performance;
- Changing the original design of the activity in light of practical contingencies; and
- Making project staff and management aware of what was happening in the field.

Project staff conducted technical monitoring for individual participants or groups involved in activities requiring innovative practical skills and know-how, such as on-farm experiments and CPR management initiatives. This was done through participatory extension visits during which farmers and staff conducted field observations and measurements, discussed this information in light of previous records, and agreed upon actions to be taken to correct technical errors and deal with subsequent stages of the activity or experiment.

Participatory Evaluation

Participatory evaluation aims at extracting the lessons from the implementation experience. Participants and staff through interactive techniques conduct it. It strives to identify both the positive and negative aspects of the work completed and to provide suggestions for future plans. In the project, participatory evaluation proved to be essential in refining the content and methods of participatory implementation. Furthermore, by linking past experience and future planning, it contributed to ensuring continuity and 'vision' to overall participatory process.

Evaluation exercises focusing on the technical quality of the work sought to answer the basic question: "How good are the results of our work?" These evaluations considered physical work (such as building infrastructure, establishing soil conservation measures and reforestation works) and assessed whether the work was done according to the recommended technical specifications and whether these specifications were sound. These exercises, jointly carried out by the farmers concerned and the project staff, often contributed to identifying immediate actions for upgrading the work and led to suggestions for improving the technical design of similar works in the future.

Progressive Withdrawal of Support

Major efforts were made to allow user groups and VDCs to carry out different activities with less project support (i.e. decreased methodological, technical and financial assistance). Progressive withdrawal of support to local actors in participatory planning, implementation and evaluation had a dual objective:

- Promoting the self-reliance of VDCs in managing their own decision-making and action-taking process; and
- Decreasing the costs of support in each site, so that resources could be used to extend the process to other communities.

Promoting Forums for Collaborative Watershed Management

Participatory and integrated watershed management cannot become collaborative if it is not strongly supported by local governments. VDCs must have the option of referring to an administrative authority with whom they can negotiate. Development and conservation initiatives should be conducted within a local strategic planning process, coordinated by the government bodies of the district. These considerations are especially important in view of the decentralization process and the establishment of bottom-up local planning structure. The project helped local governments to strengthen their planning strategies and include in these strategies elements of participatory and integrated watershed management. The main activities the project carried out in this area included:

- Sensitizing local institutions: Project carried out activities to sensitize local administrators and professional to the participatory and integrated approach to natural resource management and human development, and to improve their competence in this area.
- Promoting inter-sectoral collaboration among local line agencies and NGOs: Cooperation among line agencies and sectoral NGOs is essential to local planning and is also key element in participatory and integrated watershed management. In selected sub-watersheds, technical assistance was provided to the planning bodies of local governments to coordinate inter-sectoral actions with the demands of grassroots organizations and with non-governmental initiatives in general.
- Involving the private sector: In areas where the private sector was found to be a significant stakeholder, representatives of corporations and companies were asked to join the participatory and integrated watershed management process.
- Legal recognition of VDCs and user groups: In accordance with local laws and regulations, the project supported user groups and VDCs in acquiring legal status. However, in actual practice, none of these organizations got themselves registered.
- Conflict management: During the implementation of the above activities, at times, the project teams had to mediate conflicts among communities and institutions or within the local administration.

METHODOLOGICAL ASPECTS OF PEOPLE'S PARTICIPATION

The government polices began to emphasize people's participation in development programmes during early 1990s. The *panchayati raj* institutions were given a constitutional status with the 73rd and the 74th constitutional amendments. The *panchayats* are given many more powers, apart from the setting up of reservations for the disadvantaged sections of women and low castes. In order to ensure local people' participation in development programmes; the most readily available institution was the *gram panchayat*. Nonetheless, IWDP (Hills-II), Jammu and Kashmir have formed separate committees in order to operationalized participatory watershed management. The *panchayats* are statutory bodies, whereas the VDCs are informal bodies and have no legal standing. After the *panchayats* were given a constitutional status, various development polices gave directions about involving the *panchayats* at different stage. However, when actually seen in the field, the link is weak and in many cases, as good as non-existent. Many users committees have a very poor relationship with the *panchayats*. Sometimes villagers see *panchayats* and users committees as competing alternative forms of organization. At the initial stages of the phase second of IWDP (Hills), the *panchayats* were non-existent in the State of Jammu and Kashmir. The elections for *panchayats* took place later in the year 2001. Besides, the view of the project functionaries towards *panchayats* is also very clear. The *panchayats* are elected bodies, and although supposed to be representative of the people, don't remain so in reality. The *panchayats* are supposed to be non-political, but these are influenced by party politics, which does not allow for a fair representation of village interest. Hence, the need for the formation of separate watershed committees at village level has been emphasized. A structure parallel to the *panchayats* is also necessary, as it gives scope to the project functionaries to have parameters of their own choice in the membership and the executive body, addressing issues of equity in particular. Not only this, the government departments do not have any control over the expenditure of funds made by *panchayats*. Not trusting the *panchayat* system, the project officials want to have control over the disbursement and expenditure of money by creating parallel participatory institutions.

VDC Formation and People's Participation

The first step in organizing collective action is defining the boundaries of the CPRs and specifying those authorized to use it (Ostrom, 1990). However, simply defining the resource boundaries and identifying the users is not enough. It is possible for a limited number of the appropriators to increase the intensity of resource use, so that they may totally destroy the resource. As such, the role of user groups in resource management is not overemphasized. PRAs reveals that IWDP (Hills-II) project functionaries along with local resource users conducted a walk through survey of the villages included in the sub-watersheds to identify the degraded village common lands (VCLs), forestlands, and grazing lands to be enclosed for protection and regeneration. Special attention has been paid to include the women, landless, and disadvantaged in the user groups. The rural poor have a greater dependency on CPRs for livelihoods and thus their representation on user groups has been assured through positive discrimination. There have been wide variations in the size of villages (72 to 500 households) in which VDCs have been formed. VDC comprises a group of about 10-25 villagers from each of the villages. The participatory social development functionaries in the presence of village communities have formed the VDCs through adopting the process of selection of the members of VDCs, keeping mandatory gender and disadvantaged groups' representation into account. In none of the case, the VDC has been formed either by election or by nomination of the project functionaries.

The members of the VDC represent specific socio-economic classes within the community. The family background, experience, sobriety of behaviour, sense of tolerance, quality of impartiality and proven honesty appears to be the predominant consideration for the selection of VDC members across the selected sub-watersheds. In some cases, the possession of landholding has also been given consideration for selection of members to VDCs. Even though in a large number of cases, the relatively elderly persons or middle-aged persons were selected to VDCs, in some cases, although not many in number, persons of relatively younger ages were returned to these committees. In the forested sub-watershed of Ramnagar, more than 70 per cent of the VDC members reported that experience, family background, sobriety of behaviour, sense of tolerance and quality of honesty were the dominating factors in their selection on VDC. Whereas in agricultural sub-watershed, more than 80 per cent of the members of VDC reported that the qualities of tolerance, impartiality, honesty and sobriety of behaviour were taken into consideration in getting the members selected to the VDC. Similarly, in Akhnoor sub-watershed, honesty (in case of more than 80 per cent) of the member was given main consideration in selecting a member to VDC, in both the forested as well as agricultural watersheds.

During PRAs, it was found that all the members of VDCs were not perceptible and receptive of project activities. The young members bubble with excitement and in many times tend to ignore the virtues of humility and healthy cooperation with development functionaries. The user groups, which had been formed, were not necessarily represented in the VDC. Thus, they may be unable to directly influence the decision making process in terms of the activities of the VDC. However, the experience gained in-group formation is very valuable. They should be the building blocks of the farmers' organization or VDC. More than 77 per cent of the VDC members across the selected sub-watersheds reported that membership procedure was rigid. Guidelines need to be framed for enabling those who remained non-members in the beginning to join later on. While the early members have every right to demand that late joiners should pay in some form of the effort the former have already invested, VDC should be discouraged from believing that they can permanently exclude non-members from access to the resources. It is ironical to note that not even a single VDC has been registered. However, the lists of the members exist on the records of VDCs and IWDP officials. The main reasons for non-registration of VDCs with appropriate authorities were reportedly lack of legal knowledge and cumbersome procedure involved in the process of registration. As a result, all of the sampled VDCs have not initiated the process of registration. A large proportion of the VDC members across the selected sub-watersheds reported that VDC has not been got registered due to lack of legal knowledge and cumbersome procedure. It is significant to note that in none of the VDC, factors like internal conflict and/or non-cooperation from the project/registration authorities have played a part in non-registration of the VDC.

The membership procedure was reportedly rigid in most of the cases in both the forested as well as agricultural sub-watersheds. It is significant to note that no change in VDCs membership has taken, once the VDCs were formed by selection. It has also been noticed that educated traders with little or no land have dominated the VDCs in forested sub-watershed of Ramnagar. However, the interested landless and poor farmers were not accommodated in VDCs, whereas, in agricultural sub-watersheds of both Ramnagar as well as Akhnoor, the landless and near landless villagers have been given representation on VDCs. During PRAs, it was revealed by some of the members in agricultural sub-watershed of Ramnagar that their names have been incorporated in VDC's list, but they don't know the purpose of such an inclusion. PRAs confirm that a new type of leadership has come to dominate the stage at the village level in the form of VDCs. The poor farmers and agricultural labourers, including the disadvantaged groups have very much come to the forefront, but the large landowners and educated traders were dominating the VDCs.

Women Participation in Watershed Management

Along with the project-implementing agency, it is the VDCs and user groups that are responsible for planning and decision-making in watershed development projects. Most of the women are unaware of the role they can play in watershed development projects. Project implementing agency efforts to involve women are often limited to awareness camps, or explanations of the concept of watershed development in women's meetings. One of the main objectives of these camps and meetings was to enforce the ban on free grazing and open access to CPRs. Given the prevailing socio-cultural constraints, it is only exceptionally strong women who are be able to make themselves heard under such circumstances. Furthermore, single woman, or even two, may find it difficult to represent the interests of all women in the village. Although women in general can be divided into two broad groups, landed and landless, there will be different needs and priorities within these groups depending on the occupations and socio-economic groups. Women from lower status-households were more interested in watershed protection activities since these directly affected their access to CPRs. Women from large landholding-households were less interested, as they were not directly affected by restrictions on CPRs.

One of the positive effects of the involvement of the users in project activities by forming VDCs is the increasing visibility of those groups who hitherto had not been expected to enter the public arena on equal terms. If we limit ourselves for the present to women, the record of VDCs is rather disappointing in active participation. The majority of the women VDC members were found in the age group of 30-45 with poor educational attainment. It is ironical to note that none of the women VDC member was educated above middle standard. Besides, their participation in VDC has been reportedly veiled and passive. Thus, their representation on VDC is reported as marginal and may be too restricted to facilitate the dynamic and vibrant participation of all the stakeholders. PRA reveals limited, often symbolic participation of women in the VDC. Out of six VDC selected for the present study, only two (one each in forested watershed (Dehari) and agricultural watershed (Chohara) had significant involvement of women, due to their participation having been facilitated by participatory social development functionaries. In other VDCs, 2-3 women have been made members of the managing committee. By and large, women remain outside the participatory process. As women are often the largest single group of resource users, their absence from VDC decision-making has often resulted in their priorities remaining unheard and the negative impact of increased labour and time required for water, fuel-wood and fodder collection. This applies particularly to women of the poorest households with no private property resources to fall back upon when forest areas are closed or certain types of usage are forbidden. It has been observed that a few women were aware about the existence of VDC in their villages and their functioning. This is not an issue of gender equality alone but may also have a critical impact on the sustainability of both the VDC and participatory watershed development and protection, without which the programme objectives will only remain rhetorical.

The mandatory requirement of at least one woman, or an adequate number of women on VDCs and user groups encourages their participation but does not demonstrate a committed effort to involve women in decision-making. For effective participation, it is essential that at least one-third, preferably one-half, of the committee should consist of women. Furthermore, female committee members need to be given specific responsibilities and made signatories to the bank account for the project, in order to emphasis the importance of their role. As a first step towards strengthening the involvement of women in decision-making, the objectives of the project implementing agency and the plans for intervention in the watershed should be made available to women in the community from the beginning of the project. Watershed plans are usually presented by the project-implementing agency through the gram sabha and it is therefore essential that full participation of women in these meetings be attained. This can be achieved by specifying that the unit of participation is the individual adult and not the household. Sensitive scheduling of the time and location of the meeting can also encourage attendance by the poorest women in the community.

Women who want to participate in community activities can do so only after completing their household duties and other work-whether it is in their own fields or outside employment. Participation in community activities therefore becomes the third work burden of women. On the one hand, we recognize that women can make significant contributions to community development and expect them to participate. At the same time, we criticize them for not attending meetings and for their inability to take off from their domestic responsibilities to participate in community activities. Facilitating women's participation begins with understanding the community in which the watershed activity is to be undertaken. Gender roles, responsibilities and gender based division of tasks in the household and community need to be analyzed before planning any development activity. Although certain socio-economic generalizations can be made, each community will be unique in terms of specific norms and relationship. Watershed development projects depend on community action and it is important, therefore, to understand each individual community before any attempt can be made to overcome social and cultural barriers.

If decisions related to ownership, access and control of productive resources remain exclusively in the hands of the men, it is unlikely that women will receive a fair share of benefits. Efforts must be made to improve intra-household distribution of benefits through community projects. CPRs not only meet daily household needs for fuel and fodder, but also provide livelihood options for women. Although CPRs remain heavily degraded in many areas, the imposition of accessing restriction on common and forestland has led to successful regeneration of the resource in watershed development areas. Women, however, rarely benefits from this regeneration, mainly because they are unable to pay, for rights to cut and carry fuel and fodder. As a result, many women have been forced to reduce or sell their livestock. Furthermore, women have to go further failed to fetch fuelwood, increasing the time spent in collection. Only women from land-owing families who have benefited from irrigation are able to use gas or kerosene stoves as a substitute for the lack of access to fuelwood. Similarly, when water becomes available, it is men-specifically landowners-who tends to assume control over the resource. Water resources developed through community enterprise should belong to the community as a whole and logically, women should be made equal partners in the management of these resources. Watershed development projects are, therefore, taking away livelihood options from women and increasing the drudgery involved in accessing fuel, fodder and water. It is important that project-implementing agency should develop mechanisms to enable the women a wider sharing of benefits. Where CPRs are concerned, women can be given rights over village common and forestland to access the resource they need for their livelihoods. The project can provide infrastructural support to record and confer user rights to these resources. It can also help set up systems-such as social fencing or rotational grazing – to ensure that women do not completely loose their access rights to CPRs.

VDCs and Decision Making in Watershed Management

Every VDC has convened monthly meeting as a routine affair to discuss the problems confronting watershed development and protection. In some cases, fortnightly meetings of Executive Committee (EC) also took place to discuss the pertinent issues. In most of the cases, the decisions have been taken by majority vote, although cases were also reported where influential members got their decision enforced. Some cases have also been reported where decisions have been taken at the behest of IWDP field functionaries, which ranges between 21 per cent to 45 per cent, the lowest in forested sub-watershed of Ramnagar to highest in agricultural sub-watershed of Akhnoor. The EC was authorized to take decisions on behalf of the VDCs' members, whereas only the general body took some pertinent decisions. However, the working of the VDCs and ECs were reportedly not very transparent. In the agricultural sub-watersheds of Ramnagar as well as Akhnoor, more than 60 per cent of the VDC members reported that the ECs have played a greater role in decision making related to watershed management. The rules entrusting duties and responsibilities to the members of VDCs and ECs have been strictly complied with in most cases. In the sub-watershed of Ramanagar, 10 per cent of the VDC members reported that important decisions regarding watershed management were taken only in the general body, whereas in Akhnoor sub-watershed, such proportion are as high as 21 per cent. The functioning of the VDCs was reportedly not very transparent.

In the forested sub-watershed of Ramnagar, 29 per cent of the VDC members reported transparent working of the VDCs, whereas in agricultural sub-watershed, only 15 per cent reported transparent working of their respective VDC. More or less similar is the situation of forested sub-watershed of Akhnoor, whereas in agricultural sub-watershed none of the VDC members reported VDCs' functioning as transparent, which is really a cause of concern. The robustness of a participatory institution is reflected in the extent of the identification of its general body of members with the institutions and the latter's capacity to take up activities of common interest. Only two (one each in forested watershed (Dehari) and agricultural watershed (Chohara)) of the 6 VDCs selected during the course of the present study had developed some self-governing traits of this nature. The livelihood needs of the disadvantaged had been over-looked, which have resulted in their further marginalization due to the less visible and subtle processes of exclusion, delegitimization of their traditional resource use patterns, and emphasis on monetary and wage incentives rather than making existing resource based livelihoods more sustainable.

Nature of Participation in Watershed Management

PRA exercises revealed that with the formation of VDCs, community involvement in watershed development and protection has increased significantly, which have resulted in social mobilization and confidence building among the rural community. VDCs have implemented demand driven activities on priorities within the framework of the project design. The participatory social development functionaries have created the much needed awareness among the local community regarding sustainable management of natural resources, for which local stakeholders have been empowered through capacity building programmes to inculcate the habits of self reliance and sustainability of the assets created through project interventions. The idea of participatory decision-making has been appreciated and operationalized by the members of VDCs for which regular meetings were organized and pertinent issues confronting the village have been discussed. None of the VDC has played any role in fund management such as checking and allocation of funds for watershed development activities. The VDCs in forested as well as agricultural watersheds in Ramnagar and Akhnoor were found actively engaged in framing and execution of the development schemes, management of CPR resources, maintenance of assets created through project interventions, solving internal conflicts, if any. For example, asset maintenance (85.18 per cent) followed by interaction with project functionaries (81.48 per cent) were reportedly the main activities of the VDC members in forested sub-watershed of Ramnagar, whereas creation of SHGs was reportedly the dominant activity of VDC members in forested sub-watershed of Akhnoor. In the agricultural sub-watersheds, asset maintenance and fund raising from the project was reportedly the dominant VDC activity in Ramnagar as well as Akhnoor. VDC members were also helping the participatory social development functionaries in resource conservation activities, for which Village Development Plan (VDP) has been framed in each of the selected village.

VDCs have been responsible for preparation of Village Development Plan (VDP) on the basis of demands and priorities put forward by the village communities. After the preparation of VDP, a general Memorandum of Understanding (MOU) has been signed with the President and Secretary of the VDC, which highlights the roles and responsibilities of the VDC as well as the project staff. The MOU has also been signed before the execution of work of each activity between VDC, user groups and the team leader (Chief Executive Officer) of the sub-watershed. The content of the MOU reads like: cost of contract, disbursement of funds/release of payments, completion time, maintenance of accounts, dispute settlements, duties and responsibilities of respective parties, details of cost sharing vis-à-vis beneficiary contribution, undertaking for maintenance and sustainability, and benefits sharing of the assets created. PRA exercises revealed that VDPs preparation has been done in a cursory manner, and the means to meet their priorities are often unrealistic. The project functionaries along with VDCs and user groups have relied heavily on available secondary data sources for the preparation of VDPs, and PRA exercises have been attempted in ritualized way with little relevance to preparation of the VDPs. No doubt, project functionaries have invested a lot to time and effort in preparing the VDPs. These are fairly lengthy and elaborate documents. The sheer length and bulkiness of the documents is, however, itself a problem. Most of the members of VDCs and local stakeholders have not been fully informed about their contents. The preparation of the plans does not include systematic consultation with diverse resource user groups. As the VDPs are prepared mostly by the inputs from the project functionaries, the approval of these is sought only from the members of the executive committee of VDCs, prior to getting them singed as a part of the MOU. This is a highly undesirable practice.

The project functionaries must ensure that all villagers, both VDC members and non-members as well as user groups, are fully aware of the VDP contents and find them acceptable, before asking a VDC representative to sign it on their behalf as a legally binding document. Non-members need to be equally informed as they also have legal rights in watershed resources. At the same time, VDP must include viable alternatives in time and space for all those currently dependent on watershed resources proposed to be closed to grazing and extraction; the resource poor groups cannot wait for several years to be able to satisfy their daily consumption needs. For this, micro planning must incorporate an analysis of the differing, often conflicting, needs and priorities of different socio-economic groups. This must include an analysis of gender differences in resource use, and must ensure that the resource needs of poorer women and disadvantaged resource users are adequately addressed. In the longer term, VDP should aim to reduce gender differences in work burden as well as access to resources.

System of Financial Management

The system of financial management adopted by the VDCs has not been reportedly satisfactory. In forested sub-watersheds in Ramnagar and agricultural watershed of Akhnoor, 81 per cent of the VDC members reported that joint bank accounts were opened, whereas in the agricultural sub-watershed of Ramnagar, only 36.84 per cent of the VDC members reported opening joint account for fund management. VDCs were supposed to maintain record of all transactions. However, the project was reportedly playing a major decisive role in utilization of funds and keeping the records of all transactions in the name of VDCs. For instance, only 42 per cent of the VDC members in agricultural sub-watershed of Ramnagar reported that their respective VDCs maintained the record of all transactions. PRAs revealed that even the President and Secretary more often were made just a signatory on financial documents and project functionaries performed the key roles. The financial matters were discussed in the VDC meeting and the system of internal audit was put into practice in a few VDCs. No attempt has been made in any of the sampled sub-watersheds for fund mobilization from the community and use of savings from the sale of usufructs share for VDCs' fund capitalization. In a few cases, there has been system of provisioning of credit to the members out of accumulated VDC fund. However, in practice, this has not been operationalized due to poor fund accumulation. VDCs have also been given autonomy in utilizing the available funds, but such provisions were impracticable due to non-availability of funds with them. The project has provided funds for resource conservation and protection activities and these funds were utilized specifically on planned activities. No diversion of funds was allowed under any circumstances, except the permission of project implementing agency at sub-watershed level.

Interaction of VDC with Project Functionaries

At present, there is lack of development of more balanced partnerships between IWDP functionaries and VDCs, which is reflected in the imbalance in power and control between IWDP and VDC. The responsibility for maintaining VDC accounts, convening its meetings and preparing the village development plans, powers of dissolution and conflict resolution is largely held by project field functionaries. Thus, instead of the VDCs' executive committees (ECs) being accountable to the general body to assure democratic and responsive functioning of the VDCs, they are instead, accountable to project field officers. This defeats the very purpose of participatory watershed management. VDCs have interacted with project functionaries with regard to the types of schemes to be undertaken, developing norms and executing project interventions. In the forested sub-watershed of Ramnagar, more than 80 per cent of the VDC members reported that they interacted with the project functionaries with regard to types of activities to be implemented and maintenance of the assets created by the project. In the agricultural sub-watershed, 84 per cent of the VDC members interacted with project functionaries regarding training. In the forested sub-watershed of Akhnoor, more than 70 per cent of the VDC members interacted with project functionaries on types of activities, manpower assistance, training, and CPR management. Whereas in agricultural sub-watershed, about 81 per cent of the VDC members interacted with project functionaries on maintenance of assets created followed by training (72.72 per cent) and executing development schemes (63.63 per cent). On the whole, VDCs have played a useful role in the maintenance of assets created, although their participation in the supervisory work might have remained limited, but these must have brought about the motivation among the villagers to collectively safeguard the assets created for their benefit. It is significant to note that some of the basic features of sustainable participatory institutions have not been taken care of and no attempt has been made to explore the ways and means for fund capitalization by the VDCs. The poor villagers have not contributed in any form to provide technical assistance in project interventions due to inadequate capacity building. However, whatever and wherever possible, the VDCs have contributed in the form of arranging local manpower support, may be hired or voluntary, but mostly, the hired labour, because of poverty, the voluntary labour contribution was very small. VDCs have also sought IWDP help in solving internal conflicts over usufruct sharing, contract assignment to members of EC, etc.

Interaction of VDCs with Other Agencies

One of the purposes of VDC creation was to encourage forward and backward linkages between the villagers and the development agencies. It, therefore, makes it imperative that the VDCs wherever existing should have cultivated healthy linkage with development agencies operational in the area. This had considerable significance in making VDCs vibrant agents in focussing on the problems faced by the beneficiaries and act as catalysts of change in the rural areas. VDCs have interacted with local NGOs on the issues like organizing, initial guidance and capacity building, which have been facilitated by the project. They never discussed the financial issues like fund raising, maintenance of proper accounts, etc. with local NGOs, which may be due to poor financial base as well as capacity building on the part of local NGOs to assist the VDCs in this regard. In forested sub-watershed of Ramnagar, more than 70 per cent of the members reported that they interacted with NGOs on the issues such as organizing, initial guidance and capacity building. Whereas, in the agricultural sub-watershed, about 63.15 per cent, 47.36 per cent and 68.42 per cent of the VDC members reportedly interacted with local NGOs on such issues. Likewise, in the forested sub-watershed of Akhnoor, about 78.78 per cent, 45.45 per cent and 39.39 per cent of the VDC members interacted with NGOs for capacity building, organizing and initial guidance. In agricultural sub-watershed, only 54.54 per cent and 36.36 per cent of the VDC members reportedly interacted with NGOs on such issues. A large proportion of the VDCs has open account in the banks to operate the development funds allocated to undertake resource conservation and protection activities. However, they never approached the banks for other purposes such as raising of the funds for income generating activities through creation of SHGs. None of the VDCs have registered themselves with appropriate authorities, as such they have not interacted with *panchayats* on issue of registration. VDCs have interacted with *panchayats* for availing of watershed development schemes and sought their assistance in this regard.

Besides above, VDCs have interacted with local traders and contractors for developing marketing links as well as repairs and construction of irrigation distribution channels, water harvesting structures, forest enclosures, soil conservation structures, etc. VDCs have also been involved in conflict resolution and forging of linkages with administrative, developmental and political agencies. In none of the case, police intervention was sought, as the conflicts are resolved at local level with the mediation of the *panchayats* and local leadership. The VDCs are often more concerned with non-resource issues, but there is no blending of the project with *panchayats* and other line departments. The present linkages between IWDP (Hills-II) and other line departments are weak and need strengthening, if the VDC have to emerge as the focal points for all round watershed development. There is urgent need to form suitable links with government and non-government agencies, which provide alternative/ supplementary social and economic programmes with the potential to provide additional or alternative incomes and livelihoods for rural people.

SOCIAL AND ECONOMIC DIMENSIONS OF PARTICIPATION: FACTOR ANALYSIS

The responses of the VDC members provide the necessary information for estimating their role and strategies at the VDC level. The random sample consists of 90 VDC members in three VDCs each selected from sub watershed of Ramnagar and Akhnoor. A factor analysis, which is a method for translating a large set of variables into a few independent choice variables, separates participatory indicators into a set of principal components, known as factors. Each factor represents an independent choice. As a rule of thumb, variables with a coefficient in absolute value above 0.5 are said to be dominating in a factor. Another rule of thumb is that all factor with a value larger than one should be used in the analysis. Table 1 shows the results for forested and agriculture watershed separately in the sample area. The factor analysis yields two factors viz. social and economic in forest and agricultural watersheds. The social and economic factors tell us the dimensionality of participation in Ramnagar and Akhnoor separately for forested and agricultural watersheds. A perusal of data presented in table 1 make it clear that in case of forested and agricultural watersheds of Ramnagar, the dominating variables in social factor are all related to VDC members' attitude towards the meeting, which explain respectively 42% and 33% of the variations. This is typically a social aspect of participation. The dominating variable in economic factor, which explain 14% and 10% of variations respectively in forested and agriculture watersheds, express contribution to and benefiting from participation as well agreement to decisions. The interest to attend the meetings and purpose it serves to the participant has again a high factor loading. While economic consideration is important, two participatory indicators related to meetings are also dominant. These two participatory indicators relate to the acceptance of the meetings, whether they are conformed themselves to the discussion in the meetings. The economic factor represents VDC members' perceived economic benefits and contribution and their acceptance of the institutional arrangement. This shows that participation in watershed management in Ramnagar consist of two dimensions.

In the case of sub watershed of Akhnoor, the dominating variables in the social factor, which explain 44% of variations in forested watershed are all, related to people's participation in evaluation and decision making which typically symbolizes social choice. It also symbolizes the acceptance of institutional arrangement in the forested watershed. The dominating variable in economic factor, which explains 11% of variation, express people's contribution to the VCL pool, which typically symbolisms an economic choice. The economic factor is also dominated by the importance of meetings and almost negatively dominated (factor loading - 0.24) by the frequency of meetings. This means that people who consider the meetings to be important also believe that the meetings are not held frequently; it is the reflection of VDC members who are quit pessimistic about the present practice of local institutional arrangements to manage watershed resources.

It is significant to note that negative indicators of participation are quite high in agricultural watershed of Akhnoor, but not dominating, supporting the statement of a negative attitude that is reflected in economic factor. This mean that the person who gain from participation and contribute more to VDC, perceive the meetings as in-frequent; they also are quite negative about their possible influence on the VDC as indicated out by negative participatory indicators. In brief, the factor analysis shows that participation in the selected sub-watersheds is two-dimensional. In the social factor, all co-efficient that are related to meetings dominates. In the economic factor, the co-efficient that are related to economic aspects of participation dominate. On the combined level, we see a clear division of the participatory choice into two components where social considerations are most important, economic considerations constitute the second main important considerations.

The following analysis explains the conditions under which a person is most likely to participate in watershed management. The links between several socio-economic variables and participation are found with the help of multiple regression analyses. The table 2 shows the general patterns for institutional settings in sub-watersheds of Ramnagar and Akhnoor. The following equation is estimated:

$$\theta = \alpha + \beta_1 \text{ RES} + \beta_2 \text{VCLDEP} + \beta_3 \text{ AVAGE} + \beta_4 \text{EDVS} + \beta_5 \text{CASGR} + \text{error}$$

Where

θ is the level of participation;

α is a constant; and

β_1 is the coefficient of a socio-economic variable.

RES: Level of resources, based on the principal component of three indicators, present quality, change in quality, and availability of resources.

VCLDEP: Dependence of village common land (VCL) – total use of VCL resources like fuelwood, fodder, etc. divided by total need for per family.

AVAGE: Average age in the family.

EDUS: Years pf schooling of the respondent.

CASGR: Caste group (higher number means a lower caste).

Besides regression, the descriptive variables are also checked for multico linearity by excluding correlated variables. For instance, the education of the respondent is strongly correlated to the average education in the family. The later is therefore excluded in all cases. The adjusted R^2 is low (< 0.17) and even negative in those cases where all considered descriptive variables are insignificant. The low R^2 is inherent to cross-section data and it is not caused by the sample size, it suffices to interpret linkages with significant t-statistics. The variables that are not significant in the regression equations can also be interpreted, namely, that they do not influence the behaviour of interest as it is described by the dependent variable. The regression outcomes are quite diverse for the institutional settings, but some general patterns are apparent. The level of resources is always positively linked to participation and significantly in eight out of twelve cases. This shows that participation is enhanced when people perceive their resource as being of a god quality. A similar conclusion can be drawn for the forest dependence. This link is also positive in al cases and significantly so in ten out of twelve cases, meaning that high VCL dependence stimulates people's participation in watershed management. Better resources and increased and increased dependency on the common resource lead to a higher level of participation. This suggests that improving levels of common resources strengthen people's participation. A higher level of dependence on common resources means that people have a higher stake in the VCLs, which is reflected, in their higher level of participation.

Table 1
Grouping of Participatory Indicators into Principal Components

Level of Participation	*Project Area with VDC*							
	Ramnagar				*Akhnoor*			
	Forested WS		*Agrl. WS*		*Forested WS*		*Agrl. WS*	
Factor	*Social*	*Economic*	*Social*	*Economic*	*Social*	*Economic*	*Social*	*Economic*
Planting in VCLs	0.376	0.054	0.138	0.034	0.284	0.082	0.324	0.018
Contribution to VCL pool	0.234	**0.572**	0.218	**0.611**	0.168	**0.631**	-0.134	**0.586**
Benefiting from VCL pool	0.091	**0.786**	0.068	**0.672**	0.082	**0.576**	0.076	**0.541**
Ability to use VCL pool	0.076	**0.682**	0.052	**0.584**	0.068	0.112	0.034	**0.518**
Benefits from using VCL pool	0.162	**0.732**	0.098	**0.682**	0.098	*0.432*	0.132	0.389
Importance of meetings	**0.632**	0.084	**0.584**	0.064	**0.092**	0.621	0.612	0.132
Agreements with decisions	0.052	**0.832**	0.048	**0.756**	**0.658**	*0.482*	0.076	*-0.448*
Attendance of meetings	**0.789**	0.162	**0.686**	0.14	**0.762**	0.184	**0.548**	-0.172
Ability to influence decisions	**0.672**	0.376	**0.598**	0.352	**0.252**	0.286	**0.732**	-0.192
Frequency of the meetings	**0.786**	0.018	**0.734**	-0.018	**0.688**	-0.024	**0.724**	-0.018
Interest in meetings	**0.681**	**0.521**	**0.612**	**0.548**	0.784	0.234	0.186	-0.098
Gain from meetings	**0.638**	**0.541**	**0.584**	**0.612**	**0.611**	0.023	**0.538**	**0.542**
Suggesting in meetings	**0.623**	0.335	**0.608**	*0.414*	***0.442***	0.372	**0.514**	0.292
Percentage of variance explained	42.20%	13.60%	32.60%	10.30%	43.60%	11.40%	36.70%	11.80%
Number of observations	27	27	19	19	33	33	11	11

Note: Numbers in bold face denote a dominating indicator (factor loading ≥ 0.5 or ≤ -0.5)
Numbers in italic face are almost dominating factor (factor loading close to 0.5 and -0.5)

The indicator of the average age in the family is only (negative) significant in three cases, which implies that younger people in forested watersheds of Ramnagar as well as Akhnoor participate most. When we look at the indicator for education of respondent, two positive significant linkages are found, namely social participation in forested watershed and overall participation in agricultural watershed in Akhnoor, which shows that when education is significant, it stimulates participation. The link between caste and social participation is significant and positive in agricultural watershed in Ramnagar and forested watershed in Akhnoor. In forested watersheds, it tends to be negative, but not significantly so. This shows that people from lower caste in agricultural watershed in Ramnagar and forested watershed in Akhnoor participate more. The same link is also found for the combined sample in forested watershed in Akhnoor. The above analysis shows under which conditions a person is most likely to choose a high level of participation. When the condition of the common resource is good and /or when people are dependent on the common resource, participation goes up. Low average levels of education in the family and high levels of education of the respondent enhance participation. A high level of people's participation facilitates the initiation of a participatory institution. Once an institution is created, a lower level of participation in needed to keep the participatory process going.

CONCLUSION

Participatory approaches have come into existence having felt the need for collective action to regenerate watershed resources to meet consumption needs, or to gain access to such resources for economic empowerment to emerge from existing social oppression. The participatory approaches have been evolved by the project and not self-initiated. The strengthening of participatory approaches in the formative years involves a great deal of efforts on the part of project implementing agency and local stakeholders including villagers, political leadership and NGOs. Over the period, participatory institutions have gained confidence and build their capabilities; they diversify their activities and have become more self-reliant. In project area, VDCs have a two-tier structure. the general body of its members, and the executive committee, which has representation from the hamlets/caste/clans/caste groups. One-third of women has been inducted which are mandatory under the existing provisions. VDCs met at least once a month. The organizational structure and practices of VDCs reflect the reverence for collective wisdom of its members as well as their democratic action and management of common resource such as village common lands (VCLs), water resources, pastures and forestland. The creation and maintenance of closures on VCLs and forestland, and repairs and maintenance of water distribution channels and water harvesting structures with the object of meeting their needs have been the primary activity of user groups and VDCs. This initial

Table 2
Links between Socio-economic Variables and Levels of Participation

Participation	*Project Area with VDC*											
	Ramnagar						*Akhnoor*					
	Forested WS			*Agrl. WS*			*Forested WS*			*Agrl. WS*		
Factor	*Social*	*Economic*	*Total*	*Social*	*Economic*	*Total*	*Social*	*Economic*	*Total*	*Social*	*Economic*	*Total*
Constant	0.448***	0.486***	0.412***	0.324*	0.136	0.196	0.402***	0.486***	0.372***	0.432**	0.512***	0.517***
	(0.100)	(0.088)	(0.105)	(0.132)	(0.152)	(0.148)	(0.044)	(0.062)	(0.050)	(0.097)	(0.138)	(0.092)
RES	0.0348	0.0042	0.0598	0.0732	0.310**	0.238**	0.136**	0.246***	0.246***	0.188*	0.324**	0.322***
	(0.0676)	(0.0634)	(0.0684)	(0.0812)	(0.100)	(0.088)	(0.043)	(0.060)	(0.050)	(0.072)	(0.112)	(0.092)
VCLDEP	0.225***	0.028	0.186**	0.196*	0.258**	0.322***	0.238***	0.026	0.188**	0.00238***	0.258**	0.328***
	(0.052)	(0.0656)	(0.058)	(0.078)	(0.089)	(0.086)	(0.058)	(0.0622)	(0.064)	(0.00130)	(0.094)	(0.090)
AVAGE	-0.00352**	-0.00038	-0.00289*	0.00106	-0.00296	-0.00126	-0.00348**	-0.00200	-0.00020	0.00044	-0.00020	0.00168
	(0.00128)	(0.00128)	(0.00138)	(0.00188)	(0.00232)	(0.00210)	(0.00132)	(0.00130)	(0.00100)	(0.00088)	(0.00200)	(0.00162)
EDUS	0.00472	-0.00258	0.00336	0.00672	-0.00586	0.00054	0.00503**	-0.00340	0.00348	0.00132	0.00343	0.00760*
	(0.00258)	(0.00284)	(0.00300)	(0.00386)	(0.00454)	(0.00416)	(0.00190)	(0.00250)	(0.00200)	(0.00238)	(0.00402)	(0.00328)
CASGR	-0.0014	-0.0098	0.0084	0.0414*	-0.0098	-0.286	0.0287***	-0.0175	0.0222**	-0.0124	0.0162	0.0272
	(0.0201)	(0.0184)	(0.0211)	(0.0152)	(0.0182)	(0.164)	(0.0072)	(0.0099)	(0.0074)	(0.0161)	(0.0234)	(0.0194)
Adjusted R^2	0.14	-0.04	0.11	0.16	0.15	0.17	0.16	0.04	0.17	0.08	0.07	0.15

Note: The value in the parenthesis is the S.D. *, P < 0.05; **, P < 0.01; * * *, P < 0.001

activity has inculcated a sense of ownership and collective responsibility among members.

In nutshell, the government takes the first move to rehabilitate the degraded CPRs because they possess the resources. The local people lack the initiative to take the first move, which does not mean that a top-down approach should be followed. The process should commence in those villages where participation is most likely to take place. For instance, the best chances for voluntary participation can be found among the villagers who depend highly on the CPRs and perceive the quality of the forest as good. After initial parformance in watershed management, the successful villages can then serve as an example for other villages to extend the process. Motivated by success in the first stages, resources can be mobilized to replicate the process in villages with less favorable circumstances, hence, the process should not be button-up either, but it should be an interaction between the state and the people, leading to a win-win situation. Transparency of the state and legal rights for the people are important aspects for success as well. In order to improve watershed management practices, people should be given more freedom to act on their own. The state should provide resources and assistance by formally allowing them a share in rehabilitated CPRs. This would enhance the development of the village and the mutual trust between villagers, so that mutual participation can be sustained, by getting closer to the optimal level of watershed development.

REFERENCES

Carney, D. and John F. (1998), *Natural Resource Management and Institutional Change*, Routledge Research/ODI Development Policy Studies, Routledge: London.

Conway, G. R. and J. A. McCracken (1990), 'Rapid rural appraisal and agro-ecosystem analysis', in M. A. Altieri and S. B. Hect (eds.), *Agroecology and Small Farm Development*, CRC Press: Florida.

Ostrom, E. (1990), *Governing the Commons: The Evolution of Institutions for Collective Action*, Cambridge University Press. Cambridge.

Sadhu, A. N. and Sudan, F. K. (1999), *'Retrospective study: IWDP (Hills-I)'*, Integrated Watershed Development Project, Hills-II, Jammu and Kashmir, World Bank and Integrated Watershed Development Project, Hills-II, Jammu and Kashmir.

Saxena N. C. (1999), 'Foreword', in John Farrington, *et al.* (ed.) *Participatory Watershed Development*: Challenges for the Twenty first Century, University Press: Oxford.

Turral, H. (1995), 'Devolution of Management in Public Irrigation Systems: Cost Shedding, Empowerment and Performance: A Review', Working Paper No. 80, Overseas Development Institute, London.

Wade, R. (1988), *Village Republics:. Economic Conditions for Collective Action in South India,* Cambridge University Press: Cambridge.

World Bank (1997), *The State in a Changing World,* Oxford University Press, New York.

World Bank (1998), *Project Concept Document: Integrated Watershed Development (Hills-II) Project,* Rural Development Sector Unit, South Asia Region, New Delhi.

World Bank (1999), *Project Appraisal Document: Integrated Watershed Development (Hills-II) Project,* Rural Development Sector Unit, South Asia Region, New Delhi.

2

FOSTERING RESEARCH ON WETLAND MANAGEMENT IN BNGLADESH: A POD OF ECOLOGICAL RESOURCE DEVELOPMENT

M. Zulfiquar Ali Islam

INTRODUCTION

The ecosystem of wetland provides a variety of plant and animal species with a unique habitat. In respect of its multifarious values and eco-friendly home to support their survival, other ecosystem has not what the wetland has. The wetland improves the natural water quality through filtering, controls flood as natural sponges, protects shorelines and steam banks against erosion, provides home with threatened and endangered species, contributes natural products to our economy, and also at the same time, it helps storing recreational, historical, scientific, and preserving cultural values. It is one of the most productive ecosystems in the world. Furthermore, wetland directly helps to moderate global climate conditions for us by storing carbon within their plant communities and soil and by releasing no carbon dioxide to the atmosphere. In spite of such values and utilities of wetland, it is disappointing but fact that the wetland in Bangladesh to date, has not received any special attention either by government and/or non-government organizations or by the natural and social scientists. Consequently, their vulnerability to both man-made and natural loss and degradation is alarming in our country. We have to maintain the present status of wetland and to restore an ecosystem to a close approximation of the condition prior to their loss and degradation as well. In the present context, we need fostering natural research on the ecological values of wetland and social research on its conservation, restoration, development and management as well.

In spite of unique ecological functions and values of wetland that are very useful for human life, there has been little enthusiastic effort in research to foster the understanding of how the wetland is functioning eco-friendly for our environment. Also such effort does not help understand how we should manage, conserve, restore, and develop such ecological resource to make our environment sustainable for better survival. Bangladesh is facing critical policy issues armed with very little understanding and analytical framework of the wetland management. It is a disappointing fact that no social research on the wetland management in Bangladesh has, to date, been done and too little natural science research on this issue has been conducted by the ecologists. One of the reasons is the investment of time and energy required for ecological and social research to understand enough of an unfamiliar research area to enable serious progress in collaborative exertion.

It is noteworthy that the government and non-government organizations may gather better perspectives on the wetland management in several disciplines, such as ecology, botany, zoology, geography, hydrology, as well as sociology, anthropology, and economics. It is expected that a gathering of a group with pertinent perspicacity and experience as well as openness of the benefits of interdisciplinary analysis might pave the ways for sustainable management and development of wetland in Bangladesh. The outcome of this exploration are designed to be useful to the community people of Bangladesh, and also to endow illustrations of thriving research with and put forward attention-grabbing queries to the natural and social scientists. The policy makers and implementers of Bangladesh are benefited from the nature, findings, and utility of such interdisciplinary research.

METHOD AND DATA SOURCES

The research is based on secondary data and information of different sources. A considerable number of conceptual and theoretical literature and data sources are consulted here to finalize the paper. The paper is designed in a framework of qualitative analysis and it is enriched with the insights of the author.

Conceptualization

Wetland is perpetually or periodically flooded lands. They are found in flat vegetated areas, in depressions on the landscape, and between water and dry land along the edges of steams, rivers, lakes, and coastlines (Sipple, 2001). The recurrent or prolonged presence of water at or near the soil surface is the dominant factor determining the nature of soil development and the types of plant and animal communities living in the soil and on its surface (Cowardin *et al.*, 1979). The present paper considers tidal and non-tidal marshes, forested and shrub swamps, bogs, and fens as wetland in Bangladesh.

Both the inland and coastal wetlands are to be wrapped for their management at community and institutional levels. The concept of wetland management wraps all activities that can be conducted with, in, and around wetland, both natural and man-made. The various efforts and activities are generally designed to protect, restore, manipulate, or provide for their eco-friendly functions and values. The wetland management involves natural wetland protection, activities involving natural wetland that are specifically exempted from regulatory requirements, wetland creation and restoration, and wetland construction for water quality improvement (Watershed Academy Web, 2002). It encompasses all planning for and strategies of responses formulated by the community people and institutional sources to the protection and conservation of wetland. Also it entails to include preventive measures for reducing both the vulnerability and degradation of wetland and to include curative measures for restoring the wetland lost and degraded as well.

Ecological Values and Degradation of Wetland

The exploration of functions and values of wetlands is a recent effort through which we are now in a new horizon of knowledge. Wetlands are once considered useless and disease-ridden places and thus those are failed to draw attention of the policy makers and implementers and social scientists. Nowadays, the realization of their ecological functions and uses and multifarious benefits to human society has been incepted. The ecosystem of wetlands produce abundant food for a great variety of animal species and thus it is found in the catalog of the most productive ecosystems in the world. It is a source of substantial biodiversity that provides support of life with numerous species from microbes to mammals. The complex and dynamic relationships among the organisms inhabiting wetland environments form the base of the food web. It is found that numerous species of birds and mammals rely on wetlands for food, water, and shelter, especially while migrating and breeding (Sipple, 2001). Another ecological function of wetlands is biogeochemical cycling that involves the biologic, physical, and chemical transformations of various nutrients within the biota, soils, water, and air. It provides the conditions needed for the removal of both nitrogen and phosphorus from surface water. The scientists recently discover its additional function of atmospheric maintenance. Wetlands world-wide help to moderate global climatic conditions as they store carbon within their live and conserved (peat) plant biomass and soil instead of releasing it to the atmosphere as carbon dioxide. Also they can receive, store, and release water in various ways. This hydrologic cycle is performed through ground water and surface water, as well as biologically through transpiration by vegetation (Sipple, 2001).

The unique ecosystem of wetlands is extremely valuable and productive habitat for fish, wildlife, and plants. It is the primary home of many species that depend on it for their survival and also it is seasonal habitat for others. As a matter of fact, since some species of migratory birds use certain wetlands round the year and would become extinct if those wetlands were destroyed, an international agreement to protect wetlands of international importance was developed. It is estimated by the US Fish and Wildlife Service that up to 43 per cent of the federally threatened and endangered species rely directly or indirectly on wetlands for their survival (Sipple, 2001). The estuaries and their coastal marshes serve as important nursery areas for the young of many game and commercial fish and shellfish as they produce so much plant biomass and invertebrate life. Wetlands are valuable to us in respect of improving water quality, including that of drinking water, by intercepting surface runoff and removing or retaining inorganic nutrients, processing organic wastes, and reducing suspended sediments before they reach open water (Sipple, 2001). In performing this filtering function, they save us a great deal of money. Also they reduce environmental problems, such as algal blooms, dead zones, and fish kills that are generally associated with excess nutrient loadings. Moreover, some wetlands maintain stream flow during dry periods; others replenish groundwater. The sponge effect of wetlands helps control flooding and prevents water logging of crops. Aside from this, wetlands reduce the likelihood of flood damage to homes, businesses, and crops in agricultural areas. They lessen monetary flood damage and protect human health, safety, and welfare as well. As wetlands are located at the margins of lakes, rivers, bays, and the ocean, they help protect shorelines and steam banks against erosion. The wetland plants hold the soil in place with their roots, absorb the energy of waves, and break up the flow of stream or river currents (Sipple, 2001).

The economic benefits of wetland resources are enormous. The wealth of natural products from wetlands includes mammals and birds, fish and shellfish, wild rice, and timber. We may obtain medicines from wetland soils and plants. Wetlands are used to grow certain cash and food crops and to cultivate fish in our country. Wetlands provide us with many recreational, educational, and research opportunities. Also they are often regarded as the storehouse of historical, scientific, and cultural heritage (Islam and Akand, 2002). Much of the nature-based tourism involves birds and many of them are wetland-dependent. An inordinate amount of the recreational birding is associated with wetlands and aquatic habitats. It undoubtedly relates to the fact that birds in particular tend to gravitate towards wetlands and aquatic habitats, which in turn attracts natural history and outdoor enthusiasts. In addition to this, artists and writers capture the beauty of wetlands on canvas and paper, or through cameras, and video and sound recorders. Others appreciate wetlands by hiking, boating, and other recreational activities. Wetlands are studied in collaboration with environmental programs at the adult stage continuing education facilities and at environmental centers. Furthermore, many school systems at the grammar, middle, and high school levels use these valuable ecosystems as out-of-door laboratories for environmentally related courses in United States. It is because that they serve as excellent study sites to learn about vegetative structure (*e.g.*, the density and cover of the vegetation) and ecological functions (*e.g.*, nutrient cycling), natural ecological processes (*e.g.*, plant succession), biodiversity, and plant-animal interactions. For more advanced students, particularly those at the high school and college levels, and professionals seeking to learn more about wetlands, they serve as excellent research sites.

The wetland in Bangladesh is one of the richest and the most productive aquatic resources in the world. The wetland in Bangladesh amounts to more than 2.8 million hectares, which are the home of over 300 fish species and an immense variety of plants and animals. It is disappointing but fact that wetland is once considered useless and disease-ridden places and thus those are failed to draw attention of the policy makers and implementers and social scientists. Consequently, the hidden values and usefulness of wetland in respect of the ecological resources are not always apparent and known to our people; however, they are often filled in or dredged to make way for housing, recreation, industry, and garbage damps.

It is noteworthy that our unplanned interventions cause the modification and loss of this unique ecosystem. The plight of wetland in Bangladesh is alarming as their protection laws and regulations are like the toothless watchdogs. The objectives of the treaty adopted at the 1971 Ramsar Convention on Wetland has not yet implemented in Bangladesh. It is framed out to reduce wetland loss and to encourage the recognition of the ecological functions and social and scientific values of wetland. Wetland is frequently and dim-wittedly subjected to conversion into other uses rather than the actual usefulness of this unique productive ecosystem.

APPROACHES TO MEETING REQUIRMENTS FOR RESEARCH DATA ON WETLAND

The following approaches to meeting requirements for data needed to advance research on the wetland management in Bangladesh. It is expected that these approaches are intended as illustrations of ways to enable researchers — in concert with government and the non-government organizations — to address the need for more extensive, more timely, and new sources of data on wetland.

Making Data Available Through a Clearinghouse

The data are to be related to the ecological and socioeconomic impacts of wetland and communications are to be available to the natural and social science researchers through a clearinghouse. A clearinghouse would provide documentation and archiving of data sets on wetland management in Bangladesh. It might be necessary to develop incentives for researchers to contribute data to this archive in our country.

The research data on the various aspects of wetland have to deposit in a clearinghouse and it could be an efficient way for successor researchers to satisfy the expectation of authentic data. An explicit expectation is that the research data should be deposited in appropriate data banks and that could also be added as a condition of receiving grants from various organizational sources at the national and international levels. Furthermore, the academic and action researchers are intended to collaborate with commercial data sources on defining standards of wetland management and development in Bangladesh. A clearinghouse could encourage their enthusiastic effort in both designing a comprehensive research and its methodology. In addition to this catalog, the joint work between natural and social scientists could lead to building new kinds of data clearinghouses and they employ new tools and techniques for making use of the data deposited in carrying out successful research.

Exploring Ways for Researchers to Gain Access to Private-sector Data

The consultant, journal, and non-government organization data is another valuable resource to researchers who examine the ecological and socioeconomic values of wetland in Bangladesh. The private-sector sources of data may have a number of possible limitations: a lack of consistent definitions and methods over time and the tendency of private-sector firms to preserve only the current information. The rigorous and closer working relationships between researchers and the private sector can provide solutions in many of these cases.

Increasing Data Collection Efforts by Government

The policy makers of the government need more, not less, information to frame a comprehensive national policy guide for wetland management, conservation, restoration and construction. But the disappointing fact is that the governmental organizations of Bangladesh suffer from the quantity and availability of data on wetland. The government has no allocation of budget and planning for gathering research data for wetland management.

In fact, the government has fewer resources for analyzing data and for making the data publicly available. It is to be noted that loss of such data sources pins down natural and social science explorations of the ecological and socioeconomic values of wetland. At a minimum, government decision makers need to be aware of the valuable data and information on wetland.

Exploring the Development of New Multipurpose Data Sets by the Research Community

It may be possible to start the development of both qualitative and quantitative multipurpose data sets. The blending of qualitative and quantitative multipurpose data may pave the ways for successful research on wetland in Bangladesh. The researchers should rely much on qualitative data with the association of quantitative ones in case of their academic research and they should give much emphasis on quantitative data with the association of qualitative ones in case of action research.

Establishing Stronger Ties with Wetland-owners' Association to Facilitate Collaboration

There should have some experience in collaborative work between the wetland-owners' association and the researchers and this collaboration is needed for successful research designed for effective and comprehensive policy guide. Both the communities may allow each to be benefited from the resources of the other. It is to be noted that limited financial resources may place constraints on such collaboration in materializing their purposes.

Blending of Indigenous Knowledge and Research Data

There should have an opportunity of blending the indigenous knowledge of community people and research data to make the wetland resource development sustainable in the context of Bangladesh. It is because that the indigenous knowledge in the basis for local-level decision-making in ecological resource management (see Warren, 1991).

Exploring Values and Uses of Wetland as a Productive Ecosystem

It would be very useful to convene seminars and workshops of ecologists, social scientists, academics, development practitioners, and representatives of commercial interests to the ecological and socioeconomic functions and values of wetland in Bangladesh. The values and uses of wetland as a productive ecosystem are explored in the various sessions of conference, seminar, and workshop. The findings of such endeavor may suggest useful policy strategies for wetland management in Bangladesh.

Research on Wetland Management for Framing Policy Guide

The expedited augmentation of various human activities has markedly caused unexpected and unplanned negative influences on wetland in Bangladesh. This alarming situation has generated a demand for fostering research, planning and management on wetland resources. In order to effectively establish and maintain an efficient, cost-effective and economically safe wetland ecosystem, the policy planners, managers, and administrators require the blending of indigenous knowledge and research findings. The research on wetland in Bangladesh implies both academic and policy implications. The successful natural and social research findings may valuable and helpful for framing our national policy guide in the following areas.

Identification and National Inventory

The government should employ the scientists and experts those could use both the on-site and off-site identification methods in identifying wetland in Bangladesh. The government should map the wetland and assess its ecological resources every 10 years. The national wetland inventory maps should be in paper as well as digital format and magnetic tape. A national inventory should be conducted for estimating the amount of wetland and exploring their characteristic features.

Research and National Database

The government and non-government organizations should conduct baseline research on the wetland for building a national database. The phase-wise and region-wise research would be expedient and prolific in this context. The government and non-government organizations should establish a well-organized research center for developing a national database on the ecosystem of wetland in Bangladesh.

Wetland Protection Grants Programme

The government and non-government organizations should encourage and provide the natural and social scientists with research equipments and financial support to conduct required intensive and extensive research on wetland in Bangladesh. The organizations at the national and international levels may develop comprehensive wetland protection strategies under the wetland protection grants programme.

Incorporating Indigenous Knowledge on Wetland

The development organizations need to recognize the indigenous knowledge on wetland, value it and appreciate it in their interaction with the community people. Before incorporating this knowledge in their development approaches, the organizations need to understand it and critically validate it against the usefulness for development programs. Their development strategies should incorporate indigenous knowledge to make a sustainable development of wetland resources.

Legislative and Administrative Measures

The existing legislation in Bangladesh is working as toothless watchdog for protecting the unique and productive ecosystem of wetland. The government should formulate and implement adequate legislative and administrative mechanisms in this regard. The government must implement the objectives and policies framed by the Ramsar Convention 1971.

The government and non-government organizations should allocate adequate budgets, and deploy trained and skilled stuff and expertise, which should be associated with adequate monitoring and enforcement authorities. Also it was suggested by the bureaucratic personnel of different ministries of GOB in a conference[1] that the Ministry of Fish and Livestock, Ministry of Forest and Environment, Ministry of Land, and Ministry of Law should work together for managing and conserving our wetland resources in unanimous ways.

Protection, Conservation and Restoration

The government and community people collaboratively should work together for the protection of representative examples of all types of wetland in both private and public ownership. In this context, the conservation of the least modified and most ecologically productive samples of each category should be addressed on the priority basis. Also they should protect wetland habitats, which are eco-friendly home for native flora and fauna especially fish stock, and the rare and endangered species of animals.

The government, non-government organizations and community people should prop up the management of all wetland catchments so that the complex and webbed relationships within a wetland ecosystem and between a wetland and surrounding ecosystems are to be understood and espoused by all the people at the different levels.

Inclusion of Wetland in the National Textbook

The government should include wetland ecosystems as texts in the national curriculum from the primary to advanced educational levels. The learners should be well informed and be aware of the hidden values of our productive resources in wetland ecosystems through their texts within the befitting ways.

Mitigation Banking

The government and non-government organizations should establish mitigation banking for the creation, restoration, and protection of an area of functioning wetland in advance of and to offset anticipated wetland impacts within the same ecoregion.

Professional Association

There should be an association for the professionals who do research and related work on wetland in Bangladesh. This association should be organized by the natural scientists, environmental sociologists, ecological anthropologists, economists, legal advisors, environmental journalists, environmentalists, development workers, and policy makers and implementers as well.

The members of this dynamic association should be enthusiastic in organizing seminar, symposium, and conference on wetland in Bangladesh on a regular basis. They should manage available fund and encourage it researchers to conduct action research on the various aspects of wetland in Bangladesh.

Social Mobilization for Public Awareness

The policy planners and scientists should properly explore the multifarious hidden values of the unique ecosystem of wetland. The exploration should be disseminated through the mass-media, journals, national dailies, and books and be known to the people of the country. The government, non-government organizations and community people should display various banners, placards and posters showing the pictographic information about the multifarious values on the World Wetland Day on 2nd February every year.

The community people should participate in such programs and activities so that they could do the vital role in conserving and restoring both the private and public wetland resources. The scientists, social scientists, experts, legal advisors, journalists, and policy planners should write and publish articles on wetland utilization for creating a massive awareness among the educated people who could mobilize their uneducated counterparts in this respect. The government and non-government organizations should open a rich website on wetland in Bangladesh for presenting wetland sites and distributing data and information to the global level.

SUMMARY

The plight of wetland is alarming in Bangladesh, as it is suffering neglect from adequate legislative, administrative and social management. It deserves multifarious ecological values that may contribute to our various economic, aesthetic, scenic, recreational and tourism, educational, and cultural needs. Also it is more valuable for hydrological, biological, and atmospheric maintenance of our aquatic and its surrounding ecosystems. It is noteworthy that our unplanned interventions cause the modification and loss of this unique ecosystem. We must be cautious that some of its values lost may not be replaceable. The government, non-government organizations and community people should collaboratively design both preventive and curative mechanisms for the conservation and restoration of our wetland. To meet this need, we should explore various possible ways for conducting successful research from the perspectives of both the natural and social sciences.

NOTE

1. Conference on Wetlands: People, Land, Water and Fish organized by Network on Management of Wetland and Aquatic Ecosystem and Ministry of Fisheries and Livestock at the Auditorium, IDB Bhaban, Sher-e-Bangla Nagar, Dhaka on 13-14 July 2002.

REFERENCES

Cowardin, L. M., *et al.* (1979), *Classification of Wetland and Deepwater Habitats of the United States*. FWS/OBS-79/31. Washington D.C.: U.S. Fish and Wildlife Service.

Islam, M. Zulfiquar Ali and Mustafa Kamal Akand (2002), 'Community and Institutional Responses to Wetland Management in Bangladesh: A Capsule of Suggestive Policy Recommendations', a paper for publication in the proceedings of the Conference on Wetlands: People, Land, Water and Fish organized by Network on Management of Wetland and Aquatic Ecosystem and Ministry of Fisheries and Livestock at the Auditorium, IDB Bhaban, Sher-e-Bangla Nagar, Dhaka on 13-14 July.

Sipple, William S. (2001), *Wetland Functions and Values*, U.S. EPA office of Water, Watershed Academy Web. URL: http://www.epa.gov/watertrain/wetland/text.html.

Warren, D. M. (1991), 'Using Indigenous Knowledge in Agricultural Development', *World Bank Discussion Paper No. 127*. Washington, D.C.: The World Bank.

Watershed Academy Web (2002), 'Wetland Management', URL: http://www.epa.gov/watertrain/wetland/text.html.

3

NATURE, EXTENT AND DETERMINANTS OF COLLECTIVE ACTION OVER WATER AND GRAZING MANAGEMENT IN BOTSWANA

Patricia M. Makepe

INTRODUCTION

Common property resource (CPR) regimes have been under scrutiny for over three decades (Hardin, 1968; Ostrom, 1990). A CPR is one over which no single individual has control, making it susceptible to unrestricted access, overuse and degradation. For example, livestock grazing in Africa is widely practiced under arrangements where the ownership of livestock is private, but the land on which they are grazed is utilized in common. Left unchecked, overstocking causes overgrazing which can lead to land degradation. Other examples of these tragedies of the commons are overfishing and deforestation.

Several instruments such as taxes, quotas, licenses, the assignment of private property rights and the imposition and enforcement of the rules of use and access by an external agency have been suggested as measures to address the CPR overuse problem (Wade, 1986). In Botswana, privatization was implemented in 1975, and later in 1991, to address the problem of overgrazing. Privatization failed to address the overgrazing problem because of high transaction costs and the high incentives herders in Botswana have to overstock[1]. Consequently, the problem of overgrazing persists.

Recently, policymakers and scholars have acknowledged that as a resource management tool, privatization is not always superior to community resource management (Runge, 1992; Baland and Plateau, 1996) although privatization may be optimal in cases where community resource management is so poor it becomes welfare increasing to privatize land (McCarthy, et al., 2001). Various factors such as poverty, dependence on the natural resource, and environmental risks may make common property a more rational solution to problems of resource management (Runge, 1992).

Botswana's rangelands lie in a non-equilibrium environment. Non-equilibrium environments are ones in which no stable equilibrium between animal and plant population exists because of highly fluctuating rainfall and temperature both in time and space. As such, the application of the carrying capacity concept as it is used on fenced ranches in stable environments is not realistic (Behnke and Scoones, 1993). In non-equilibrium environments, there is no fixed number of animals for a particular range over time and space, so livestock mobility plays an important role in enabling herders to allow their animals to maximize on variations in range productivity (Niamir-Fuller, 2001). This practice allows livestock to utilize greener pastures while other pastures are given time to regenerate, something which is not possible with enclosure. Therefore, in non-equilibrium environments like those in Botswana, it may be welfare increasing for rangelands to be managed as common property.

Over the last decade, scholars have advocated for the restoration of indigenous common property rights structures to enable communities to manage their own resources (Ostrom, 1990; Runge, 1992; Baland and Platteau, 1996). As a result, community based resource management approaches relying on collective action, have received greater attention as plausible ways to deal with the CPR overuse problem. But the devolution of resource rights back to communities alone is not enough to ensure that resources will be used in sustainable ways. The success with which communities collectively manage their resources also depends on the quality of local level institutions of resource management; the organizations to enforce them and the ability of these institutions to govern resource use (Baland and Platteau, 1996; Rasmussen and Meinzen-Dick, 1995; Swallow and Bromley, 1995). In the pre-colonial era harsh ecological conditions were sufficient to limit herd sizes keeping pressure on the range minimal (Baland and Platteau, 1996; Poteete, 1999). Since then population pressure, mechanical technology, market and government failures and the lack of use regulations for communal rangelands have led to their overexploitation[2].

Communal grazing lands in Kgatleng District of Botswana are under severe grazing pressure. The factors, which can enhance the way cattle owners in the district, can collectively manage the communal grazing resource system need to be identified. Cattle owners have organized themselves into syndicates for purposes of drilling underground water critical for livestock production. Membership in a syndicate allows cattle owners to control access to water and the surrounding grazing lands. Traditionally, the land board made sure that boreholes were drilled 16 km apart but in some places boreholes are closer together. Grazing lands are under the jurisdiction of the Land Board, which has failed to address the problem of overgrazing. It has become necessary to explore whether grazing rights ought to be devolved to the syndicates, so that this user group can take over the management of the grazing resource (Baland and Platteau, 1996; Ostrom, 1990; 1992). Before this happens, it is vital to understand the factors underlying collective action at the syndicate level over water provision in Kgatleng district as it currently stands. In this study, collective action is defined as any action taken jointly by syndicate members (or a committee of elected syndicate members) to achieve a common objective, whose outcome depends on the interdependence and reciprocity of syndicate members. Collective action occurs whenever members take steps actively to work together in managing water and grazing resources. What are the determinants of collective action over the management of water and grazing resources in Kgatleng district? After presenting some descriptive statistics of some indicators of collective action, an index of collective action at the syndicate level has been developed. Econometric methods have also been used to examine the determinants of collective action over borehole water and grazing management. The main concern is to establish whether cattle owners in Kgatleng District actively and jointly take steps together to manage their water[3] and grazing resources. Recently, policy makers and researchers have acknowledged that a critical dimension to environmental sustainability is the success with which local communities husband collectively managed natural resources (Baland and Platteau, 1996). By looking at collective action as a possible alternative to privatization, the study contributes to the search for appropriate communal resource management strategies.

STUDY AREA

Kgatleng district, located in southeastern Botswana, covers a total area of 7,960 km^2. It was selected for the study for the following reasons. First, all the grazing land in this district is under communal tenure and it is dominated by pastoralism. Second, small and large cattle owners in this district have formed themselves into syndicates and drill boreholes to tap underground water which provides permanent water for livestock production. Finally, the area is considered to be under a lot of grazing pressure (Van Vegten, 1981; Moleele, 1994; Arntzen, 1990; Allen, 1996; Jeltsch, 1996; Ringrose, 1996).

Land use in the district is broadly divided as shown in the map (see Fig. 1). Villages are found along the Madikwe River, the railway line and the main road. The predominantly mixed farming areas are found in the southern part of the district where crop production is the dominant activity. The predominantly grazing area is found in the northeastern part of the district. Kgatleng district has a mean annual precipitation of 450 – 550 mm. Precipitation is erratic both in space and time with a seasonal rainfall variability of between 30 to 40 percent. Mean annual temperature is 18 – 20 degrees Celsius. Soils are poor in organic matter and nutrient quality (Moleele, 1994).

Figure 1
Land Use in Kgatleng District, Botswana

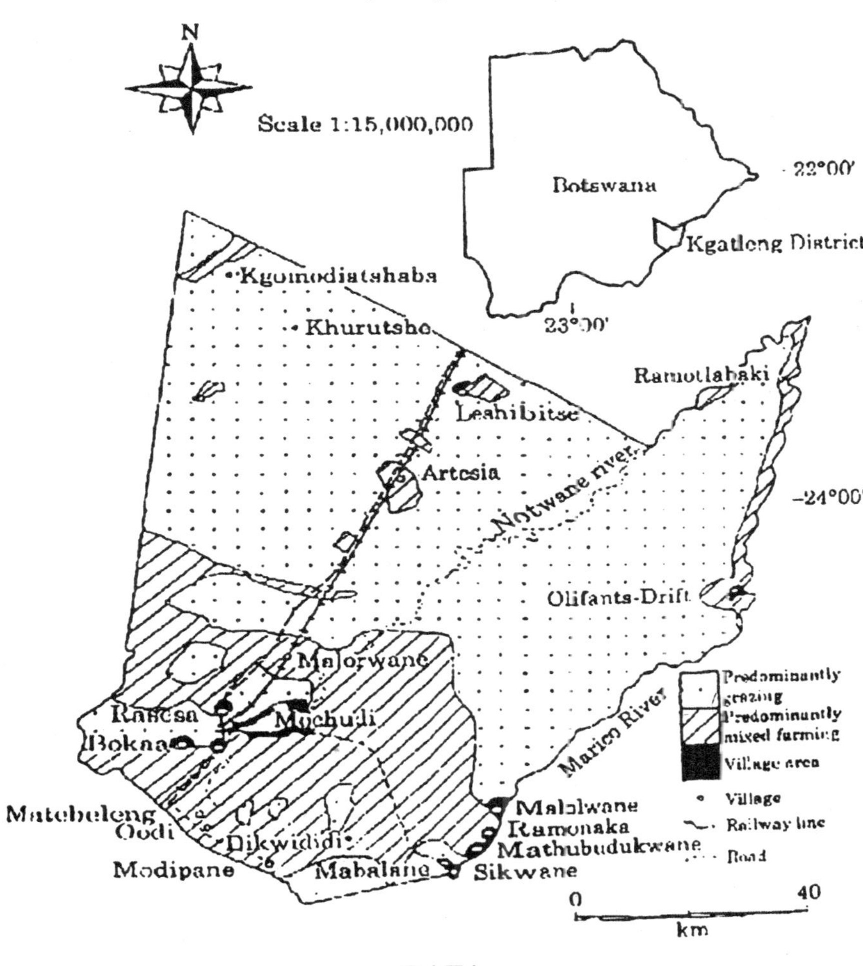

DATA

This study is based on a survey of 73 borehole syndicates conducted from February to September 2002, in Kgatleng District. A list of all the boreholes that were drilled in Kgatleng district was obtained from the Department of Water Affairs in Gaborone, Botswana. From this list, all the boreholes that were owned by syndicates in the district were selected. Of the 80 syndicates, five declined to participate in the survey and so a total of 75 syndicates were interviewed. Of these syndicates, two were outliers[4] and were dropped from the analysis. As a result, data from the 73 remaining syndicates was used in the analysis. The borehole syndicate was selected as the unit of analysis because it is the relevant institution responsible for borehole water provision and management. Therefore, it became expedient to look beyond the household as the unit of analysis. Boreholes are spread all over the northeastern part of the district, a predominantly grazing area covering more than 60 percent of the district's total land area. The average number of cattle watered in one season per borehole is 1110. This figure consists mainly of members' cattle and a small number of in-migrating cattle, which make it to the borehole because the grazing area is not fenced.

The membership in a syndicate allows members to share ownership and control over the use of the borehole, and grazing resources within the vicinity of the borehole. The average number of members in a syndicate is 18. All members share joint responsibility for the borehole's operational and maintenance costs. These include the costs of fuel, oil, services and repairs. Tasks such as maintaining the troughs and fences surrounding the holding pens are done by members with assistance from herd boys (people hired to look after cattle). If a member's access to borehole water is terminated, other members do it as a sanction for his failure to pay his share of the borehole's operational and maintenance costs. This step is taken after several attempts at persuading the member concerned to pay his share of the operating and maintenance fees.

Each syndicate has a tiered structure, which extends access to the borehole beyond its members to include dependents and hirers. A member's dependents are the people who come behind him and gain access to the borehole through his membership to the syndicate. Dependents include the children or relatives of the member of the syndicate. For example, a son or a relative whose cattle are being held by the member qualifies as a dependent. On average, syndicate members allow four dependents access to the borehole. Dependents have no independent rights to the borehole except through the member and they do not pay fees. Nevertheless, non-monetary exchanges do occur between adult dependents and members, particularly if the dependent is not a child of the member. A dependent may offer his services as a herd boy to a member in return for access to borehole water and grazing. Therefore, while there may be no monetary exchange there is some reciprocal exchange-taking place between members and dependents over time.

The next categories of cattle owners who have access to the syndicate borehole are the hirers. Hirers are people who pay members a fee for the privilege of enjoying use rights to the borehole water and grazing. Each hirer pays an amount proportional his herd size. 34 percent of syndicates have no hirers, while another 63 percent have an average of two hirers but do not intend to accept any additional hirers because of resource scarcity.[5] Neither dependents nor hirers have any say in the running of the borehole.

A structured questionnaire was administered to groups of three individuals representing the members of each syndicate. The survey collected information about the structure and function of the syndicate; the use and maintenance of water and range resources, range degradation and improvements; causes and the extent of disputes over resource use; and market accessibility. In addition, informal interviews were carried out with key informants in the Ministry of Agriculture at the district level and community elders to provide supplemental information.

COLLECTIVE ACTION OVR WATER AND GRAZING RESOURCES IN KGATLENG DISTRICT

In Kgatleng district, access to communal grazing land is solely determined by access to borehole water, which is owned and managed collectively by a syndicate. Communal tenure entitles community members to graze their cattle on land designated for grazing. But the spacing rule used by the land board to allocate land to syndicates for drilling boreholes prohibits others from drilling a borehole near the existing borehole. Hence, syndicates have *de facto* control over the grazing surrounding the borehole. The ambiguity, which exists between rights to grazing and rights to water makes managing grazing difficult (Peters, 1994). While syndicates have established rules regarding the use of borehole water the same cannot be said for grazing. But because access to grazing is determined by access to water, collective action for managing water is inextricably linked to grazing management.

Collective action for managing water occurs within and between syndicates. Each syndicate acts as a unit within itself, and also as a unit in relation to other syndicates. Within the syndicate, collective action amongst members occurs in the form of shared responsibility, joint action and reciprocity toward water provision for themselves. Between syndicates, collective action occurs in the form of reciprocal behavior toward water provision for members of other syndicates experiencing a disruption in their water supply. Assistance is sought from the nearest syndicate with a reliable water supply. Once permission is granted, grantees supply oil and fuel for running the borehole while their cattle are at the borehole. This form of collective action is a standard practice amounting to an unwritten rule, which does not require econometric analysis. Therefore, this paper focuses on collective action within the syndicate. 59 syndicates reported giving and seeking assistance from others when there is a water shortage, while the remaining 14 reported only giving assistance to others but not seeking it.

To establish how well members work together, respondents were asked about the causes and extent of disputes over resource use. 40 percent of the syndicates had a dispute in the last year. Disputes arise mainly over problems related to fee collection. 79 percent of the syndicates had problems with the payment of fees. In the spirit of working together said to be the cornerstone of syndicate organization (Peters, 1983) if there is a dispute, members first talk it over before resorting to punishment.

To establish whether members worked together better now than in the past, respondents were asked to determine whether the incidence of disputes had reduced (or increased) and whether their resolution was easier (or more difficult) now than in the past ten years. 58 percent of the syndicates reported that there were fewer disputes. Conversely, 42 percent of the syndicates reported that there were more disputes. However, 70 percent of the syndicates stated that it was easier to resolve disputes. The remaining 30 percent had difficulties in resolving disputes.

Table 1 indicates the various reasons given for the above observations. Having members who are relatives enhances their ability to work together, an important factor in reducing the incidence of disputes. When accompanied by the application of strict punishment rules it became easier for syndicates to resolve disputes. In contrast, having too many boreholes in the area, high cattle numbers and opportunistic behavior on the part of members with small herd sizes contributed to the increase in disputes[6]. These factors and the disregard for syndicate rules plus the inability for members to work together made it more difficult for some syndicates to resolve their disputes now than in the past.

Despite the incidence of disputes, syndicate membership is fairly stable. 63 percent of the syndicates had no members join or leave the syndicate in the last five years. Based on the 61 percent level of attendance at meetings, members do actively contribute their labor and time to activities related to water provision most of the time. Meetings are held six times a year on average to discuss matters pertaining borehole management. Although nearly half the syndicates stated that the incidence of disputes had risen, more than half reported that it was easier to resolve them. The above assessment shows that syndicates in the district do work together in water provision.

MEASURING COLLECTIVE ACTION AT BOREHOLE SYNDICATE LEVEL

The process of measuring collective action at the syndicate level was done in three stages. First, five indicators of collective action were selected from the survey data. Second, these indicators were used to develop two indices related to collective action using factor analysis. Finally, these indices were used to analyze the underlying determinants of collective action within syndicates over the management of water and grazing in the district.

A series of five variables have been used because there is no single ideal indicator of collective action. The five variables were chosen based on observable syndicate features representing several aspects of collective action, namely: how well members get along; members' provision of labor, time and capital, and the use of underlying institutions to punish violators of syndicate rules for water provision. Using this criterion, the following five indicators of collective action were chosen: (1) whether disputes had

Table 1
Reasons for the Extent of Conflicts and Conflict Resolution over Water and Grazing in Botswana's Kgatleng District

Reasons why there are more conflicts now than in the past ten years	*Frequency*	*Percentage*
Too many boreholes and cattle congestion	11	36.7
Arable land encroachment	1	3.3
Generation gap between young and old members	2	6.7
Member's relaxed attitude toward paying fees and low punishment	4	13.3
Members with small herd sizes taking advantage of those with larger herds	12	40.0
Total	30	100
Reason there are less conflicts now than in the past ten years	*Frequency*	*Percentage*
Members are related to each other and work together well	19	48.7
Good leadership accompanied by strict punishment rules	5	12.8
Better educated members	1	2.6
Syndicate is still new	2	5.1
Few members	5	12.8
Few boreholes in the area	2	5.1
Herd boys look after cattle	1	2.6
Drift fence separates arable land from grazing	4	10.3
Total	39	100
Reason it is easier to resolve conflicts now than in the past ten years	*Frequency*	*Percentage*
Most of the members are related to each other	2	4.0
Members work together well and sort out problems	29	58.0
Few in number	2	4.0
Strict punishment rules and using the penalty system	15	3.0
Herd boys take good care of cattle	1	2.0
Syndicate is still new	1	2.0
Total	50	100
Reason it is more difficult to resolve conflicts now than in the past ten years	*Frequency*	*Percentage*
High demand for grazing	1	5.9
Members do not work together well	11	64.6
Financial constraints	2	11.8
Some members do not abide by the rules of the syndicate	2	11.8
We have no control over members	1	5.9
Total	17	100

Source: Survey data.

occurred in the last year (2) the number of meetings held in the last year, (3) the percentage attendance at meetings, (4) the presence of defaulters, and (5) the severity of punishment for defaulters. Similar approaches have been used in empirical studies on collective action (McCarthy, et al., 2001) in Ethiopia, (Meinzen-Dick, et al., 2000) in India and (Dayton-Johnson and Bardhan, 2001) in Mexico and South India. The selection of these indicators of collective action was constrained by data availability. The syndicate level interviews made it difficult to collect data on sensitive and private information, such as data on the main sources of income of members.

Whether there were any disputes in the last year indicates the extent of violations regarding water use, allocation and borehole maintenance and how well members get along. Disputes often arise amongst members because of the allocation of costs amongst the group. If a member uses his own vehicle to transport fuel to the borehole a dispute may arise over the proper allocation of the cost amongst members. On the other hand, if hired transport was used it is easier to divide the transportation cost equally among members. Using the absence of disputes in the last year as an indicator of collective action has been used before (McCarthy, et al., 2001; Dayton-Johnson and Bardhan, 2001).

The number of meetings held in the last year and the percentage attendance at meetings were both chosen as indicators of collective action because they represent individuals' active and joint supply of labor and time in the provision of the water. Members regularly hold meetings throughout the year to discuss and take action on matters pertaining to the rules of use, operation and maintenance of the borehole. On average syndicates met six times during the year, with an average attendance rate of 61 percent.

Meetings proceed when enough members are present to constitute a quorum. The types of decisions reached at these meetings include decisions related to the management and maintenance of the borehole, planned improvements for the borehole, payments and rules of use, and penalties to be meted out on defaulters. Decisions and actions are taken jointly and made democratically after members have reached a consensus; no single member dominates the meeting. Even though syndicates appointed a chairman, his main purpose was to chair the meetings and keep a record of the syndicate's affairs. He did not make unilateral decisions. Others have used the actual amount of labor time spent in the construction, development and maintenance of a water source by members of a user group as a dependent variable (Dayton-Johnson and Bardhan, 2001). Unfortunately, this data was not available. To address this issue, percentage attendance at meetings as an indicator of collective action has been used. McCarthy et al., (2001) also used this variable as an indicator of collective action.

The presence of defaulters as an indicator of collective action has been used because this variable represents the provision of individuals' capital in providing water. If there were defaulters this indicated a lack of collective action. Defaulting on fees could indicate a lack of funds and not a lack of collective action. To address this issue, respondents were asked why they thought other members had not paid fees. Only 20 percent of the syndicates indicated a lack of funds to be the main reason for not paying fees. 44 percent of the syndicates reported that some members could pay fees but they would not pay because they knew that other syndicate members would meet the borehole's operational and maintenance costs; evidence that some members free ride on the contribution of others.

Members share jointly the responsibility for the borehole's operational and maintenance costs. 88 percent of the syndicates used a flat rate system, with each member contributing an equal amount towards maintaining the borehole. 11 percent of the syndicates had members pay proportionately according to herd size. Only one syndicate used money from hirers for borehole maintenance[7]. In rare cases where the problem had to be fixed urgently, some members paid more than their share of the monetary cost and collected the money from other members later. Patterns of reciprocity come into play when it comes to those who cannot pay at some point in time. If a member cannot pay today, he may do something in return for another member who may pay his share.

What happens to members who do not pay maintenance fees has been used as an indicator of collective action because this variable reflects the way in which members deal with those who fail to cooperate. Traditionally, members have developed a graduated penalty system of dealing with defaulters. Initially, a defaulter is persuaded to pay. Then he is taken to the customary court. When this fails he is taken to the high court. If the problem persists he is evicted. Members take eviction as a final resort frustrated by those who continuously violate syndicate rules. As observed by Peters (1983): "This penalty is rarely imposed, since it indicates a failure of the syndicate's major internal sanctions - verbal reminders and recriminations—and a rupture in the valued solidarity of the organization. Members are reluctant to have recourse to such a final sanction, since it constitutes public acknowledgement that the mutual understanding or *kutlwano* said to underlie syndicate organization has broken down" (p.156). Nevertheless, nine syndicates indicated that they had evicted some defaulters. Using the establishment of a penalty system and graduated sanctions to deal with defaulters as an indicator of collective action has been used before (McCarthy, et. al., 2001).

Given the context and manner in which syndicates provide and manage water provision, the approach used in this study to measure collective action made it possible to capture sufficiently the most suitable and appropriate indicators of collective action for the following reasons. The syndicate level interviews made it possible to obtain accurate data on the indicators that were subsequently selected to represent collective action because respondents were able to recall together, facts about their syndicate. Members know each other very well from many years of interaction. Therefore, they were able to state with confidence whether or not members of their syndicate were cooperating, based on the indicators selected above. The interviews included follow up questions to verify and clarify certain points and this served to enrich the data collected. Nevertheless, the syndicate level approach made it difficult to obtain accurate data on sensitive individual level information. Respondents were reluctant to reveal the main sources of income of other members. Such questions were regarded as sensitive and private information. Further, there were no resources available to conduct a household level survey to get such sensitive and disaggregated data. Despite these shortcomings, group level surveys are the most commonly used approach to study collective action.

DEVELOPING INDEX OF COLLECTIVE ACTION

Using STATA 7 software, a principal components factor analysis on all five indicators of collective action has been performed. This yielded a list of eigenvalues for each variable. The results of the factor analysis are in Table 2. The eigenvalues for the first and second principal components were 1.84 and 1.14 respectively. Together they accounted for 60 percent of the variation. Only these two factors had eigenvalues values greater than one. Therefore, using Kaiser's Rule[8] these two factors were retained for further analysis. The scree plot test[9] has been performed to ascertain this finding, and it supported this conclusion. The results of the scree plot test are on Figure 2. A sharp bend occurred after the second component indicating a noticeable drop in the variance explained by each factor additional to the second. The factor loadings[10] from the two retained factors were then subjected to a varimax rotation, forcing the principal components to be orthogonal. These orthogonal component loadings were then used to construct the principal component scores for each observation in the survey data set. These principal component scores were then normalized to facilitate interpretation.

The first factor is highly and positively correlated for variables associated with disputes, nonpayment and punishment. Specifically, the incidence of a dispute in the last year, the presence of defaulters and the severity of punishment all weighed heavily onto component one and weakly onto component two. This index has been taken to represent the combination of these indicators. However, its meaning is ambiguous because while nonpayment suggests a lack of collective action, taking strong steps to punish defaulters suggests collective action. This anomaly arises from the possibility that a higher incidence of disputes represents a situation of reporting disputes quickly so that they can be resolved.

The second factor is highly and positively correlated with variables associated with members actively participating in matters related to the management of the borehole. Specifically, the number of meetings held in the last year and the percentage attendance at these meetings all weighed heavily onto the second component and weakly onto component one. This has been taken as an index of members taking active steps together in matters related to the operation and management of the borehole.

Table 2
Factor Analysis Results on Five Indicators of Collective Action

Eigenvalue decomposition of the indicators of collective action

Factor	*Eigenvalue*	*Difference*	*Proportion*	*Cumulative*
1	1.843	0.6973	0.3686	0.3686
2	1.146	0.1931	0.2292	0.5978
3	0.953	0.2763	0.1906	0.7884
4	0.677	0.2954	0.1353	0.9237
5	0.381	0.0763	1.0000	

Principal component factor loadings for indicators of collective action

Variable	*Component 1*	*Component 2*	*Uniqueness*
Disputes	0.6653	-0.2665	0.4864
# Meetings	-0.0588	0.7621	0.4157
% Attendance	-0.1221	0.6468	0.5667
Nonpayment	0.8463	0.1148	0.2705
Punishment	0.8160	0.2499	0.2717

Principal component loadings for the orthogonal solution

Variable	*Component 1*	*Component 2*	*Uniqueness*
Disputes	0.6164	-0.3656	0.4864
# Meetings	0.0590	0.7621	0.4157
% Attendance	-0.0212	0.6579	0.5667
Nonpayment	0.8540	-0.0017	0.2705
Punishment	0.8445	0.1215	0.2717

Principal component score coefficients based on rotated factors

Variable	*Scoring coefficient 1*	*Scoring coefficient 2*
Disputes	0.3209	-0.2858
# Meetings	0.0707	0.6621
% Attendance	0.0213	0.5680
Nonpayment	0.4691	0.0284
Punishment	0.3210	0.1474

Source: Survey data.

GENERAL MODELS

Multiple regression analysis was performed to examine the determinants of collective action using as dependent variables the index indicating members not following rules and the index indicating greater activity in meetings, and also the five individual indicators (disputes, nonpayment, punishment, number of meetings, and percentage attendance at meetings). The analysis has been done both ways to ascertain and enrich the analysis because the indices are not without some difficulty in interpretation. Ordinary least squares models were used to explain the two indices related to collective action. A Tobit model was used to explain the percentage attendance at meetings because this variable is bounded between zero and one. Binary probit models were used to explain the presence of members who do not pay fees and whether there were any disputes in the last year. An ordered probit model was used to explain what happens to those who do not pay fees because responses took on a graduated order from one to five, with one being the least severe penalty (doing nothing) and five indicating the most severe penalty for violation (eviction from the syndicate). Multiple regression analysis was done on the number of meetings held in the last year. In selecting the explanatory variables for the models it was not possible based on the available survey data to distinguish between people's willingness and capacity to cooperate.

Figure 2
Scree Plot of Eigenvalues from Preliminary Principal Components Analysis is on the Indicators of Collective Action

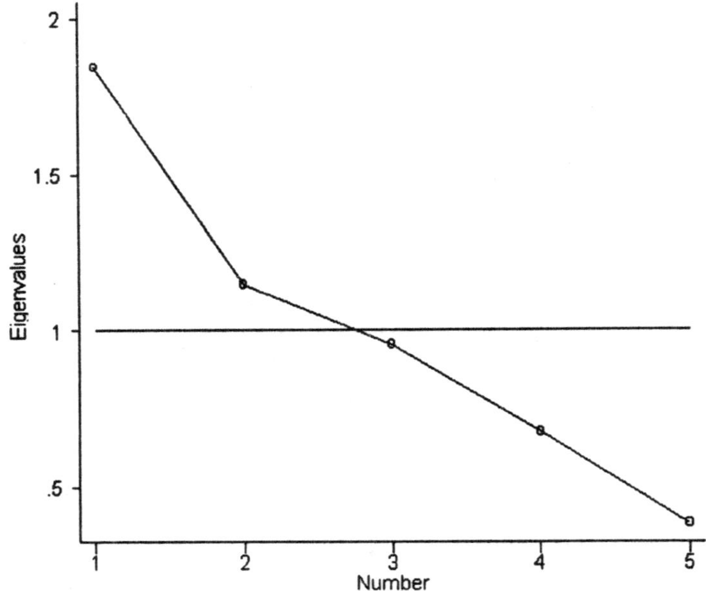

EXPLANATORY VARIABLES AND HYPOTHESIZED RELATIONSHIPS

In this section, the explanatory variables and the hypothesized relationships based on existing literature on collective action have been explained. The variables used to explain collective action within syndicates include the number of households who are members of a syndicate, heterogeneity in cattle ownership, the average herd size per member, total distance to the major market where cattle are sold, membership in other village organizations, age of the syndicate, years of schooling of the chairman, the presence of hirers, percentage of women members, the major source of water in the wet season, the presence of a water shortage during the dry season and the turnover ratio in membership. All these variables are common to all the models. Table 3 shows the descriptive statistics of all variables used in the analysis. In what follows, a description of the hypothesized relationships based on extant literature on collective action has been presented.

Table 3
Descriptive Statistics of Variables Used in the Analysis

Variable	*Obs*	*Mean*	*Std. Dev*	*Min*	*Max*
Disputes in the last year	73	0.40	0.49	0	1
Number of meetings held last year	73	6.31	8.28	1	52
Percentage attendance at meetings	73	0.61	0.26	0.14	1
Presence of defaulters	73	0.79	0.40	0	1
Severity of punishment	73	2.55	1.19	1	5
Index of members not following rules	73	0.52	0.28	0.08	0.97
Index of activity in meetings	73	0.60	0.09	0.48	0.99
Age of the syndicate	73	35.60	19.6	4	77
Number of households	73	17.97	10.55	3	49
Years schooling of chairman	73	7.45	4.01	0	17
Membership other village organizations	73	0.52	0.50	0	1
Total number of cattle watered at the borehole	73	1110.51	1103.62	200	6000
Average herd size per member	73	63.70	69.70	4	440
Main source of water during the wet season	73	0.75	0.43	0	1
Total distance to major market	73	173.52	54.74	35	283
Heterogeneity in cattle ownership	73	0.21	0.10	0	0.67
Number of hirers	73	1.93	3.57	0	15
Percentage women members	73	0.12	0.13	0	0.82
Water shortage during the dry season	73	0.37	0.49	0	1
Turnover ratio	73	0.23	0.51	0	3

Source: Survey data, Makepe (2002).

People work better in smaller groups (Olson, 1965; Wade, 1986; Baland and Plateau, 1996). In small groups the common knowledge assumption of models of strategic decisions is more valid. Norms and beliefs are shared, and patterns of reciprocity are common, especially if the group is socially and culturally homogenous as is the case here (Peters, 1983). In small groups, it is also easier to reach a consensus in decision-making and in enforcing social sanctions through reputation mechanisms and existing relationships (Bardhan, 1993; Meinzen-Dick et al., 1997). Group size also represents a balance between increasing transaction costs and economies of scale. In a small syndicate, the demand for collective action over water provision may be low while the transaction costs of achieving it are high because of high fixed costs. But a larger syndicate may be able to pool its resources and meet its costs. As syndicate size grows and resources become scarce, it raises the stakes for members to remain actively engaged in collective action. Therefore, there are some positive economies of scale to be gained by larger syndicates when it comes to the pooling of resources. But if the syndicate gets too large, it can be detrimental for collective action because the gains from collective action ultimately exceed incentive problems associated with it. Rising scarcity raises the benefits from free riding on the contributions of others and so collective action may work better in groups of intermediate size (Pender, 1999). It has been hypothesized that there is an inverse U-shaped relationship between intermediate syndicate size and collective action.

The effect of heterogeneity on collective action is ambiguous. Communities may be heterogeneous according to socio-cultural background, interest and endowment (Baland and Platteau; 1996; 1999). These aspects may affect collective action differently. The population in the district is ethnically and culturally homogeneous (Peters, 1983; Poteete, 1999). Therefore, a variable for cultural heterogeneity has not been used. The wealth heterogeneity in terms of cattle ownership amongst members has been considered. The heterogeneity has been measured by the coefficient of variation of the distribution of the proportion of households with a certain number of cattle based on the following categories: less than 20 cattle, 20 to 40 cattle, 40 to 60 cattle, 60 to 100 cattle, 100 to 200, 200 to 300 and more than 400 cattle.

In Botswana, government policy strongly supports cattle producers through high prices and input subsidies for cattle. As the average herd size per member increases, so does the incentive for collective action over the provision of water critical for cattle keeping. It is expected that the collective action be enhanced by the average herd size per member because it increases the profitability of providing and managing water (McCarthy, et. al., 2001). Therefore, it has been hypothesized that average herd size per member has a positive impact on collective action.

Total distance to market represents members' accessibility to cattle markets. It also represents members' accessibility to markets for their labor. Hence, the effect of market access on collective action is ambiguous. In the study area market access varies within the sample. Roads get poorer as syndicates get further away from major towns. But generally the district has relatively good access to markets compared to other places. Once cattle reach a major town, the marketing infrastructure is in good condition. Better access to markets increases the value of resources and the value of managing resources well. Both factors favor collective action. Fujita, Hayami and Kikuchi (1999) have shown that in rural communities with poor access to markets, members have incentives to cooperate because they expect to continue their interaction indefinitely. Better access to markets may undermine an individual's incentive to engage in collective action by increasing the opportunity cost of labor or by providing more exit options, making it more difficult to punish those who fail to cooperate (Baland and Plateau, 1996; McCarthy, et al., 2001).

It has been hypothesized that membership in other village organizations has a positive impact on collective action. Possible learning effects and experience acquired from being in other organizations reduces the costs and increases members' ability to enforce agreements (Wade, 1986; Rasmussen and Meinzen-Dick, 1995; Baland and Plateau, 1996). Different village organizations such as farmer organizations, revolving funds and burial societies, exist in the study area. Experience gained from membership in these organizations has a positive spillover effect on the running of the syndicate. It has also been hypothesized that the age of the syndicate has a positive impact on collective action. Syndicates with members who have been around longer and have more experience working together are likely to cooperate more than in syndicates that are fairly new. The average age of syndicates in the study is 36 years and membership is fairly stable. Years schooling of the chairman was selected as a proxy for the management ability of the syndicate chairman (Kajisa, 1999). It has been hypothesized that years of schooling of the syndicate chairman has a positive effect on collective action. Schooling is likely to provide the chairman with better management skill and leadership in matters pertaining to running the syndicate. Syndicates located near rivers or seasonal pans (low-lying areas which become submerged with water for part of the year) shut down the borehole in the wet season and use these alternative water sources. This reduces the borehole's operational and maintenance costs. The main source of water in the wet season is a dummy variable, which is zero if the syndicate used its borehole during the wet season, or one if a river or seasonal pan was used. It has been hypothesized that reliance on alternative water sources in the wet season impacts negatively on collective action because it lowers members' stakes to work together to secure water on a permanent basis.

Whether the syndicate experienced a water shortage during the dry season is used as an explanatory variable to explain collective action. During the dry season all syndicates rely mainly on borehole water. Therefore, it has been hypothesized that the presence of water shortages during the dry season has a positive impact on collective action because reliance on the borehole during this period raises the stakes for members to act collectively to ensure that water is available on a permanent basis. It has also been hypothesized the presence of hirers in the syndicate to impact negatively on collective action. The presence of hirers raises the difficulty of collecting fees and enforcing rules. Similarly, it has been hypothesized that the percentage of women members in the syndicate has a negative impact on collective action. Cattle keeping is a male dominated enterprise. Women members are represented by a male relative in decision-making at the syndicate level. Therefore, they have no direct say in how the syndicate is run. This contributes to disagreements amongst members when decisions are made which they are not party to. Finally, turnover ratio has been selected as an explanatory variable. The turnover ratio is the number of people joining the syndicate plus the number of people leaving the syndicate divided by the syndicate size. Turnover ratio affects borehole management because people are not in the syndicate long enough to develop an ability to work together. It has been hypothesized that the turnover ratio has a negative impact on collective action within the syndicate.

ECONOMETRIC RESULTS

Robust results of the econometric analysis first with the indices of collective action and second with the indicators of collective action as dependent variables are shown in Table 4 and Table 5 respectively. There are few significant variables, making it difficult to establish and identify measurable relationships that determine collective action. This is consistent with other findings in the empirical literature, which also point to this difficulty. Nevertheless, the regressions do yield some significant relationships, which deserve mention. Looking at each explanatory variable, the study find evidence of an inverse U-shaped relationship between syndicate size and collective action. For percentage attendance at meetings, both the number of households and its squared term are significant at the one and ten percent levels respectively (Table 5). This supports the hypothesis that collective action works better in syndicates of intermediate size, consistent with the extant literature on collective action (Olson, 1965; Wade, 1986; Bardhan, 1993; Baland and Platteau, 1996; Meinzen-Dick, et. al., 1997; Pender, 1999; McCarthy, et. al., 2001). Syndicate size represents a balance between increasing transaction costs and economies of scale. Intermediate syndicate size raises the likelihood that members attend meetings, have few defaulters and use more severe penalties by enabling members to organize themselves more effectively in providing labor, time and capital for water provision.

Table 4
Determinants of Collective Action over Borehole Water Management in Botswana's Kgatleng District (With Indices of Collective Action)

	Members not following rules index	*Greater activity in meetings index*
Number of households	0.0065	-0.0068
	(0.48)	(1.64)
HouseholdsSquared	-0.0002	0.0001
	(0.65)	(0.74)
Wealth heterogeneity	-0.0730	-0.0387
	(0.21)	(0.37)
Average herd size per member	-.0014	-.00002
	(2.52)***	(0.09)
Distance to cattle market	0.0012	-0.0002
	(1.77)*	(1.14)
Membership in other village organizations	0.1660	-0.0147
	(2.45)**	(0.72)
Age syndicate	-0.0012	0.0007
	(0.70)	(1.44)
Years schooling - chairman	0.0058	0.0003
	(0.0.67)	(0.13)
Hirers	-0.0345	-0.0111
	(0.50)	(0.53)
Percentage of women households	0.4012	0.1940
	(1.67)*	(2.71)***
Water source wet season	-0.0312	0.0003
	(0.29)	(0.001)
Water shortage dry season	0.1223	0.0343
	(1.79)*	(1.68)*
Turnover ratio	0.0923	-0.0599
	(1.32)	(0.29)
Constant	0.2319	0.6918
	(0.93)	(9.32)***
Observations	73	73
R^2	0.32	0.43
F test	2.14	3..45
	0.02	0.00

Source: Survey data
Numbers in parenthesis are absolute value of *t* statistics. *significant at 10% level **significant at 5% level ***significant at 1% level

Average herd size per member impacts positively on collective action. It has a negative effect on the index for members not following rules at the one percent level of significance. This supports the hypothesis that factors, which increase the profitability of water provision, enhance collective action. McCarthy et al., (2001) report a similar finding. The result is supported by negative and significant effect average herd size per member has on the following indicators of collective action: presence of defaulters and severity of punishment. In both models, average herd size per member is negative and significant at the one and five percent levels respectively.

Table 5
Determinants of Collective Action over Borehole Water Management in Botswana's Kgatleng District (With Indicators of Collective Action)

	Number of Meetings last year	*Percentage attending meetings*	*Disputes last year*	*Presence of Defaulters*	*Severity of punishment*
Number of	-0.1178	-0.03799	0.0236	0.0104	0.0578
households	(0.28)	(3.11)***	[0.31]	[0.13]	[0.97]
Household	-0.0006	0.0005	-0.0004	-0.0007	-0.0012
Squared	(0.07)	(2.22)**	[0.26]	[0.44]	[1.01]
Wealth	-1.4713	-0.1668	-0.4391	-0.8446	0.1081
heterogeneity	(0.14)	(0.57)	[0.24]	[0.34]	[0.07]
Average herd	-0.0063	0.0005	-0.0019	-0.0089	-0.0059
size per member	(0.37)	(1.11)	[0.62]	[2.63]***	[2.21]**
Distance to	-0.0059	-0.0006	0.0056	0.0053	0.0037
cattle market	(0.29)	(1.16)	[1.65]	[1.25]	[1.33]
Membership	-1.6682	0.0078	0.7522	0.4129	0.8548
village orgs.	(0.80)	(0.14)	[2.02]**	[0.91]	[2.84]***
Age syndicate	0.1235	-0.0016	-0.0149	0.0021	-0.0065
	(2.30)**	(1.09)	[1.52]	[0.18]	[0.87]
Years schooling	-0.1542	0.0109	0.0324	-0.0447	0.0469
- chairman	(0.58)	(1.53)	[0.68]	[0.72]	[1.23]
Hirers	-1.0169	0.0441	0.3832	0.1545	-0.5458
	(0.47)	(0.78)	[0.99]	[0.32]	[1.78]*
% Women	20.6881	0.3198	1.7693	1.5253	1.5562
households	(2.80)**	(1.62)	[1.43]	[0.74]	[1.52]
Water source	-1.1724	-0.0110	-0.6357	0.1641	-0.8332
wet season	(0.47)	(0.17)	[1.45]	[0.34]	[0.24]
Water shortage	3.8814	0.0223	0.2866	0.9107	0.4441
dry season	(1.84)*	(0.40)	[0.78]	[1.55]	[1.50]
Turnover ratio	-0.9312	0.0559	0.2423	0.3998	0.4112
	(0.43)	(0.75)	[0.64]	[0.82]	[1.37]
Constant	5.8382	1.0750	-1.5838	0.2583	
	(0.76)	(5.28)	[1.22]	[0.18]	
Observations	73	73	73	73	73
R^2	0.27				
F test	1.72				
	0.08				
Wald		16.62	19.43	25.49	
LR		55.22			
x^2 statistic		0.00	0.22	0.11	0.02
Pseudo R^2		1.02	0.17	0.26	0.13
Log likelihood		0.62	-40.73	-27.36	-86.99

Source: Survey data, Makepe (2002).
Numbers in parenthesis are absolute value of t statistics. Numbers in brackets are z statistics. *significant at 10% level ** significant at 5% level ***significant at 1% level

For the index capturing members not following rules, it has been found that membership in other village organizations has a positive and significant effect. The ambiguity of this index made it necessary to examine the coefficient on membership in other village organizations in the regressions on the following indicators of collective action: disputes in the last year and severity of punishment. The membership in other village organizations also increases the likelihood that disputes occurred in the last year and severity of punishment at the five and one percent levels of significance respectively.

As mentioned earlier, higher incidence of disputes may capture higher reporting of disputes, possibly indicating that experience in other organizations increases the awareness of the benefits of addressing disputes quickly in order to resolve them. The positive relationship with the severity of punishment is consistent with the extant literature on collective action, which states that when individuals participate in other organizations, learning takes place and social capital is created. Both these aspects serve to reduce the costs and increase the ability of members to enforce agreements (Wade, 1986; Rasmussen and Meinzen-Dick, 1995; Baland and Plateau, 1996).

The age of the syndicate has a positive impact on collective action based on the number of meetings at the five percent level of significance. This supports the conjecture that syndicate members with more experience working together get along better. The presence of hirers in the syndicate reduces the likelihood the penalty system is used at the ten percent level of significance. Increasing resource scarcity means that hirers face a real and credible threat of exclusion if they default. Therefore, they must abide by syndicate rules. That there is increasing resource scarcity in the area is supported by the limits placed on the number of hirers accepted and the refusal by other syndicates to accept hirers at all.

The percentage of women members in the syndicate has a positive and significant impact on the index of members' not following rules at the ten percent level of significance. The same variable also has a positive and significant effect on the index for number of meetings and percentage attendance at meetings and on the number of meetings at the one percent level of significance. This supports the conjecture about the problem of direct representation faced by women members. Cattle production is a male dominated enterprise and women become members when the husband passes away. However, in matters related to decision making at the syndicate level they are represented by a male relative. This lack of direct representation increases the incidence of disputes. Women are more likely to attend meetings to have these disputes resolved.

Whether there was a water shortage in the dry season has a positive and significant effect on the index of members' not following rules as well as the index of greater activity in meetings at the ten percent level. The same variable was positive and significant for the number of meetings that were held in the last year at the ten percent level. The finding supports the hypothesis that the presence of a water shortage in the dry season raises the stakes for members to be actively involved in collective action over water provision because a permanent water supply is critical for cattle production. The remaining explanatory variables – heterogeneity in cattle ownership, total distance to market, years schooling of the chairman and source of water during the wet season and turnover ratio – are not significant although they have the expected signs.

Putting the results from the analyses together, significant relationships has been found between some indicators of collective action within syndicates and the following explanatory variables: syndicate size, average herd size per member, hirers, percentage women members and water shortage during the dry season; social capital as represented by membership in other organizations and age of the syndicate. These results are borne out in the literature on cooperation and collective action (Olson, 1965; Wade, 1986; Bardhan, 1993; Baland and Platteau, 1996; Meinzen-Dick et al., 1997, McCarthy, et al., 2001). The study found that collective action in Kgatleng district is enhanced by factors, which reduce the costs of water provision and increase its profitability. It has also been found that collective action is hindered by factors, which increase the costs of water provision, and factors, which reduce its profitability.

Multicolinearity could be responsible for making some explanatory variables used in the models statistically weak. The multicollinearity has been tested in the model of the index representing members' not following rules and the model representing greater activity in meetings. There were minor but manageable problems between number of households and its squared term.[11] These variables were kept, as they were needed to test the inverted U-shaped relationship between the size of the syndicate and collective action.

CONCLUSION

Using factor analysis, two indices related to collective action were developed and used to examine the underlying factors, which determine collective action within syndicates in Kgatleng district of Botswana. To strengthen the analysis, multivariate econometric analysis was also done directly on all five indicators thought to represent collective action. The analysis clarified whether and how syndicate-specific characteristics affect collective action in the district.

The results show that the following syndicate level characteristics had significant effects on collective action: physical factors (syndicate size, average herd size per member, percentage women members and whether there was a shortage of water during the dry season). Two factors which representing members' experience in working in groups (membership in other village organization and age of the syndicate) also have a significant effect on collective action.

Collective action is considered necessary to manage common property resources, and this study takes the preliminary step of investigating the extent to which collective action is found in Kgatleng district. While it is difficult to quantify exact relationships and findings are not entirely conclusive, this study provides some evidence that collective action is likely to occur in syndicates with intermediate size, which are located in areas where members have experience working in groups. The study also provides evidence that members actively take steps to work together over managing water. Nevertheless, the management of water and grazing is inextricably linked because access to grazing is solely determined by access to water in Kgatleng. Finally, collective action to manage communal grazing is probably a necessary but not sufficient condition for conservation.

NOTES

1. A detailed discussion of the factors which led to the failure of privatization as a policy to address the problem of overgrazing is given in Makepe (2003).
2. A detailed discussion of these effects is given in Makepe (2003).
3. In this study it, was not possible based on the available survey data to distinguish between people's willingness and capacity to act collectively. Inaction on the part of a member simply indicates a lack of collective action.
4. In one syndicate there were only two members because all others had been evicted. Of the remaining members, one complained that the other member was not contributing to the running of the borehole at all. In fact, he was in the process of getting him evicted from the syndicate. The second syndicate had 100 members which was unusually high.
5. Increasing resource scarcity is evident in two ways: first, in the way syndicates turned down applications from cattle owners who wanted to be members of existing borehole syndicates, and second, in the practice of limiting the number of dependents the syndicate could have.
6. Syndicate members with large herds have a higher stake in ensuring that borehole water is available on a permanent basis and accused those with smaller herds of free riding on their contributions. This contributes to the incidence of disputes among syndicate members.
7. In this particular case the hirer had negotiated access to the borehole with the full members of the syndicate who had given him permission to use the borehole water over a five-year period provided that he invested money towards the development of the water source.
8. According to Kaiser's rule only those principal components with eigenvalues greater than one can be retained for further analysis. This rule is based on the rational that any principal component, which is a

measure of common variance should account for more variance than any single variable in the standardized score space (Greene, 1978)

9. The scree plot test is used to determine how many components can be retained. The method involves plotting the variance accounted for by each component in the order of extraction and then looking for an elbow in the curve.

10. The factor loadings are the correlations between the original variables and the new principal components.

11. Using STATA I calculated the Variable Inflation Factor (VIF), which helps flag multicollinearity. The VIF values for number of households and its squared term are 23 and 21, respectively, which are below the rule of thumb cut-off value of 30.

REFERENCES

Allen, J. (1996), A Study of the Spatial Distribution of Livestock Pressure in the Oliphant's Drift Area, South East Botswana using Landsat TM, Unpublished Minor Field Studies No. 6, Swedish University of Agricultural Sciences: Uppsala.

Arntzen, J. W. (1990), Economic Policies and Rangeland Degradation in Botswana. *Journal of International Development*. 2(4): 471-499.

Baland, J. P. and J. P. Platteau (1996), *Halting Degradation of Natural Resources: Is There a Role for Communities?*, Oxford University Press.

Baland J. P. and J. P. Platteau (1999), The Ambiguous Impact of Inequality on Local Resource Management, *World Development*, 27(5): 773-788.

Bardhan, P. (1993), Analytics of the Institutions of Informal Cooperation in Rural Development, *World Development*, 21(4): 633-639.

Behnke, R.; Scoones, I. and C. Kerven (1993), *Range Ecology at Disequilibrium: New Models of Natural Variability and Pastoral Adaptation in African Savannas.* ODI: London.

Dayton-Johnson, J. and P. Bardhan (2001), Inequality and the Governance of Water Resources in Mexico and South India, Workshop Paper: Economic Inequality, Collective Action and Environmental Sustainability, Santa Fe Institute, September.

Fujita, M., Y. Hayami and M. Kikuchi, (1999), The Conditions of Collective Action for Local Commons Management: The Case of the Philippines, Paper Presented at World Bank Seminar : Washington DC: September 1.

Greene, W. H. (1978), *Econometric Analysis*, Upper Saddle River, N.J: Prentice Hall.

Hardin, G. (1968), The Tragedy of the Commons, *Science*, 162: 1243-1248.

Jeltsch, F. Milton, S. J; Dean W. R. J. Rooyen-van, N. Harcombe, P. and M. Palmer (1997), Simulated Pattern Formation around Artificial Waterholes in the Semi-Arid Khalahari, *Journal of Vegetation Science*, 8(2): 177-188.

Kajisa, K. (1999), Contract Theory and its Application to Groundwater Markets in India, Ph.D Dissertation, Michigan State University.

Makepe, Patricia (2003), The Evolution of Institutions and Rules Governing Communal Grazing Lands in Botswana, Paper 1, Ph.D Dissertation, Department of Resource Development, Michigan State University.

Meinzen-Dick, R. S. and A. Knox (1997), Collective Action, Property Rights and Devolution of Natural Resource Management: A Conceptual Framework, Paper Presented at Workshop on Collective Action, Property Rights and Devolution of Natural Resource: Puerto Azul, Philippines, June 21-24.

Meinzen-Dick, R.; A. Knox and M. Di Gregorio (2001), Collective Action, Property Rights and Devolution of Natural Resource Management: Exchange of Knowledge and Implications for Policy, Proceedings of the International Conference: Philippines, 21-25 June.

Moleele, N. (1994), Ecological Change and Piospheres: Can the Classical Range Succession Model and its Modifications Explain Changes in Vegetation and Soil Around Boreholes in Eastern Botswana, M. Sc. Thesis, University of Canberra.

McCarthy, N; Kamara, A and Kirk, M. (2001), The Effect of Environmental Variability on Livestock and Land Use Management: The Borana Plateau, Southern Ethiopia, EPTD Discussion Paper No. 75, IFPRI: Washington, D.C.

Niamir-Fuller, M. (2001), Managing Mobility in African Range Lands, In N. McCarthy, B. Swallow, M. Kirk and P. Hazel (eds), *Property Rights, Risk and Livestock Development in Africa,* IFPRI: Washington, D.C.

Olson, M. (1965), *The Logic of Collective Action,* Harvard University Press: Cambridge.

Ostrom, O. (1990), *Governing the Commons: The Evolution of Institutions for Collective Action,* Cambridge University Press: Cambridge.

Ostrom, E. (1999), *Crafting the Institutions for Self-Governing Irrigation Systems,* Institute for Contemporary Studies. San-Francisco California.

Pender, J. (1999), Rural Population Growth, Agricultural Change and Natural Resource Management in Developing Countries: A Review of Hypotheses and Some Evidence from Honduras, EPTD Discussion Paper 48, IFPRI: Washington D.C.

Peters, P (1983), Cattlemen, Borehole Syndicate and Privatization in Kgatleng District of Botswana: An Anthropological History of the Transformation of a Commons, Ph.D Dissertation, University of Boston.

Peters, P. (1994), *Dividing the Commons: Politics, Policy, and Culture in Botswana,* University Press to Virginia: Charlottesville and London.

Poteete, A. R. (1999), Disaggregating State and Society: Accounting for Patterns of Tenure Change in Botswana 1975-1996, Ph.D Dissertation, Duke University.

Rasmussen, L. N. and R. Meinzen-Dick (1995), Local Organizations for Natural Resource Management: Lessons from Theoretical and Empirical Literature, Environment and Production Technology Division Discussion Paper No. 11, IFPRI: Washington D.C.

Ringrose, S. (1996), The Use of Integrated Remotely Sensed and GIS Data to Determine Causes of Vegetation Cover Change in Southern Botswana. *Applied Geography*, 16(3): 225-242.

Runge, C. F. (1992), Common Property and Collective Action in Economic Development, In Bromley (ed), *Making the Commons Work: Theory, Practice and Policy*, pp. 41-62, International Center for Self Governance: San Francisco.

Swallow, B. W. and D. Bromley (1995), Institutions, Governance and Incentives in Common Property Regimes for African Rangelands, *Environment and Resource Economics*, 6: 99-118.

Van Vegten, (1981), Man-made Vegetation Changes: An Example from Botswana's Savanna, Gaborone: University of Botswana.

Wade, R. (1986), The Management of Common Property Resources: Collective Action as an Alternative to Privatisation or State Regulation, *Cambridge Journal of Economics*, 11: 95-106.

4

SUSTAINABLE LIVESTOCK DEVELOPMENT IN SEMI-ARID WATERSHEDS OF SHIVALIKS, JAMMU AND KASHMIR

Falendra K. Sudan

INTRODUCTION

Integrated watershed management is the process of formulating and implementing a course of action involving natural and human resources in a watershed, taking into account the social, political, economic and institutional factors, operating within the watershed and the surrounding river basin and other relevant regions to achieve special social and environmental objectives. Typically, this process would include: (i) establishing watershed-management objectives; (ii) formulating and evaluating alternative resource management actions involving various implementation tools and institutional arrangements; (iii) choosing and implementing a preferred course of action; and (iv) thorough monitoring of activities and outcomes, evaluating performance in terms of degrees of achievement of the specified objectives (Easter, Dixon and Hufschmidt, 1986). Watershed-management practices are those changes in land use, vegetative cover, and other non-structural and structural actions that are taken in a watershed to achieve watershed-management objectives (Pereira, 1989). The watershed approach is the application of integrated water management in the planning and implementation of resource management and rural development projects or as part of planning for specific resource sectors such as agricultural, forestry, or mining. Imbedded in this approach is the linkage between uplands and lowlands in both biophysical and socio-economic contexts.

Watershed management in India has been defined as a rational utilization of land and water resources for optimum and sustained production with minimum hazards to natural resources. Increased public participation of individuals and groups in the design and implementation of projects would increase the returns-both environmental and economic.

The *World Development Report* (1992) stresses the importance of participation in creating effective programmes and viable resource-management institutions. An FAO Conservation Guide (Bochet, 1983) focuses on the role of mountain communities in the design and implementation of watershed management programmes in upland areas. National Policy on Watershed Development had recently been characterized by a concern that decisions on rehabilitation and subsequent rights and responsibilities should be taken in ways, which support the livelihood of poorer groups, and are institutionally sustainable. The Guidelines on Watershed Development (Government of India, 1994) marked a significant step towards approaches that are participatory and involve a high degree of decentralized decision taking and allocation of funding. People's participation in watershed management reduces the cost of project interventions, increases the benefits to people participating in the programme, correct the mistakes made by the project functionaries in designing various structures, increase the level of political awareness of the people, decrease the perpetual dependence of the people on government and thereby making the programme self sustaining. Mobilisation of local resources and project implementation becomes easier and smooth as a result of participation of local people (Singhal, 1999).

Institutionally sustainable rehabilitation and management of watershed development project is possible only where stakeholders/ beneficiaries act jointly to manage the resources. Watershed development essentially a resource-base approach to environmental protection and livelihood enhancement. Unless adequate safeguards can be built in, the danger is that, as the commons become more productive, better-off farmers are tempted to take control of them and customary access rights of the poor are denied (Farrington et. al., 1999). Watershed development project is being affected by three broad groups of factors: How the project is designed and implemented, how the project is organized and what external agencies are operating at watershed level. Community awareness, knowledge, support and public participation in the decisions that affect the environment and economic outcomes are equally important. This is, best secured by creating centralized systems of regulation, use and management of resources upon which local communities depend and giving these communities an effective control over the use of these resources (WCED, 1987). To the extent over exploitation of natural resources is due to lack of inappropriate structure of property rights, creating and enforcing appropriate property rights in common property resources could be an instrument of averting their tragedy and improving the environmental and economic gains (Singh, 1997). India's experience with various watershed development projects shows that the participatory approach could ensure sustainable use of renewable common pool resources (Singh, 1995).

In India, watershed development programmes were conceived as tools for correcting the regional imbalances in agricultural development created by green revolution, through investments in soil and water conservation and natural resource management in rainfed areas. The watershed development programmes have evolved from being purely technically oriented soil and water conservation programmes to more integrated and participatory programmes aiming at natural resource management with organisation of beneficiaries and more recently targeting livelihood improvement. Though the rationale for watershed development programmes has been to promote inter-regional equity, the impact of watershed investments on inter-household equity has turned out to be mixed (Puskur et. al., 2004). The overall impact of watershed development programmes has been positive and significant (Rao, 2000), increase in physical and economic access to groundwater (Chandrakanth et. al., 2004), landless and marginal households hardly benefited from watershed development, while the better off households located in downstream areas and with access to irrigation have benefited most (Farrington et. al., 1999). The availability of irrigation is central to increased cropping intensity, crop productivity, hence labour absorption and reduced migration (Shah, 2001). With the focus of most watershed development projects on increased crop productivity, the inequities embedded in the distribution of agricultural land were only reinforced. Recent evidence points out that in many watersheds inequities increased, since for non-land owning and well-owning households access to drinking water, grazing lands and other natural resources decreased (Kerr, 2002; Batchelor et. al., 2003).

It is increasingly being recognised that to increase the equity of watershed development, more attention has to be paid to livestock production, which is a major livelihood activity in most rainfed areas. In addition to providing income through animal and dairy sales, it has an important function for crop production by providing manure and draught power. Especially in the ecologically fragile areas of semi-arid watersheds, livestock makes an important contribution to the survival of the economically weaker sections (India Task Force, 1987), small ruminants in particular playing an important role in ensuring rural livelihoods against drought (Pasha, 2000). Water and fodder are the most critical constraints for livestock development in semi-arid areas. Although in some cases watershed development programmes have tried to improve biomass production on commons, in most cases livestock production has insufficiently been accounted for, missing out on important opportunities to improve the livelihood of the watershed poor.

Watershed development appears to have even negatively affected livestock production: due to grazing restrictions on common lands, households dependent on small ruminants have in some cases been forced to migrate with their livestock and/or to sell it off. For large ruminant owners the effects were not so adverse: as these species usually depend more on crop residues and stall-feeding. The predominant farming system in most semi-arid watersheds is the 'mixed crop-livestock farming system' under rainfed conditions, supported with limited irrigation from existing water harvesting structures, ponds, and gravity based irrigation channels. Traditionally, livestock production played an important role in the rural economy of India's semi-arid zones (Walker and Ryan, 1990). On an average, 15 per cent of household income was derived from livestock production. For poor households, like landless shepherds and women, the importance of livestock production is much larger: two-third of the rural poor have livestock for income and drought insurance. Whereas landless, marginal landowners and nomadic and semi-nomadic groups would mainly depend on goat and sheep, landowners mostly held large ruminants, as the costs of maintenance were high due to fodder scarcity (Walker and Ryan, 1990).

A desk study of livestock-environment interactions in watersheds by Mangurkar et. al. (2001) highlighted some important issues. They found support for the hypothesis that watershed development programmes tend to change the livestock composition (numbers, proportion of different breeds and species in their total population) depending on the restrictions on access to feed/fodder resources and, quantitative and qualitative change in the types and availability of fodder. Watershed development programmes with integrated livestock component strongly strengthen the crop livestock integrated mixed farming system, with a positive impact on the environment. The introduction of practices such as stall feeding, prohibited access to grazing resources and declining traditional arrangements in watershed development programmes do influence the promotion of natural regeneration, but tend to adversely affect the livelihoods of the resource-poor livestock owners. Lack of appropriate policies and approaches, adequate poverty focus and convergence of related sectors in natural resource manegement programmes are limiting the promotion and effective utilisation of positive livestock-environment interactions. Kurup (2003) focused on not only the inter-linkages between livestock production, resource endowment and market conditions, but also on the management of resources and policy environment: since sustainable livestock production depends strongly on community organisation and their capacity to manage the common resources. Thus, the present study is of contemporary relevance and examines the impact of livestock development intervention on increasing the livestock production potential in semi-arid watersheds of *Shivaliks* in the context of Integrated Watershed Development Project (IWDP), Hills-II, Jammu and Kashmir.

MATERIALS AND METHODS

Study Area

The state of Jammu and Kashmir with an area of 1,38,214 km^2 comprising the uppermost drainage of Indus, Jhelum, Middle Chenab and Western Rabi exhibits great contrast in relief features, climate, soil and vegetation within comparatively narrow geographical spread. The climate of lower plains of Jammu and Kashmir that merges with plains of Punjab is sub-tropical. The climate is predominantly sub-tropical in the main *Shivaliks*. The climate in the region covering upper reaches of Pir-Panjal forming the main mountain backdrop changes from sub-tropical in the lower reaches to moist-temperate in upper reaches. The climate of Kashmir Valley (bordered between the Great Himalayas and the Pir-Panjal is dry temperate. The great Himalayan range: the innermost line of high mountains with Ladakh situated in trans-Himalayan zone is a cold desert and experiences sub-zero winter temperature. *Shivaliks,* locally known, as *'Kandi'* comprise of piedmont deposits made of boulder and pebbles, gravel and sand with minor clays mixed in varying proportions. One of the characteristic features of the area is the stream called *choes* or *khads,* which remain dry for most of the year. Though total rainfall is high (average 1000-1500 mm.), but its distribution is very erratic resulting in frequent droughts. This area is subject to soil erosion due to undulating topography, steep slopes, poor vegetative cover (scrub forest) and coarse to medium texture of the sedimentary material. *Shivalik* hills consist of highly erodible sand stones, conglomerates, siltstones and shales. The uncontrolled deforestation and overgrazing in the past has resulted in the reduction of vegetative cover and accelerated erosion in the *Shivaliks*. The area experiences paucity of water for plant growth due to excessive run-off, high evaporation especially during summer, and unpredictable and erratic rainfall. The eroded material from *Shivalik* hills, brought down by the seasonal rivulets *(choes)* is deposited in the sloping piedmont plain and also in the area around *choes*. The repeated deposition of coarse sediments renders these areas comparatively low in agricultural productivity. Due to lack of irrigation, subsistence rainfed agriculture is the prevalent production system is selected sub-watersheds. In terms of physiographic features, there may not be 100 per cent similarity between the two sub-watersheds but, certainly, in term of accessibility, natural resource endowment, infrastructure development and general index of socio-economic awareness, the selected sub-watersheds are comparable.

IWDP, Hills-I was started in 1990, and concentrated on integrated development of high priority zones of rainfed areas. The total area of *Shivaliks* and *Karewas* is 9.45 lakh Ha. (*Shivaliks*: 7.50 lakh Ha and *Karewas*: 1.95 lakh Ha); and about 5.9 lakh Ha. is identified as problem area, out of which, 0.52 lakh Ha. (8.81 percent of the problem area), have been treated under IWDP (Hills-I). This includes the sub-watersheds of Devak and Ramkote in *Shivalik* hills (Jammu region) and Dudhganga in *Karewas* (Kashmir Valley). IWDP (Hills-II) targets another 0.61 lakh Ha. (10.33 percent of the problem area), leaving balance of 4.77 lakh Ha (80·86 percent of problem area) still untouched. The principal lessons learnt from IWDP (Hills-I) were: the need to develop an integrated approach to the *Shivalik* Watersheds; the need for a coordinated approach to the use of available funds from all sources; the need to involve stakeholders in planning, implementation and maintenance; the need to increase stakeholder awareness of environmental and socio-economic considerations in articulating communities' needs; the need to focus on measures to arrest soil erosion and promote *in-situ* moisture conservation from ridge to valley; the need to assess marketing prospects, especially of horticultural products; the need to improve infrastructure within watersheds, particularly rural links to markets, water harvesting and drinking water; the need to develop monitorable project implementation and development objectives' indicators, and the need to involve local NGOs in the project as facilitators and trainers in the planning and implementation of the project (World Bank, 1998).

IWDP, Hills-II, Jammu and Kashmir covers two sub-watersheds in *Shivaliks*, viz. Ramnagar and Akhnoor and two sub-watersheds in *Karewas*, viz. Rajwar and Rambiyara. The proposed study will be confined to two sub-watersheds of Akhnoor (Jammu district) and Ramnagar (Udhampur district). While Ramnagar is in the inner *Shivaliks*, Akhnoor is on its outer part. Ramnagar sub-watershed is the catchment area of Ramnagarwali *Khad* (gorge) in the middle catchment of river Tawi. It has an area of 32,630 Ha. and is subdivided into 39 micro-watersheds. Akhnoor sub-watershed with an area of 42350 Ha. is sub divided into 37 micro-watersheds. It covers about a dozen rivulets (*nallahs*) and a large number of small nallahs originating from *Kalidhar* ridge and draining into Chinab river on the western portion of *Kalidhar* ridge, drains join Manawar Tawi which in turn also meets the Chenab river. Following are the objectives of IWDP (Hills-II), Jammu and Kashmir: to restore on a sustainable basis the productive potential of the region and improving the quality of life; to reduce soil erosion and improve availability of water; to help increase production and income, grain crops, horticulture, fodder, fibre, fuelwood, livestock and household based products; to promote holistic and sustainable agro-ecological development involving people's participation; to strengthen community participation, and to develop local level institutions to enhance the sustainability of the model (World Bank, 1999).

Methodology

During the Khon Kaen Conference (1985), different types of Rapid Rural Appraisal (RRA) techniques were identified, one of which was labeled "participatory". The orientation of a participatory rural appraisal (PRA) was to facilitate or stimulate community awareness and capability regarding a problem or issue. Particular attention was given to enabling local people to conduct their own analysis of problems, and to share their findings. The role of the outsider became one of catalyst, rather than one of expert. Stimulation of community awareness and capability was also intended to reduce the extractive nature of RRA, and to help local people to empower themselves. In that regard, PRA is consistent with some of the basic aspects of sustainable development (local empowerment, equity, social justice) (Conway and McCracken, 1990). The primary data has been collected using well-structured and pre-tested questionnaires, participatory rural appraisal (PRA) techniques and group meetings. Whenever possible and required, secondary data sources have also been utilized.

Through participatory observations and focus group discussions, it is possible to describe what goes on, who and what is involved, when and where things happen, how they occur, and why-at least from the stand point of participants-things happens as they do in particular situations. The methodology of participatory observation and focus group discussions has been used to ascertain important differences between the views of insiders as opposed to outsiders. The participant observation inquiry has been helpful in an open-ended logic and process of enquiry as well as qualitative description of the phenomenon under study. Both the unfocused and focused observation techniques have been used. The unfocused initial has been use to become increasingly familiar with the insider's world to refine and focus subsequent observation and data collection. All the observations have been recorded on site and misunderstanding, if any has been corrected thereof. In all the focus-group observations, 10-12 stakeholders have been recruited from different settings. Highly formal interviews have been conducted using structured interview schedules. Besides, study has been supplemented by utilizing potentially rich sources of secondary data and research materials. The content analysis technique has been used to analyze the data and information qualitatively and quantitatively (using descriptive statistics). The content analysis technique has been supplemented by use of code and labels, field notes, sorting, shifting, constructing and reconstructing these materials. The household survey has been drawn on a stratified random sample of village household. The questionnaire was pre-tested before the main survey with small 'focus' group assembled to discuss their reactions to questionnaire prior to detailed survey.

For the collection of primary data and information on livestock production and development at household level, three villages each in project area with village development committee (VDC) and project area without VDC from the two sub-watersheds of Akhnoor and Ramnagar, where a maximum number of people are known to have benefited from livestock development intervention have been selected. In the non-beneficiary category, three sample villages each have been selected on random basis, each 20-25 km away from the sampled micro-watersheds villages (*i.e.*, project area). Therefore, a total of eighteen villages (six each from 'project area' and three each from 'non-project area') have been selected. The stratified sampling technique has been used to select villages. During the stratification, care has been taken to include both "forested watersheds' villages and agricultural watersheds' villages" in order to makes a comparative study. A purposive sample of approximately 20 per cent of household level respondents in 'project' and 'non-project' area has been selected as the base of enquiry. The sample size was restricted to 376 and 404 households in sub-watersheds of Ramnagar and Akhnoor. Out of which, 185, 119 and 72 households; 158, 72 and 174 households belong to project area with VDC, project area without VDC and non-project area respectively in Ramnagar and Akhnoor.

RESULTS AND DISCUSSION

Livestock Institutional Support

The cattle population in the State of Jammu and Kashmir mainly consists of cows, buffaloes, sheep, goat, horses and camels. The camel and horses are few in numbers in the *Shivaliks* and are used for transport purposes. The cows and buffaloes are reared for dairy, whereas, sheep and goat are reared for meat and wool. The cattle population in the project area is 2.38 lacs, spread over an area of 1.11 lacs hectares, out of which the cows are 1.23 lacs, buffaloes 0.35 lacs, and sheep and goats are 0.80 lacs. The nomadic cattle population increases with their arrival during summer in *Karewas* and winter in *Shivalik* hills. The livestock predominantly consists of local breeds with low yield of milk and wool. PRA exercises were carried out by the project functionaries to ascertain the problems in livestock development in the project area, which identified the problems like (a) absence of effective health care system; (b) low availability of fodder; and (c) inadequate artificial insemination facility for breed improvement.

IWDP (Hills-II), Jammu and Kashmir has established 3 veterinary hospitals and 7 veterinary dispensaries in the project area. It is significant to note that all the veterinary hospitals have been established in the sub-watershed of Ramnagar. The veterinary facilities created across the project area are catering to the requirement of the cattle population. More specifically, attention have been given on (i) mass vaccination against foot-mouth diseases, haemoharrahic speticimia etc.; (ii) dosing against intestinal worms and liver fluke; (iii) artificial insemination; (iv) natural services by way of providing breeding cow bulls and rams; (v) castration of scrub bulls; (vi) dipping of sheep against ecto parasite; (vii) distribution

Table 1
Livestock and Genetic Improvement in Selected Sub-Watersheds (Rs. in lacs)

Livestock: Genetic Improvement		*1999-2000*						*2001-2002*						*Ramnagar Change 1999-2000 to 2001-2002*		*Akhnoor Change 1999-2000 to 2001-2002*	
		Ramnagar			*Akhnoor*			*Ramnagar*			*Akhnoor*						
	Unit	*Phy.*	*Fin.*	*Fin. %*	*Phy.*	*Fin.*	*Fin. %*	*Phy.*	*Fin.*	*Fin. %*	*Phy.*	*Fin.*	*Fin. %*	*Phy. %*	*Fin.*	*Phy. %*	*Fin. %*
Cryocans	No.	0	0	0	0	0.42	3.09	0	0	0	0	0	0	0	0	0	-100
A.I Centre	Rs.	0	0.3	2.19	0	0.27	1.98	0	0.3	3.958	0	0.19	1.482	0	0	0	-29.63
Dairy Unit	No.	0	1	7.3	10	2.04	15	10	3.38	44.59	25	5	39	1000	238	150	145.1
Veterinary Hospital	No.	1	6.45	47.1	0	2.37	17.4	0	2	26.39	0	1.93	15.05	0	-68.99	0	-18.57
Veterinary Dispensary	No.	2	2.25	16.4	0	6.94	51	0	1.9	25.07	0	1.95	15.21	0	-15.55	0	-71.9
Equipments	Rs.	0	1.3	9.49	0	0.71	5.22	0	0	0	0	0	0	0	-100	0	-100
Poultry Unit	No.	0	0	0	0	0	0	0	0	0	0	2.49	19.42	0	0	0	249
Pregnant Cows and Calves Feed	No.	0	0	0	0	0	0	0	0	0	0	1	7.8	0	0	0	100
Sterility Control	Rs.	0	0.6	4.38	0	0.29	2.13	0	0	0	0	0	0	0	-100	0	-100
Castrators	No.	0	0.6	4.38	0	0.57	4.19	145	0	0	4	0.26	2.028	0	-100	400	-54.39
Feeding Stall	No.	0	0.61	4.45	0	0	0	0	0	0	0	0	0	0	-100	0	0
Bull/Buffalo	No.	0	0.3	2.19	0	0	0	0	0	0	0	0	0	0	-100	0	0
Maintenance of Livestock	Rs.	0	0.29	2.12	0	0	0	0	0	0	0	0	0	0	-100	0	0
Total		3	13.7		10	13.6		155	7.58		29	12.8		0	-44.67	190	-15.81

Source: Physical and Financial Achievements, 1999-2000 and 2001-2002, IWDP, Hills-II, Jammu and Kashmir, 2003.

of cross-bred ram bouliet/marino rams; (viii) health cover for different diseases; and (ix) feed for calf rearing, pregnant eves and lambs born.

The data presented in table 1 makes it clear that under the component of genetic improvement of livestock, artificial insemination centres have been established in both the sub-watersheds. However, the impact of artificial insemination programme is reportedly modest in project area, which needs to be propagated by the field functionaries. In order to improve the livestock health and productivity, veterinary hospitals and dispensary have been started with project interventions. The dairy and poultry units have also been given assistance, however, very marginally. The problem of poverty, unemployment and underemployment is very acute in the project area and dairy and poultry rearing could be alternative viable income generating activities (IGAs) for the people living in the sub-watersheds, if undertaken on group basis. The products of these micro-enterprises have ready market in the local villages and nearby townships. It is ironical to note that livestock development infrastructure has been created, but the maintenance of the infrastructure has not been given adequate attention. A significant proportion of the population in the project area is rearing of sheep and goat as a source of livelihood. The transhumance pastoral and semi-pastoral communities depend heavily on sheep and goat rearing, for which provision has been made for migratory health-cover facilities (See table 2). New modified techniques have also been propagated under sheep husbandry to improve the quality of sheep and goats reared by the transhumance pastoral and semi-pastoral communities in the selected sub-watersheds. However, the sheep husbandry has not been promoted in a desired way, keeping in view the problems being faced by transhumance and local residents.

Besides above, the project has also been providing facilities in the villages, where veterinary hospital or dispensary is not available. The veterinary camps have been organized in each village of the selected sub-watersheds. On the whole, 2251 artificial insemination has been done, 178 veterinary camps has been organized and 104, 877 and 182 of dairy, poultry and sheep units were set up across the project area over the period ending September, 2001. The concerted efforts of the project have resulted in a significant increase in the yield of milk and wool. For instance, the yield of cow milk has increased from 2.75 litre per day to 4.25 litre per day in case of local cow and from 6.75 litre per day to 8.60 litre per day in case of crossbred cow after project intervention. Similarly, over the project period, the yield of buffalo milk has increased from 3.50 litre per day to 5.25 litre per day in case of local buffalo and from 8.60 litre per day to 10.30 litre in case of cross-bred buffaloes. Likewise, the yield of sheep wool has increased significantly from a base line yield of 2.25 Kg. per annum to 3.40 Kg. per annum (IWDP, Hills-II, 2001). On the whole, the livestock development intervention has improved significantly the production and productivity

Table 2
Sheep Husbandry Intervention in Selected Sub-Watersheds (Rs. in lacs)

Sheep Husbandry		1999-2000 Ramnagar			1999-2000 Akhnoor			2001-2002 Ramnagar			2001-2002 Akhnoor			Ramnagar Change 1999-2000 to 2001-2002		Akhnoor Change 1999-2000 to 2001-2002	
	Unit	*Phy.*	*Fin.*	*Fin. %*	*Phy.*	*Fin.*	*Fin. %*	*Phy.*	*Fin.*	*Fin. %*	*Phy.*	*Fin.*	*Fin. %*	*Phy. %*	*Fin.*	*Phy. %*	*Fin. %*
Rams	No.	0	1	20.6	0	0	0	359	0.14	3.867	0	0	0	359	-714.1	0	0
Sheep Units	No.	0	0	0	0	0	0	0	0	0	0	2	28.78	0	0	0	2
Health Cover (Migratory)	Ls.	0	2.55	52.6	0	0.74	25.2	15	2.5	69.06	0	1.99	28.63	15	-99.5	0	-35.2
Rams Feed	No.	0	0.5	10.3	0	0.4	13.6	0	0	0	0	0	0	0	-100	0	-100
Pregnant Eves Feed	No.	0	0	0	0	0	0	0	0.98	27.07	0	1.5	21.58	0	0.98	0	1.5
Dipping Vats/Tubs	No.	0	0.8	16.5	0	1.8	61.2	5360	0	0	4.4	1	14.39	5360	-100	4.4	-179
Dipping Troughs	No.	0	0	0	0	0	0	0	0	0	0	0.46	6.619	0	0	0	0.46
Total	0	4.85	0	2.94	5734	3.62	4.4	6.95	5734	-130.4	4.4-35.35						

Source: Physical and Financial Achievements, 1999-2000 and 2001-2002, IWDP, Hills-II, Jammu and Kashmir, 2003.

of livestock and helped improved the livelihoods of the poor, which will be evident from the following discussion.

Livestock Production and Development

The project has promoted livestock development by introducing artificial insemination techniques, which has resulted in hybridization of livestock with improved yield and consequent reduction in number of animals. This has significant reduced the grazing pressure on watershed. The system of stall-feeding has also been propagated, as most of the common property resources are enclosed for rehabilitation and regeneration. The project has also popularized the use of improved cooking stoves and LPG gas by demonstrating the benefits to the farming households. With more use of these alternative sources of energy, the pressure for on common property resources has been reduced significantly.

The artificial insemination programme has made a modest dent on crossbreeding. The semen bank and deep-freezing of bovine semen, as a part of this project, was an important landmark in the field of cattle breeding in rainfed area of Jammu and Kashmir. However, artificial insemination and frozen semen technology has been restricted to few villages, the cattle sheds for good quality bulls have not been properly managed and the bulls are underfed. On the whole, the breed improvement programme has been only moderately successful. There has been significant increase in milk yield, which may be attributed to higher productivity of crossbred animals. The yield rates of local milch animals remained stagnant, whereas, the yield rates of crossbred animals continued to increase at a rapid rate initially. The initiation of protection of village common lands and forestlands by creating enclosures and the poor usufruct sharing arrangements has resulted in marginal decline in the yield of even crossbred animals lately.

The growing economic opportunities for undertaking dairy farming as a commercial proposition combined with the interaction among the number of factors in agrarian rainfed economy, households have been progressively reducing their holding of drought animals and increasing their stock of milch animals. It has been reported that the requirement of work animals in the project area has been declining rapidly. The factors underlying this process are decline in the average size of cultivated holding, shift in cropping pattern, increase in the cost of rearing work animals, less availability of village common lands and common grazing lands, protection and closures of village common lands and forests. On the whole, the decline in the requirement of work animals and increase in the requirement of milch animals has resulted in significant changes in the composition of cattle population. While the adult male cattle population showed a sharply falling trend, the adult female cattle population has shown an increasing trend. Consequently, sex ratio of adult cattle has shifted in favour of females; the population of buffaloes has shown a declining trend. At the same time the rise in the profitability of milk production has resulted in an intensive selection process in the rearing and maintenance of cows for milk. Thus, while attempts are being made to rear the best young female to adulthood, the unproductive and low productive animals are eliminated from the herd by disposing them off in the market.

Table 3
Livestock and Yield (per household)

Item	*Unit*	*Project area with VDC*		*Project area without VDC*		*Non-project area*	
		Forested WS	*Agrl. WS*	*Forested WS*	*Agrl. WS*	*Forested WS*	*Agrl. WS*
Milch Animals	No.	2.8	3	3.2	3.6	4.2	4
Milch Yield							
Cow	Litre/per Animal	4.25	4	3.75	3.5	3.25	2.5
Buffalo	"	6.5	5.25	4.5	4	4	3.5
Sheep	No.	4.8	3.6	3.2	2.8	2	1.2
Wool Yield	Kg./per Sheep	3.1	2.4	1.5	1.3	1	0.8

The average number of milch animals per household and milk yield is given in table 3. The data presented in table makes it evident that average number of milch animals was reportedly more in forested watershed than in agricultural watershed across the sample area. In the non-project area, the average number of milch animals were reportedly higher than project area with VDC and without VDC, which clearly reflects that households in project area were rearing lesser number of milch animals than households in project area without VDC and non-project area. The differences in average number of milch animals in project area with VDC and without VDC were quite small than project area with VDC and non-project area. For instance, in the forested watershed, the difference in average number of milch animals in project area with VDC and without VDC was – 0.4, whereas in agricultural watershed, the difference in project area without VDC and non-project area was – 4. The table also reveals that average milk yield per animal across sample area. A mere perusal of the data given in table 6 makes it evident that milk yield was comparatively high in project area with VDC than project area without VDC and non-project area (see table 4).

The average number of sheep per household was comparatively high in forested watershed across the sample area. It is significant to note that average number of sheep were more in project area with VDC than project area without VDC and non-project area. Similarly, wool yield was as high as 3.1 kg per sheep in forested watershed and 2.4 kg per sheep in agricultural watershed in project area with VDC. In project area without

VDC, it was reportedly 1.5 kg and 1.3 kg respectively in forested and agricultural watershed, whereas, in non-project area, it was comparatively low and stood at 1 kg and 0.8 kg respectively (see table 4).

Table 5
Livestock Production Objectives by Sample Population

	Project area with VDC		*Project area without VDC*		*Non-project area*	
Production Objectives	*Forested WS*	*Agrl. WS*	*Forested WS*	*Agrl. WS*	*Forested WS*	*Agrl. WS*
Milk	No.	%	No.	%	No.	%
No	%	No.	%	No.	%	
Self	147	67.74	69	72.63	34	56.66
84	70.59	103	88.79	91	92.8	
Both (M & S)	70	32.25	26	27.37	26	43.33
35	29.41	13	11.20	7	7.14	
Total	217	95	60	119	116	98
Milk-Based Product						
Self	187	86.17	75	78.95	50	83.33
95	79.83	85	73.27	80	81.63	
Both (M & S)	30	13.82	20	21.05	10	16.66
24	20.17	31	26.72	18	18.36	
Total	217	95	60	119	116	98
Agriculture						
Self	230	100	84	100	53	100
119	100	116	100	92	100	
Both (M & S)	0	0	0	0	0	0
0	0	0	0	0	0	
Total	230	84	53	119	116	92

Note: M & S: Market and Self

The indigenous livestock breeds of the forested watersheds in *Shivaliks* have been exposed to natural selection for a very long time and are thus well adapted to harsh environmental conditions. IWDP (Hills-II) focus on new breeds under insemination programme at subsidized price. These changes weakened indigenous breeding system that could improve livestock hardiness. The crossbred livestock of better quality is replacing the local cattle, sheep and goat. Thus, after project intervention indigenous livestock diversity has declined precariously more in the forested sub-watershed of Ramnagar than Akhnoor and the decline is reportedly more in those of the villages where the VDCs are taking all types of decisions related to watershed development and protection. It is, thus, urgently needed to speed up the development of sustainable and long-term crossbreeding programmes.

The sampled population reared the milch animals to supplement household nutrition and earnings through the sale of milk and milk products in the village nearby townships. In the project area with VDC, a higher proportion of the milk is consumed within household and more than one-third and one-fourth of the milk is sold respectively in forested and agricultural sub-watersheds. The milk-based products such as cheese, ghee, kalari, etc. are sold in the market, but often at very low price due to undeveloped rural markets. The data given in table 5 also reveals that sampled population in non-project area sell a small proportion of the milk and milk-based products in market, due to low milk yields of the local milch animals kept by them. It has been reported that the project has distributed improved grass varieties for plantation in fields and field bunds, which have resulted in improved fodder availability. The creation of user groups have facilitated in usufruct sharing on more or less equitable basis and ensured better availability of fodder. The non-project area is reportedly scarce in fodder and village common lands and forestlands are severely degraded, however, with project intervention regeneration of the enclosed village common lands and forestlands has taken place in project area, which has increased the availability of feed and fodder.

Fodder is a very crucial biomass needed for maintaining the livestock. The immediate livelihood impact of creation of enclosures on village common lands, forestlands and grazing lands has been felt with regard to the grazing practices and the availability of fodder from common property resources. The enclosed common property resources are forbidden for open grazing. The cost of restrictions imposed on open grazing in the enclosed area is to be compensated by increased grass production to some extent; the major concern remains one of ensuring fair or equal distribution. However, the usufruct sharing is reportedly inequitable. Most households in fact facing fodder scarcity and resort to one or the other mean to fill the deficit. Under IWDP (Hills-II), Jammu and Kashmir attempts has been made to protect the village common lands, forestlands and grazing lands so that common property resources should be conserved and rehabilitated. Such change has been noticed in both the project area as well as non-project area (See table 6). However, the incidence of animal grazing on common property resources has declined comparatively more rapidly in project area than project area without VDC, and more decline in animal grazing has been reported in project area than non-project area. For instance, before project intervention, on average more than 7 animals per household were grazed on common property resources in forested watershed and about 6 animals per household in agricultural watershed. After project intervention, users groups have been created to protect and maintain the enclosed common property resources. Consequently, the average number of livestock grazed on common property resources declined to less than 4, and such a decline is in the range of 38 per cent to 52 per cent, the lowest in agricultural watershed and highest in forested watershed.

Table 6
Animal Grazing on Common Property Resources by Sample Population

	Project area with VDC				*Project area without VDC*				*Non-project area*			
	Forested WS		*Agrl. WS*		*Forested WS*		*Agrl. WS*		*Forested WS*		*Agrl.-WS*	
Grazing on CPRs	No.	% Change	No.	% Change	No.	% Change	No.	% Change	No.	% Change	No.	% Change
Before	7.65		6.0364		8.04167		4.1597		5.9297		5.605	
After	3.7	-51.63	3.7364	-38.10	3.77778	-53.02	3.1513	-24.24	3.3672	-43.21	3.6303	-35.23
Difference	-3.95		-2.3		-4.2639		-1.0084		-2.563		-1.975	
Grazing Days												
Before	365		365		365		365		365		365	
After	180	-50.68	190	-47.94	210	-42.46	220	-39.72	320	-12.32	245	-32.87
Difference	-185		-175		-155		-145		-45		-120	

Not only the number of animals grazed on common property resources has declined, but the number of animal grazing days on common property resources has also declined significantly. For instance, before project intervention, due to open access to common property resources, the animals were grazed throughout the year. Whereas, after the project intervention, the decline in number of animal grazing has been reported in the range of 48 per cent to 51 days in project area with institutional arrangement and 40 days to 42 days in project area without institutional arrangement. Similarly, a decline in animal grazing in the range of 12 per cent to 33 per cent has been recorded in non-project area. It is to be noted that recently the forest department has started enclosing the forests for regeneration and animal grazing has been banned. The creation of closures has resulted in heavy grazing pressure on non-protected and non-enclosed common property resources, which are facing severe degradation. Thus, there is urgent need to introduce new high yielding varieties of grasses and fodder, which could be planted in enclosed common property resources as well as private lands. Besides, more and more user groups need to be created so that equitable usufruct sharing mechanism should be established. The user groups would also be helpful in introducing the system of stall-feeding on wider scale.

The creation of closures on village common lands and forestlands has resulted in significant decline in the availability of fodder, which were freely available before project intervention. After the project intervention, the regeneration of common property resources has taken place, which has improved the availability of fodder significantly in forested watersheds than the agricultural watershed (See table 7). The villagers, who are unable to meet their fodder, small timber, opt for a deficit strategy. The data makes it clear that rotational use of forest resources has been in use only in forested watershed, where institutional arrangements has been made through formation of user groups. In other cases, use of buffer stock and illicit use from the local forests are the dominant deficit strategy adopted. It is to be noted that even after the project intervention, illicit use of fodder is most common in agricultural watersheds, where the villagers reportedly intrude the forests in the vicinity to cut the grass to meet the fodder requirements.

Contribution to Income

IWDP (Hills-II), Jammu and Kashmir intends to rehabilitate the degraded natural resources as a potential strategy for poverty alleviation and livelihood improvement. Besides, environmental rehabilitation and improving the productive potential of the *Shivaliks*, the project also intends to meet the fodder and other requirements of livestock. Thus, the objective of poverty alleviation emerges very clearly as a concomitant of the objective of environmental rehabilitation. There are wide variations in average

Table 7
Availability of Fodder and Deficit Strategy in Sample Sub-Watersheds

	Project area with VDC				*Project area without VDC*				*Non-project area*			
Fodder Availability	*Forested WS*		*Agrl. WS*		*Forested WS*		*Agrl. WS*		*Forested WS*		*Agrl. WS*	
in Month (s)	*No.*	*%*	*No.*	*%*	*No.*	*%*	*No.*	*%*	*No.*	*%*	*No.*	*%*
Below 2	70	30.04	15	13.63	16	22.22	30	25.21	29	22.66	46	38.98
Between 2-4	64	27.46	21	19.09	23	31.94	38	31.93	31	24.22	25	21.18
Between 4-6	58	24.89	73	66.36	8	11.11	44	36.97	33	25.78	10	8.47
Between 6-8	23	9.87	1	0.90	4	5.55	0	0	13	10.16	1	0.84
Between 8-10	7	3.00	0	0	2	2.77	0	0	1	0.78	0	0
Above 10	2	0.85	0	0	6	8.33	0	0	0	0	0	0
Deficit Strategy												
Purchase	22	9.44	0	0	2	2.77	32	26.89	1	0.78	0	0
Buffer stock	141	60.51	110	100	72	100	119	100	127	99.22	112	94.91
Illicit Use	137	58.79	110	100	5	6.94	119	100	127	99.22	112	94.91
Rotational Use	8	3.43	0	0	0	0	0	0	0	0	0	0

monthly income from farm and non-farm occupation across the selected sub-watersheds. The project has facilitated in operationalizing non-farm activities by provisioning the necessary training, skills and inputs to the beneficiaries. After the project interventions, the farm households have significantly improved their livelihood options and a large proportion of them are reportedly above the poverty line (APL).

Table 8
Annual Income from Farm and Non-farm Occupation in Sample Sub-Watersheds (Rs.)

	Project area with VDC		*Project area without VDC*		*Non-project area*	
Farm Occupation	*Forested WS*	*Agrl. WS*	*Forested WS*	*Agrl. WS*	*Forested WS*	*Agrl. WS*
Marginal Farmer	10668	11208	8808	9744	8100	8544
Small Farmer	15660	17304	13848	13872	11832	9888
Semi-medium Farmer	17340	18408	15864	14952	12240	10512
Non-Farm Occupation						
Dairy	5472	6420	2620	4020	1512	1032
Poultry	4140	3072	1740	1488	1152	672
Goatry	2820	2100	1152	2580	648	408
Bee-keeping	3048	3312	540	540	648	420
Other Occupation	10380	15360	10380	5184	4140	5088

The Jammu and Kashmir Government, which carried out the BPL census in 1997 and used the cut off for defining a household appearing as below the poverty line (BPL) of Rs 11,000 per household per annum. This was also the figure used by the Planning Commission in the Eighth Plan. This figure has been revised upwards by taking the Consumer Price Index (CPI) for Agricultural Labourers for Jammu and Kashmir (Monthly Abstract, Jan.-March 2001; December 2001; June, 1999). The average index for 1997-98 (twelve months) is 263. The average for 2001-02, for which we have income data, the average index (nine months) is 329.56 (1986-87 = 100). In other words, there was a 25.31 per cent increase in the price index. The adjusted income figure comes to Rs, 13,784.10 that have been used as the cut off for poverty line in 2001-02. Another problem concerns the average size of the household. The average size of the household in the sampled households is above 5 persons, which is higher than the national average. Since the Jammu and Kashmir government has used the income figure of the Eighth Plan for defining the cut off for the poverty line, the relevant household size is that of the national average rather than that of Jammu and Kashmir. In other words, an adjustment has to be made, as the household size in sample area is larger than the national household average. The figure of Rs 13,784.10 has been revised upwards to the same extent, as the household size in sample area is larger than the national average household size. This will give us the second estimate of the number crossing the poverty line. The second revised figure for the household crossing the poverty line is Rs 15,686.31 in 2001-02, which has been used in the present study.

It is to be noted that the project has provided various types of assistance for farm and non-farm activities mainly to the members of BPL families identified by the VDCs, panchayats and/or local leadership. The average annual income of the marginal farmers from farming occupation alone is comparatively less than the cut-off level of BPL across the selected sub-watersheds (See table 8). In the project area with VDC, semi-medium farmers in forested watershed and small and semi-medium farmers in agricultural watershed have reportedly crossed the poverty line after the project intervention. In the project area without VDC, the semi-medium farmers in forested watershed have improved their livelihood status after project intervention and are found in the category of APL. The small and marginal farmers are still living in poverty, if we assume that they are deriving their earning solely from farm activities. The poverty is widespread in non-project area. It is to be noted that care has been taken to include those of the BPL population who are beneficiaries of one or the other anti-poverty scheme operational in the area. However, as mentioned earlier, the project has benefited the poor by providing wage employment on watershed development and protection activities as well as in upgrading traditional non-farm activities such as dairy, poultry, bee-keeping, etc. The exact level of income from non-farm activities by farmers' category is not available. However, on rough approximation, the income from non-farming activities taken together with farm earnings ease the poverty situation, more specifically in those villages where participatory approaches has been evolved. For instance, in project area with VDC, dairy supplement the average household income to the tune of Rs. 6421 per annum and average income from other non-farm sources (predominantly wage employment through project intervention) is as high as Rs. 15,360 per year in agricultural watershed. Besides, poultry and goat rearing are also important non-farm income yielding activities in project area. Thus, the poor farmers are engaged in multiple activities to eke out a livelihood. One thing which is quite evident from the above is that poverty scenario have improved though modestly in the project area without VDC and significantly in project area with VDC.

Consumption of Fodder and Manure

The table 9 gives the data on average annual household consumption of fodder from common property resources. A decline has been reported in fodder collection after project intervention and the decline in fodder collection is reportedly very sharp in project area with VDC. In non-project area too, a marginal decline has been reported in fodder collection. It is to be noted that collection of fodder from CPRs has declined significantly over the period in non-project area, whereas a positive trend was visible for project area with VDC and without VDC. Over the period, the manure collection from CPRs has declined across the sample area; however, the decline is markedly high in non-project area than project area with VDC and without VDC. Thus, it can be inferred that with project interventions, the productivity of CPRs has increased very rapidly in project area with VDC followed by project area without VDC compared to non-project area.

Table 9
Average Annual Household Consumption and CPRs Products Collected by Sample Population

CPRs	*Average Annual Household Consumption (Rs.)*											
	Project area with VDC				*Project area without VDC*				*Non-project area*			
Products Collected	*Forested WS*		*Agrl. WS*		*Forested WS*		*Agrl. WS*		*Forested WS*		*Agrl. WS*	
Fodder												
Before	5399		4685		5320		5643		5434		4558	
After	6656	23.28	7215	54.00	5960	12.03	6756	19.72	4535	-16.52	4020	-11.80
Difference	1257		2530		640		1113		-898		-538	
Manure												
Before	400		435		356		170		249		286	
After	254	-36.5	234	-46.20	143	-59.83	67	-60.58	56	-77.51	79	-72.37
Difference	-146		-201		-213		-103		-193		-207	

Livestock Productivity and Yield

The table 10 is devoted to analyze the impact of the project on livestock productivity and yield. Except number of milch animals per household, all other items shows a positive difference between project and non-project area. This clearly reflects that over the project period, the livestock population has declined significantly in project area. Similarly, a significant differential in milk yield has been recorded between project area and non-project area. The difference in milk yield of cows and buffalo stood at 1.0 kg and 1.31 kg respectively. The difference in number of sheep per household and wool yield was also quite significant between project area and non-project area and stood at 2.5 sheep and 1.18 kg of wool respectively. Besides above, the fodder collection from common property resources has also significantly high both in monetary as well as volume terms.

Table 10
Livestock Productivity and Yield Project and Non- Project Area

Livestock Productivity/Yield	*Project Area*	*Non-Project Area*	*Difference*
Milch Animals (No./per HH)	3.15	4.1	-0.95
Milk Yield (kg) - Cows	3.88	2.88	1
Milk Yield (kg) - Buffalo	5.06	3.75	1.31
Sheep (per HH)	3.6	1.1	2.5
Wool Yield (kg.)	2.075	0.9	1.18
Fodder Collection (Rs.)	4446.75	4278	168.75
Green Fodder Collection (kg.)	371.25	216.5	154.75
Dry Fodder Collection (kg.)	35	18.3	16.7

Cost-Benefit Ratio

The benefit-cost ratio of livestock interventions is presented in table 10. The benefit-cost ratio is estimated at a high level of 23.32 in case of milch animals. The annual benefit accruing to the individual beneficiary owning milch animals was to the tune of Rs. 3798 per annum, whereas the project cost on livestock intervention was reportedly Rs. 162.86. The sheep rearing has been taken by transhumance as well as settled tribal population in a significant way and the livestock intervention in sheep rearing has resulted in a benefit to the tune of Rs. 236 per beneficiary. In case of sheep rearing, the benefit-cost ratio is estimated at 1.66. The other important intervention with significant benefit-cost ratio is on-farm fodder production. On the whole, the benefit-cost ratio of livestock interventions is estimated at 6.36, which is very significant. However, when we take into account the total average transaction costs incurred by the project implementing agency and individual beneficiary, the overall benefit-cost ratio falls. Thus, the minimization of transaction costs has a positive influence on enhancing the benefits from participatory watershed development.

Table 11
Benefit - Cost Ratio of Selected Project Interventions

Intervention	*Cost (Rs.)*	*Benefit (Rs.)*	*B/C*
Livestock			
Milch Animals	162.86	3798	23.32
Sheep rearing	141.89	236	1.66
On -form Fodder Production	415.88	547.75	1.31
Total	720.63	4581.75	6.36

Perceptions of the Farmers

Over the period, with the creation of village common lands and forest closures, and plantation of vetiver grasses on the private lands, fodder availability has increased, though modestly in the project area. The villagers have started replacing local cows and buffaloes with crossbred animals, which are stall fed and thus reducing the grazing pressure. The farmers perceived that with increase in fodder availability, off-seasonal income of those of the farmers, who derive some income from the sale of the fodder, has increased. Increased fodder availability also provided the farmers a cushion during drought period, which is more frequent in the *Shivaliks*. With the increased fodder availability from village common lands and forest closure, additional crop activities have been carried on agricultural lands, which were earlier kept for fodder production. Some of the farmers have also started rearing additional cattle, mostly cross-bred and some of them have taken to petty trading and engaged in handicrafts. However, such cases are few and far between. On the whole the cattle population has declined and the crossbred animals are replacing the local cattle (See table 12).

The project has also protected and enclosed the degraded village common lands and forests, which have resulted in decrease in free grazing area. The access to protected common lands has also been restricted, which has resulted in an increase in dry forage production. In the project area with VDC, the respondents reported that project interventions have reduced the free grazing area and free access to community land has curtailed, because the degraded common land has been protected and enclosed for regeneration. The protection and subsequent regeneration of common lands is reportedly very high in project area with VDC than in project area without VDC. In non-project area, the grazing lands as well as village common lands and forestlands are subjected to severe grazing pressure due to open access and their productivity has declined very significantly. It is to be noted that during the recent past, government agencies in non-project area have enclosed village common lands and state common lands for regeneration, however, due to non-participation of the local community in regeneration programme, these protected resources are even now more or less open access. The increase in dry forage production and more availability of green fodder has contributed significantly in improving the livestock health and productivity, which is reflected in higher milk yields, saving of time and introduction of system of stall feeding in the project area with VDC.

One-fourth of the respondents in forested watershed in project area with VDC perceived an improvement in livelihood related to natural resources after community participation, and 8 per cent perceived significant improvement in livelihood and 7 per cent were indifferent. Similarly, in agricultural watershed in project area with VDC, 39 per cent perceived significant improvement, 14 per cent perceived improvement and 35 per cent of the respondents perceive no improvement at all. It is to be noted that attempts has been made in non-project area also to involve local community in resource protection and conservation activities by forest department, which have improved the livelihood based on natural

Table 12
Perceptions Regarding Fodder Availability by Sample Population

Perceptions Regarding	Project area with VDC				Project area without VDC				Non-project area			
	Forested WS		*Agrl. WS*		*Forested WS*		*Agrl. WS*		*Forested WS*		*Agrl. WS*	
	No.	%	No.	%	No.	%	No.	%	No.	%	No.	%
Increase in Fodder Availability												
Yes	198	84.97	69	62.72	53	73.61	46	38.65	24	18.75	7	5.93
No	35	15.02	41	37.27	19	26.39	73	61.34	104	81.25	111	94.07
Total	233		110		72		119		128		118	
If Yes												
Significant	112	56.56	13	18.84	12	22.64	0	0	0	0	0	0
Moderate	47	23.73	22	31.88	23	43.4	6	13.04	9	37.5	2	28.57
Marginal	39	19.69	34	49.27	18	33.96	40	86.95	15	62.5	5	71.43
Benefits From Increased Fodder Availability												
Off-Seasonal Income	156	78.78	59	85.50	37	69.81	23	50	13	54.17	5	71.43
Drought Period Sustenance	163	82.32	52	75.36	43	81.13	33	71.73	8	33.33	3	42.86
Additional Crop Activities	107	54.04	65	94.20	36	67.89	40	86.95	11	45.83	5	71.43
Additional Cattle	167	84.34	36	52.17	37	69.81	24	52.17	7	29.17	4	57.14
Petty Trading/ Handicraft	23	11.61	9	13.04	8	15.09	4	8.69	2	8.33	3	42.86

Table 13
Perceptions Regarding Livelihood Improvement due to Participatory Natural Resource Management

Perceptions Regarding	*Project area with VDC*				*Project area without VDC*				*Non-project area*			
	Forested WS		*Agrl. WS*		*Forested WS*		*Agrl. WS*		*Forested WS*		*Agrl. WS*	
	No.	%	No.	%	No.	%	No.	%	No.	%	No.	%
Main Livelihood Related to Natural Resources After Community Participation												
Improved	60	25.75	16	14.55	11	15.27	12	10.08	0	0	2	1.69
Significant Improved	19	8.15	43	39.09	17	23.61	5	4.20	5	3.90	2	1.69
Not Improved	137	58.79	39	35.45	24	33.33	73	61.34	93	72.66	87	73.73
Can't Say	17	7.29	12	10.91	20	27.77	29	24.37	30	23.44	27	22.88
Total	233		110		72		119		128		118	
Change in Household Livelihood Strategy After Community Participation												
Yes	74	31.75	18	16.36	30	41.66	25	21.01	2	1.56	2	1.69
No	144	61.80	80	72.73	33	45.83	68	57.14	105	82.03	94	79.66
Uncertain	15	6.43	12	10.91	9	12.5	26	21.85	21	16.41	22	18.64
Total	233		110		72		119		128		118	
Household Level Awareness of Participatory Process												
Yes	178	76.39	75	68.18	44	61.11	58	48.74	8	6.25	14	11.76
No	36	15.45	33	30	26	36.11	53	44.54	110	85.94	87	73.11
Uncertain	19	8.15	2	1.81	2	2.77	8	6.72	10	7.81	17	14.29
Total	233		110		72		119		128		118	

resources though modestly. Besides improvement in livelihood related to natural resource management, about one-third of the respondents in forested watershed in project area with VDC perceived a change in livelihood strategy after community participation, whereas, only 16 per cent perceived such change in agricultural watershed. In project area without VDC, 41 per cent and 21 per cent of the respondents respectively from forested and agricultural watersheds perceived a change in livelihood strategy (See table 13). In non-project area, as very little efforts, if any, have been made for community participation in natural resource management, negligible change in livelihood strategy has been reported. The respondent's perceptions regarding livestock based income-generating possibilities have also been ascertained. In project area with VDC, poultry rearing followed by animal husbandry has been perceived as the high potential on-farm income generating activities in forested watershed. The animal husbandry and poultry rearing have been perceived as high potential income generating activities in agricultural watershed in project area with VDC and without VDC.

CONCLUSION AND RECOMMENDATIONS

The livestock intensification and diversification has occurred with watershed project interventions through technological transfers, access to animal husbandry extension services, improved access to local markets and subsidized inputs. The vegetation has increased modestly and the forestland cover has also increased. Vegetation management practices included planting trees in the enclosed forestland, village common lands and private agricultural and community land. In most of the cases, only degraded forests have been handed over to the user groups, as there is an informal rule within the IWDP (Hills-II) not to hand over well-stocked forests. Consequently, access to many community forests has been restricted temporarily in order to allow these forests to recover. As a result, many user group members have to rely on unprotected forest to meet their basic needs. In agricultural sub-watersheds in project area, there is severe fodder scarcity. Grazing cattle in community forest is prohibited to defined periods during the year. As a result, people rely heavily on unprotected forests and village common lands for fodder, which further add to the forest degradation. Livestock composition has changed substantially, however, the number of crossbred cows increased, but other livestock fell in number. An increase in milk production and productivity has been reported in project area. The decline in the sheep population was particularly steep, which was due to the fact that mostly landless and marginal farmers were keeping sheep and goats, which had free access to common grazing lands and wastelands. As a result of the project intervention, grazing was stopped on village common lands under regeneration, reducing the grazing area and depleting the forage potential.

In spite of a reduction in total area available for grazing, dry forage production increased significantly compared to the pre- project phase. The increase in dry forage production is mainly due to reseeding of field funds and hill slopes with grasses and legumes, and regenerating old rootstocks on hill slopes. The average dry forage productivity went up during the post-project phase.

The evidence on poverty alleviation effects is ambiguous: while all members of a user group benefit in the long run, the poor are more severely affected by restrictions on forest use in the short run. However, this is compensated by provisioning of wage employment on repair/construction activities on various project interventions. The creation of closures on village common lands and forestland has generated much needed wage employment though for a shorter duration. Furthermore, there is an equity problem between members and non-member of a user group. In the long run, every member of a forest user group benefits from project intervention in the form of closures on village common lands and forestland. However, in area where participatory institutions have not been created, the open access forest would decrease further. The degraded village common lands and forestland, which are temporarily closed and effectively protected, will yield more forest-produce in the future. In this sense, community forestry leads to a Pareto improvement for the members of a user group. In the short run, however, the poor have to suffer the most as the village common lands and forestland has been closed temporarily. The opposition to village common lands and forest closures has been noticed during the field visits. On the other hand, the better-off factions of a user group, e.g. land owners with fodder and forage on their own lands, do not oppose community forestry because they have alternative to forest use. At the user groups level, the heterogeneity and fragility of land resources along with the variable rainfall made it difficult to fully harness the potential of common property resources and adequately meet the environmental risks through private resources based on crop farming alone. The balancing of intensive (by cropping) and extensive (by pasture forest) use of land, as required by the resource characteristics became a part of collective strategies for risk management and production enhancement.

On the whole, lack of appropriate policies has constrained both the benefits and equity of livestock outcomes of watershed development project as well as their contribution to poverty alleviation. Additional equitable benefits may be achieved through adopting the following measures: (i) Inclusion of the crucial dimensions of livelihood dependence of poor on the livestock sector and the value of livestock assets in poverty reduction in the strategy for watershed development; (ii) Consideration of the access of poor and the remote backward villages to the infrastructure on livestock health services, marketing, breed improvement in watershed development planning and decision-making; (iii) Emphasis on developing village common lands, common property resources, private lands and productivity enhancement activities including diversification of crop patterns, fodder production in such a way that they pave way for high value crop-livestock integrated mixed farming as an IGA; (iv) Increased effort for controlled grazing and community management of village common lands and common property resources with clear usufruct rights regarding investments in biomass; (v) Assessment and inclusion of livestock fodder needs in watershed development planning and decision-making; (vi) Assessment of livestock drinking water needs and incorporating livestock watering ponds/points into the construction activities planned under the watershed development programmes; (vii) Inclusion of livestock holders using the water harvesting structures for drinking water for their livestock in water user's associations (WUAs) and making clear provision for drinking water for animals; (viii) Resource scarcity may introduce tradeoff between livestock and crop development, between upstream and downstream users, or between free grazing and stall-fed livestock, thus, adequate consideration must be given to the sustainability of land and water use practices in order to ensure inter-generational equity; (ix) Strengthening of panchayati raj institutions to play more effective role in watershed development and protection; and (x) Delineation of clear methodologies for action planning and sound and appropriate administrative procedures for effective and efficient planning, decision-making, monitoring, etc. to ensure sustainability.

REFERENCES

Batchelor, C., Rama Mohan Rao and Manohar Rao (2003), Watershed Development: A Solution to Water Shortages in Semi-Arid India or Part of the Problem?, *LUWRR*, (3): 1-10.

Bochet, J. J. (1983), *Management of Upland Watersheds: Participation of the Mountain Communities*, FAO Conservation Guide No. 8, FAO: Rome.

Chandrakanth, M. G., Bisrat Alemu and Mahadev G. Bhat (2004), Combating Negative Externalities of Drought — Groundwater Recharge Through Watershed Development Programme, *Economic and Political Weekly*, March (13): 1164-1170.

Conway, G. R. and J. A. McCracken (1990), Rapid Rural Appraisal and Agroecosystem Analysis, in M.A. Altieri and S.B. Hect (eds.), *Agroecology and Small Farm Development*, CRC Press: Florida.

Easter, K. W., J. A. Dixon, and M. M. Hufschmidt (eds.) (1986), *Watershed Resources Management: An Integrated Framework with Studies from Asia and the Pacific*, Westview Press: Boulder, Colo.

Farrington, J., C. Turton and A. J. James (eds.) (1999), *Participatory Watershed Development*, Oxford University Press: New Delhi.

Government of India (1994), *Guidelines for Watershed Development, Department of Wastelands Development*, Ministry of Rural Areas and Employment, New Delhi.

India Task Force (1987), Report of the Task Force to Evaluate Impact of Sheep and Goat Rearing in Ecologically Fragile Zone, Ministry of Agriculture, Department of Agriculture and Cooperation, New Delhi.

Integrated Watershed Development Project (IWDP), Hills-II (2001), *Physical and Financial Achievements, 1999-2000*, IWDP, Hills-II, Jammu and Kashmir.

Integrated Watershed Development Project (IWDP), Hills-II (2003), *Physical and Financial Achievements 2001-2002*, IWDP, Hills-II, Jammu and Kashmir.

Kerr, J. (2002), Watershed Development, Environmental Services and Poverty Alleviation in India, *World Development*, 30(8): 1387-1400.

Kurup, M. P. G. (2003), *Livestock Policy Synthesis*, Unpublished Report Prepared for the LEAD India Study, IWMI, Hyderabad.

Mangurkar, B. R. and C. Ravi Kumar (2001), Livestock-Environment Interactions in Watersheds in India, CALPI/SDC, New Delhi.

Pasha, S. M. (2000), *Economic and Ecological Dimensions of Livestock Economy*, Commonwealth Publishers: New Delhi.

Pereira, H. C. (1989), *Policy and Practice in the Management of Tropical Watersheds*, Westview Press: Boulder, Colo.

Puskur, R., Jetske B., and Christopher S. (2004), Sustainable Livestock Production in Semi-Arid Watersheds, *Economic and Political Weekly*, July 31.

Rao, Hanumantha C. H. (2000), Watershed Development in India: Recent Experiences and Emerging Issues, *Economic and Political Weekly*, November 4: 3943-3947.

Shah, Amita (2001), Water Scarcity Induced Migration — Can Watershed Projects Help?, *Economic and Political Weekly*, 36(35), September 1: 3405-3409.

Singh, Katar (1995), The Watershed Management Approach to Sustainability of Renewable Common Pool Natural Resources: Lessons of India's Experience, *Research Paper* 14, Institute of Rural Management, Anand.

Singh Katar (1997), Property Rights and Tenures in Natural Resources, in John M. Kerr *et. al.* (eds.) *Natural Resource Economics: Theory and Application in India,*: Oxford and IBH Publishing Co: New Delhi.

Singhal, C. S. (1999), Community Participation in Watershed Management, *Kurukshetra*, April, New Delhi.

Walker, T. and J. Ryan (1990), *Village and Household Economies in India's Semi-Arid Tropics*, John Hopkins University Press: Maryland, USA.

World Bank (1992), *World Development Report, 1992*, The World Bank: Washington.

World Bank (1998), *Project Concept Document: Integrated Watershed Development (Hills-II) Project*, Rural Development Sector Unit, South Asia Region, New Delhi.

World Bank (1999), *Project Appraisal Document: Integrated Watershed Development (Hills-II) Project,* Rural Development Sector Unit, South Asia Region, New Delhi.

World Commission on Environment and Development (WCED) (1987), *Our Common Future,* Oxford University Press: Oxford.

5

BIOPHYSICAL GAINS OF PARTICIPATORY INTEGRATED WATERSHED DEVELOPMENT: EVIDENCE FROM JAMMU AND KASHMIR#

Falendra K. Sudan

INTRODUCTION

Integrated watershed development is the process of formulating and implementing a course of action involving natural and human resources, which includes establishing watershed-management objectives; formulating and evaluating alternative resource management actions involving various implementation tools and institutional arrangements; choosing and implementing a preferred course of action; and thorough monitoring of activities and outcomes, evaluating performance in terms of degrees of achievement of the specified objectives (Easter, Dixon and Hufschmidt, 1986). Watershed development practices are changes in land use, vegetative cover, and other non-structural and structural actions to achieve well defined objectives (Pereira, 1989). The *World Development Report* (1992) stresses the importance of participation in creating effective programmes and viable resource-management institutions.

The Guidelines on Watershed Development (Government of India, 1994) marked a significant step towards approaches that are participatory and involve a high degree of decentralized decision taking and allocation of funding. Institutionally sustainable rehabilitation and management of watershed development project is possible only where stakeholders/ beneficiaries act jointly to manage the resources. Watershed development

---

This paper is a part of the research project titled "Participatory Approaches and Environmental and Economic Impact: With Special Reference to IWDP, Hills-II, Jammu and Kashmir, India", completed in May 2003 and sponsored by Ministry of Environment and Forests, New Delhi and World Bank funded "Environmental Management Capacity Building Technical Assistance (EMCaB) Project" managed by Environmental Economics Research Committee (EERC), Indira Gandhi Institute of Development Research (IGIDR), Mumbai.

is essentially a resource-base approach to biophysical and livelihood enhancement. Unless adequate safeguards can be built in, the danger is that, as the commons become more productive, better-off farmers are tempted to take control of them and customary access rights of the poor are denied (Farrington *et al.*, 1999). India's experience with various watershed development projects shows that the participatory approach could ensure sustainable use of renewable common pool natural resources (Singh, 1995).

Integrated watershed development is a top priority in the agenda of the Government of India as well as the World Bank to address the issues of resource conservation, poverty alleviation, equity, gender issues, environmental protection, and overall improvement in livelihood conditions of the rural population. Long-term solutions to rain-fed agriculture lie in the conservation and efficient utilization of watershed resources (Jain, 1997). Some efforts have been made in this direction under the World Bank funded Integrated Watershed Development Project (IWDP), Hills-II, Jammu and Kashmir. The project intends to rehabilitate the degraded common property resources (CPRs) as a potential strategy for poverty alleviation and livelihood improvement. Besides, biophysical rehabilitation and improving the productive potential of the *Shivaliks*, the project also intend to meet the fuel wood, fodder, minor forest produce and other requirements of the rural poor inhabiting the project area. Thus, the objective of poverty alleviation emerges very clearly as a concomitant of the objective of biophysical rehabilitation. Keeping above in view, in the present paper, an attempt has been made to assess the biophysical gains of participatory integrated watershed development in the context of IWDP (Hills-II), Jammu and Kashmir.

STUDY AREA

The state of Jammu and Kashmir with an area of 1,38,214 km2 comprising the uppermost drainage of Indus, Jhelum, Middle Chinab and Western Rabi exhibits great contrast in relief features, climate, soil and vegetation within comparatively narrow geographical spread. The climate of lower plains of Jammu and Kashmir that merges with plains of Punjab is sub-tropical. The climate is predominantly sub-tropical in the main *Shivaliks*. The climate in the region covering upper reaches of Pir-Panjal forming the main mountain backdrop changes from sub-tropical in the lower reaches to moist-temperate in upper reaches. The climate of Kashmir Valley (bordered between the Great Himalayas and the Pir-Panjal is dry temperate. The great Himalayan range: the innermost line of high mountains with Ladakh situated in trans-Himalayan zone is a cold desert and experiences sub-zero winter temperature.

IWDP, Hills-I was started in 1990, and concentrated on integrated development of high priority zones of rainfed areas. The total area of *Shivaliks* and *Karewas* is 9.45 lakh Ha. (*Shivaliks*: 7.50 lakh Ha and *Karewas*:

1.95 lakh Ha); and about 5.9 lakh Ha. is identified as problem area, out of which, 0.52 lakh Ha. (8.81 per cent of the problem area), have been treated under IWDP (Hills-I). This includes the sub-watersheds of Devak and Ramkote in *Shivalik* hills (Jammu region) and Dudhganga in *Karewas* (Kashmir Valley). IWDP (Hills-II) targets another 0.61 lakh Ha. (10.33 per cent of the problem area), leaving balance of 4.77 lakh Ha (80·86 per cent of problem area) still untouched. The principal lessons learnt from IWDP (Hills-I) were: the need to develop an integrated approach to the *Shivalik* Watersheds; the need for a coordinated approach to the use of available funds from all sources; the need to involve stakeholders in planning, implementation and maintenance; the need to increase stakeholder awareness of environmental and socio-economic considerations in articulating communities' needs; the need to focus on measures to arrest soil erosion and promote *in-situ* moisture conservation from ridge to valley; the need to assess marketing prospects, especially of horticultural products; the need to improve infrastructure within watersheds, particularly rural links to markets, water harvesting and drinking water; the need to develop monitorable project implementation and development objectives' indicators, and the need to involve local NGOs in the project as facilitators and trainers in the planning and implementation of the project (World Bank, 1998).

IWDP, Hills-II, Jammu and Kashmir covers two sub-watersheds in *Shivaliks*, viz. Ramnagar and Akhnoor and two sub-watersheds in *Karewas*, viz. Rajwar and Rambiyara. The proposed study will be confined to two sub-watersheds of Akhnoor (Jammu district) and Ramnagar (Udhampur district). While Ramnagar is in the inner *Shivaliks*, Akhnoor is on its outer part. Ramnagar sub-watershed is the catchment area of Ramnagarwali *Khad* (gorge) in the middle catchment of river Tawi. It has an area of 32,630 Ha. and is subdivided into 39 micro-watersheds. Akhnoor sub-watershed with an area of 42350 Ha. is sub divided into 37 micro-watersheds. It covers about a dozen rivulets (*nallahs*) and a large number of small nallahs originating from *Kalidhar* ridge and draining into Chinab river on the western portion of *Kalidhar* ridge, drains join Manawar Tawi which in turn

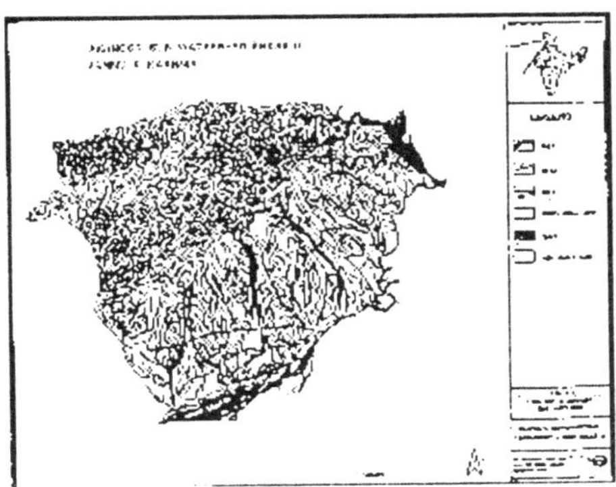

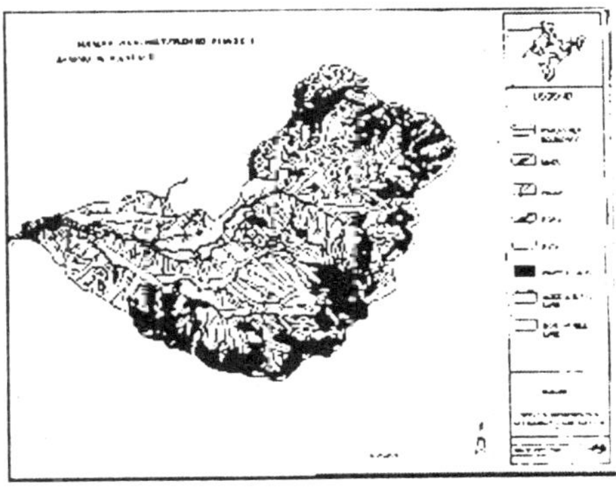

also meets the Chinab river. Following are the objectives of IWDP (Hills-II), Jammu and Kashmir: to restore on a sustainable basis the productive potential of the region and improving the quality of life; to reduce soil erosion and improve availability of water; to help increase production and income, grain crops, horticulture, fodder, fibre, fuelwood, livestock and household based products; to promote holistic and sustainable agro-ecological development involving people's participation; to strengthen community participation, and to develop local level institutions to enhance the sustainability of the model (World Bank, 1999).

The *Shivaliks,* locally known, as *'Kandi'* comprise of piedmont deposits made of boulder and pebbles, gravel and sand with minor clays mixed in varying proportions. One of the characteristic features of the area is the stream called *choes* or *khads,* which remain dry for most of the year. Though total rainfall is high (average 1000-1500 mm.), but its distribution is very erratic resulting in frequent droughts. This area is subject to soil erosion due to undulating topography, steep slopes, poor vegetative cover (scrub forest) and coarse to medium texture of the sedimentary material. *Shivalik* hills consist of highly erodible sand stones, conglomerates, siltstones and shales.

The uncontrolled deforestation and overgrazing in the past has resulted in the reduction of vegetative cover and accelerated erosion in the *Shivaliks.* The area experiences paucity of water for plant growth due to excessive run-off, high evaporation especially during summer, and unpredictable and erratic rainfall. The eroded material from *Shivalik* hills, brought down by the seasonal rivulets *(choes)* is deposited in the sloping piedmont plain and also in the area around *choes.* The repeated deposition of coarse sediments renders these areas comparatively low in agricultural productivity. Due to lack of irrigation, subsistence rainfed agriculture is the prevalent production system is selected sub-watersheds. In terms of physiographic features, there may not be 100 per cent similarity between the two sub-watersheds but, certainly, in term of accessibility, natural resource endowment, infrastructure development and general index of socio-economic awareness, two sub-watersheds are comparable. IWDP (Hills-II) is spread over predominantly rural areas. The indigenous population in the sub-watersheds is Scheduled Tribes (STs). The Scheduled Castes (SCs) are other disadvantaged social groups. In total, the project area is dominant in terms of SC population and ST population is negligible. Other social groups in the project area are nomads. The predominant economic base of the project area is primary sector activities. Agriculture is the mainstay of the rural economy, as over two-third of the population is engaged in agriculture and allied activities.

METHODOLOGY

During the Khon Kaen Conference (1985), different types of Rapid Rural Appraisal (RRA) techniques were identified, one of which was labeled "participatory". The orientation of a participatory rural appraisal (PRA)was to facilitate or stimulate community awareness and capability regarding a problem or issue. Particular attention was given to enabling local people to conduct their own analysis of problems, and to share their findings. The role of the outsider became one of catalyst, rather than one of expert. Stimulation of community awareness and capability was also intended to reduce the extractive nature of RRA, and to help local people to empower themselves. In that regard, PRA is consistent with some of the basic aspects of sustainable development (local empowerment, equity, social justice) (Conway and McCracken, 1990). The primary data has been collected using well-structured and pre-tested questionnaires, participatory rural appraisal (PRA) techniques and group meetings. Whenever possible and required, secondary data sources have also been utilized.

Through participatory observations and focus group discussions, it is possible to describe what goes on, who and what is involved, when and where things happen, how they occur, and why-at least from the stand point of participants-things happens as they do in particular situations. The methodology of participatory observation and focus group discussions has been used to ascertain important differences between the views of insiders as opposed to outsiders. The participant observation inquiry has been helpful in an open-ended logic and process of enquiry as well as qualitative description of the phenomenon under study. Both the unfocused and focused observation techniques have been used. The unfocused initial has been use to become increasingly familiar with the insider's world to refine and focus subsequent observation and data collection. All the observations have been recorded on site and misunderstanding, if any has been corrected thereof. In all the focus-group observations, 10-12 stakeholders have been recruited from different settings. Highly formal interviews have been conducted using structured interview schedules. Besides, study has been supplemented by utilizing potentially rich sources of secondary data and research materials. The content analysis technique has been used to analyze the data and information qualitatively and quantitatively (using descriptive statistics). The content analysis technique has been supplemented by use of code and labels, field notes, sorting, shifting, constructing and reconstructing these materials. Before entrusting the task of primary data collection, the research investigators have been given proper orientation in PRA techniques, focus group discussion and the various aspects of the problem under investigation. The household survey has been drawn on a stratified random sample of village household. The questionnaire was pre-tested before the main survey with small 'focus' group assembled to discuss their reactions to questionnaire prior to detailed survey.

For the collection of primary data and information on biophysical aspects at household level, three villages each in project area with VDC and project area without VDC from the two sub-watersheds of Akhnoor and Ramnagar, where a maximum number of people are known to have benefited from each project intervention have been selected. In the non-beneficiary category, three sample villages each have been selected on random basis, each 20-25 km away from the sampled micro-watersheds villages (*i.e.*, project area). Therefore, a total of eighteen villages (six each from 'project area' and three each from 'non-project area') have been selected. The stratified sampling technique has been used to select villages. During the stratification, care has been taken to include both "forested watersheds' villages and agricultural watersheds' villages" in order to makes a comparative study. A purposive sample of approximately 20 per cent of household level respondents in 'project' and 'non-project' area has been selected as the base of enquiry. The sample size was restricted to 376 and 404 households in sub-watersheds of Ramnagar and Akhnoor. Out of which, 185, 119 and 72 households; 158, 72 and 174 households belong to project area with VDC, project area without VDC and non-project area respectively in Ramnagar and Akhnoor.

LITERATURE REVIEW

As watershed issues involve long-term benefits and trans-boundary problems, the success of watershed development programmes depend on factors influencing the adoption of soil conservation practices (Anderson and Thampapillai, 1990; White 1992). The factors influencing the adoption of soil conservation are type of the technologies, knowledge of the farmers and policy incentives. Farmer's knowledge about various technical and economic relationships is crucial for adoption of conservation practices. Further, as the benefits are not immediate and only specific to site, their cooperation is extremely important. Therefore, apart from the physical linkages of soil, water and biomass, the socio-economic and political conditions play an important role in the success of watershed programmes.

Besides physical linkages, issues relating to land tenure, economic and social institutions and culture of watershed inhabitants become crucial in designing watershed projects. Brooks *et al.* (1990) state that though physical linkages cannot be excluded "the practical means of achieving sustainable projects in watershed management, conversely, can not ignore land tenure, institutions and the culture of watershed inhabitants". Blackie (1983 and 1985) has also emphasized the importance of social factors such as institutions, political and economic. He argues that the politics of the state and society (agrarian relations) are central to soil conservation. An understanding of how and how far the state will implement soil conservation policies require a careful analysis of whose interests are represented in the various institutions involved. Secondly, the analysis of the politics of agrarian society is another essential ingredient into a realistic appraisal of what is possible and what is not. White (1992) also argues that watershed should be considered as sets of vested interests (and social relations). Actors with vested interests within watershed are interdependent because water-flows across private as well as political boundaries. The achievement of watershed management, therefore, is a question of social relations, and cooperation between vested interests. These vested interests are separated by political boundaries or institutional arrangements, which rarely coincide with the topographic limits of watershed.

Studies emphasized participation of these vested interests as crucial to the success of soil conservation efforts in watershed. White and Quinn (1992) demonstrated that peasant participation in Integrated Watershed Management Project was economically viable in Haiti. White (1992) argued that for soil conservation to be adopted and sustained, it must be an extension and incremental transformation of existing cultural and technical behaviours. This transformation cannot be forced by external agents but can be achieved by dialogue with peasants. Numerous studies were undertaken to examine the biophysical impact of watershed development programme in India. Singh (1989) reviewed the experience of four watershed projects: Mittermari and PIDOW in Karnataka, Ralegaon-Siddhi in Maharashtra, and Sukhomajri in Haryana. The review reiterates that programme interventions seeking to enhance the expected benefits to people, or reduce the expected costs, are likely to elicit stronger people's participation. Other determinants are organisation of people into small groups (as in PIDOW) and leadership (as in Maharashtra), equitable sharing of benefits from collective action (as in Sukhomajri), and availability of complementary investment from government.

Chopra *et al.* (1990) reported that in the lower *Shivalik* region of northern India, the use of participatory approaches resulted in moderate to high rates of return to watershed management, greater than 12 per cent, the cut-off social rate of discount usually adopted by the Indian Planning Commission. Further, the incremental income accrued to different set of beneficiaries. Between households and village society, the project region shared about 71-82 per cent of net benefits. Externalities accruing to the government comprised 18-29 per cent of net benefits. This estimation of benefits to the project region has allowed for the direct and indirect impacts of participatory institutions on incomes of the people as well as of government agencies. Singh *et al.* (1991) studied the *Kandi* watershed, an economically backward tract of Punjab. The different components of this project and their associated rates of return were forestry (12 per cent), animal husbandry (9 per cent), soil conservation (10 per cent), horticulture (22 per cent), and irrigation (2 per cent). The overall project internal rate of return of 8 per cent represents a substantial shortfall from the earlier expectation of 12-20 per cent according to the feasibility studies. However, the internal rate of return and benefit cost ratios are 38 per cent and 2.23 for *Kinnow* (a type of citrus fruit), 26 per cent (2.48) for mango and 44 per cent (2.30) for guava. An increase in cultivated area is reported from 80.3 to 99.3 per cent, while the irrigated area increases from 1.4 to 48.9 per cent.

Deshpande and Ratna Reddy (1991) selected three micro-watersheds falling in three regions of Maharahtra to study the locational aspects and the consequent difference in technologies adopted. The watershed in the scarcity zone had land development, horticulture, contour bunding, key line formation and *nala* training as the main soil and water conservation activities. In the second watershed with modest rainfall, contour and graded bunding, *nala* bunding and training, afforestation and grass planting received prominence among the components of watershed management. In the last category of watershed with assured rainfall, drainage, water logging and waterways receive prominence. However, in the watershed treated areas, there is shift towards pulses and oil seeds. The significance of commercial crops also remains in all three locations. The analysis suggests that irrigation through water harvesting had an impact on total value productivity in both scarcity and moderate rainfall regions. They argue that development of each project is unique in itself, though certain traits are common. The study also made a comparative analysis of the cases of active beneficiary participation as against passive participation. It was noted that participatory process acts as a powerful catalyst for the programme, a result supported by Chopra *et al.* (1989).

Rajagopalan (1991) contended the integrated watershed development is a system combining erosion and run-off management with land management and the digging of irrigation wells for lifting ground water on a sustainable basis such that the amount of water withdrawn is less than or equal to the annual recharge of groundwater. The impact of integrated watershed development can be seen in improvement in resource productivity, increase in employment, better crops and crop system which ensure continuous sales and regular cash flow, additional area under sustained irrigation and cropping and reduced production risks. He argues that investment in watershed management is an appropriate development intervention, which warrants top priority as 70 % of net sown area suffers from neglect and poverty. Ramanna (1991) contended that the farmers were convinced of the benefits of programmes of watershed management both on arable and non-arable lands in the form of rise in water table level in wells and reduced siltation of tanks. Establishment of micro-watershed *sanghas* in Karnataka has demonstrated how an NGO can fruitfully involve them in watershed development.

Shah (1997) has examined watershed programmes in western Gujarat. A yield gain of 20-30 per cent is expected with an average investment of about Rs. 6000 per hectare. In terms of the acceptability of different components of watershed technology, she reports that in dynamic rainfed region returns from capital-intensive measures such as land leveling, water harvesting and commercial plantations are likely to be higher and are also likely to evoke more interest and participation. *In situ* conservation measures by themselves (including vegetative barriers) may not succeed unless accompanied by measures that are perceived to be income enhancing. She also maintains that farmers need appropriate technology, even if it is not low cost. Shah (1998) drawing on three case studies (all benefiting from special management funding conditions) doubts the efficacy of watershed-based approaches unless topography and climate are appropriate.

Kerr *et al.* (1998) reporting on a major survey of projects (AGY, NWDPRA, DPAP and IGWDP, etc.) came to the conclusion that the benefits of watershed development have been negligible, except in a few cases in which participatory approaches have been pursued effectively. The projects- NWDPRA and DPAP follow strict guidelines and are evaluated on the physical targets achieved, which limit the scope for participation by inhabitants of the project area. Of course, the DPAP and World Bank have already changed their approaches. However, Kolavalli (1998) examined the World Bank projects in Rajasthan and Orrisa, and identified continued problems with their performance. In particular, the approach to the choice of technology is still too inflexible, with little or no room for input by farmers, and there is insufficient attention to developing and strengthening community institutions to support collective action. The DPAP underwent a radical change in 1995, but the state and district level administrations have found it very difficult to implement the new guidelines. In fact, few if any, new DPAP projects had made any progress in Maharashtra.

Sadhu and Sudan (1999) have evaluated the impact of IWDP (Hills-I), Jammu and Kashmir. Over the project period, the yield per acre of various crops has increased due to greater use of improved inputs; however, their use is constraints by availability of credit. The development work related to construction of bunds and crate-work has evolved all round criticism. The selection has not been need-based but influence based and resulted into areas prone to soil erosion to degrade further. Due to increase in fodder and water availability, there was considerable increase in milch cattle and thereby increase in dairy activity. The increased agricultural activity, non-farming and horticultural plantation have resulted in reducing labour migration. With the formulation of a few Micro-Watershed Committees (MWCs), the system of people's participation in watershed management has been introduced during the later years of the project. The record of MWCs' functioning is rather disappointing. The women participation is low, veiled and passive.

Shah (2000) showed that the incidence of drought was much less severe in watershed villages when compared to the adjoining non-watershed or 'control' villages. The study concluded that the "overall impact is not only positive but also impressive .... to a large extent the participatory watershed scheme launched in 1995-96 by Ministry of Rural Development has been found to mitigate the impact of drought. However, the study pointed out that the activities and enterprises, which specially affect women such as drinking water and dairying tended to receive less attention. Further, the study observed that the expenditure on drought relief being incurred now be spent more productively by allocating it for watershed development with very little extra expenditure.

Hazara (2000) argued that the entire area of Kharaiya Nala watershed, both arable and non-arable lands were treated with appropriate soil and water conservation measures. Appropriate crop production technology was adopted for arable land, whereas, the pasture, silvipasture and afforestation programmes were taken up in the hills and plateau lands. This has resulted in bringing back about 317 hectare of wastelands into cultivation and helped in arresting run-off and soil loss from 48 per cent to 18 per cent and 20.5 tonnes/hectare to 0.9 tonnes/hectare, respectively, which resulted in generating irrigation sources from initial 10 per cent to 83 per cent cropped area under irrigation. All these have adequately reflected in terms of crop productivity, an increase by 6 times over initial level, 50 per cent shortage of fodder initially to 16 per cent surplus, doubled the milk yield, increased fuel wood availability to 85 per cent of requirement from 11 per cent availability and significant increase in farm income over initial level. Micro planning on watershed basis had helped in creating awareness and involvement/participation by villagers through creation of village level institutions in the overall management of regenerated community resources.

Wankhede and Mahalle (2000) emphasized that the success of the watershed development programmes lies in the "bottom up approach". Special efforts are made to involve local leaders, villagers and developments in "watershed literacy compaign". The impact of the programme is significant in terms of soil conservation and soil forestry activities. The extensive soil conservation and minor irrigation activities have created additional irrigation potential, which helped farmers in changing the cropping pattern in favour of cash crops. Soil conservation works have contributed in conserving soil and re-charging groundwater. Also, groundwater table has increased substantially. Consequently lands, which are higher and used to remain barren, were able to grow minor millets and pulses. Rajendran and Mashar (2000) contended that under the programme "Rural Infrastructure Development Fund, Phase-I", implemented by NABARD in the state of Kerala, soil and water conservation measures have brought about perceptible improvements in productivity of crops leading to increases in overall production. Consequently, the income generated has also shown substantial improvements. Due to policy restriction in providing assistance to farmers depending on income/area limitation, the entire area of the watershed could not perform as expected. Positive improvements have been recorded for cropping intensity, crop productivity and income.

WATERSHED DEVELOPMENT AND PROTECTION: ACHIEVEMENTS

The watershed development and protection component of the project aim at promoting proven locally adopted technologies and mechanical structures through active beneficiary involvement to conserve water and reduce soil erosion. Broadly, the treatments undertaken in the project components are: (i) Activities for arable land including contour vegetative barriers, terrace repairs and vegetative reinforcement, vegetative field boundaries, silvi-pasture, rainfed horticulture, farm forestry, on-farm fodder production, rainfed horticultural demonstrations and rainfed cropping system demonstrations. (ii) Activities for private non-arable, common and forestland including vegetative/shrub barriers in contour trenches, pasture development, silvi-pasture, afforestation, drainage lining, gully stabilization, with wire crates with vegetative support, stream bank protection, water harvesting structures and village ponds, roadside erosion control and landslide treatment. (iii) Other activities including livestock improvement and animal husbandry practices, rural infrastructure development (roads, buildings and marketing centre). All these activities are being promoted by using a variety of methods accomplished by strong stakeholders' participation including a number of mechanical and vegetative measures for watershed development and protection, improved livestock management, rural road rehabilitation and institutional strengthening (IWDP, Hills-II, 2001).

The project is being financed with the World Bank assistance of US $ 39.80 million equivalent to Rs. 198.00 crores. Accordingly annual action plans for the year 1999-2000 and 2000-2001 were formulated and approved by the Project Steering Committee to the tune of Rs. 25.00 crores and 40.00 crores respectively. Against the cumulative releases and beneficiary share the amount comes to Rs. 69.29 crores ending March 2001, out of which an amount of Rs. 66.67 crores stands incurred inclusive of beneficiary contribution, on broad range of activities undertaken. More emphasis has been paid on the cost-effective vegetative technologies under various sectors like agriculture, rainfed horticulture, silvi-pasture, afforestation in forest and village common lands (VLCs), and drainage line treatment. During 2001-2002, Rs 60.00 crore has been released and a target of 17836 hectares of land treatment under arable and non-arable category has been envisaged. Besides, drainage line treatment, erosion control measures, rural roads, building, potable water supply, livestock development, institutional support and institutional strengthening programmes have also been initiated (IWDP, Hills-II, 2001).

Arable Land Treatment

The state of Jammu and Kashmir is primarily an agricultural economy. About 61 per cent of its population is dependent on agriculture to eke out their livelihood directly or indirectly, contributing over 40 per cent to State Domestic Product (SDP). The State is mostly hilly, crisscrossed by lofty mountains and having low industrial base. The State has large proportion of population living in remote, inaccessible, isolated and economically backward region. The complete dependence on natural resources has forced the people to over-utilize the watershed resources, thus resulting in large-scale deforestation and ecological degradation. Watershed development and protection is the immediate need to restore the productive potential of the ecologically fragile *Shivaliks* and *Karewas*. The agriculture component of IWDP (Hills-II) is an important aspect in consistence with the objective of the project, contributing largely in achieving the goal of poverty alleviation through ecological improvement in the selected sub-watersheds.

The technical and non-technical inputs provided under the project includes: (i) provisioning of fertilizers, seeds and pesticides for rainfed demonstrations on cost-sharing basis; (ii) training and demonstration specifically organized for women; (iii) continuous surveillance by extension staff; and (iv) supply of improved implements, seed bins and containers to the beneficiaries. Most of the farmers living in the selected sub-watersheds have received input at least once on farm rainfed demonstration. These demonstrations have successfully promoted the introduction of improved varieties, seed treatments, appropriate application of fertilizers and pesticides agriculture implements and other improved agronomic practices to ensure moisture conservation.

Likewise, horticulture plays an important role in alleviating the social and economic conditions of the people living in hilly and rainfed areas of *Shivaliks*. The horticulture sector provides better potential to increase productivity and employment and thus rural livelihood diversification is the ultimate outcome. Under IWDP (Hills-II), the activities initiated in this sector includes: rainfed horticulture demonstration; supply of plant material, which are drought resistant having versatile adaptability; introduction of new varieties and species of fruit plants; step-up of rejuvenation programme and improvement of top-working, budding and grafting of existing fruit plants of indigenous type; training and demonstration for treatment against diseases; and supply of spraying and pruning equipment on cost-sharing basis.

The performance of interventions made under the component of arable land treatment in agriculture and horticulture sectors is highlighted in tables 3.1 and 3.2. Prior to project intervention, the small and marginal farmers were reluctant to use improved technical know-how. Large proportions of farmers were using conventional agricultural production methods. The major impediment in the introduction of high yielding varieties (HYV) of crops was non-availability of sufficient moisture at the time of sowing. The implementation of project interventions related to soil and moisture conservation has facilitated the farmers in timely sowing and using adequate doses of chemical fertilizers. Various training camps have been organized to disseminate the scientific technology from laboratory to the farmers' fields. These training camps have been organized in the selected sub-watersheds in collaboration with agricultural universities in the State of Jammu and Kashmir. The exposure visits have also been organized and the farmers drawn from across the selected sub-watersheds were exposed to latest agricultural technologies in *Krishi Mela* at New Delhi, where these farmers had ample opportunities to interact with farmers from other parts of the country.

During 1999-2000, 180 hectares and 160 hectares of arable land have been exposed to rain fed crop (R. F. C.) demonstration respectively in the sub-watersheds of Ramnagar and Akhnoor. In the year 2001-2002, a very rapid increase has been recorded in R. F. C. demonstration, which stood at 280.6 per cent in Ramnagar and 87.5 per cent in Akhnoor. The off-farm fodder (O. F. F.) production has received modest attention initially in Ramnagar, where 60 hectares of arable land has been covered, whereas, in Akhnoor O. F. F. production has received scanty attention, which may be due to better availability of fodder from existing sources/farmers' own land. However, O. F. F. production has received no attention in Ramnagar and very negligible attention in Akhnoor in the year 2001-2002 (see table 1). The vegetative field bunding has received the much-needed attention of the project. Initially, 100 hectares of arable land has been treated for vegetative field bunding in Ramnagar sub-watershed and the figure for the sub-watershed of Akhnoor stood at 160 hectares. In the year 2001-2002, the coverage has been doubled in both the sub-watersheds, which have contributed significantly in improving soil moisture quality and protected the erosion of agricultural field bunding. The terrace repair has also been given adequate attention. The system of agro-forestry has been implemented in both sub-watersheds, however, greater attention has been paid in Akhnoor, as it is predominantly agricultural watershed, where common property resources such as village common lands (VCLs) and other grazing and pasture lands were severely degraded before project intervention.

The silvi-pasture activities have been implemented initially, but not laterally. The system of mushroom cultivation has been introduced in the sub-watershed of Akhnoor in a significant way, whereas the climate of Ramnagar is more conducive for mushroom development, where it has not been attended in desired way. Besides above, improved tools and implements have also been provided to the poor farmers on cost-sharing basis in the sub-watershed of Akhnoor, whereas the provisioning of modern agricultural tools and implements has not been made in Ramnagar. Actually the sub-watershed of Ramnagar is predominantly forested and the poor farmers mostly use self-made wooden implements in agricultural

Table 1
Achievements of Watershed Development and Protection (Arable Land: Agriculture) in Sample Sub-Watersheds (Rs. in lacs)

		1999-2000						2001-2002						Ramnagar Change 1999-2000 to 2001-2002		Akhnoor Change 1999-2000 to 2001-2002	
		Ramnagar			Akhnoor			Ramnagar			Akhnoor						
Component	*Unit*	*Phy.*	*Fin.*	*Fin. %*	*Phy.*	*Fin.*	*Fin. %*	*Phy.*	*Fin.*	*Fin. %*	*Phy.*	*Fin.*	*Fin. %*	*Phy. %*	*Fin.*	*Phy. %*	*Fin. %*
Arable Land																	
Agriculture																	
R.F.C Demonstration	Ha.	180	9.22	27.939	160	4.72	13.88	685	16.9	65.8	300	24.78	58.224	280.6	82.75	87.5	425
O.F.F Production	Ha.	60	1.18	3.5758	5	1.12	3.294	0	0		5	0	0	-100	-101.7	0	-100
Vegetative Field bunding	Ha.	100	0	0	160	3.84	11.29	200	5.25	20.5	329	8.98	21.1	100	525	106.63	133.85
Terrace Repair	Ha.	198	2.6	7.8788	315	5.97	17.56	200	3.49	13.6	353	7.68	18.045	1.01	34.23	12.063	28.64
Agro-Forestory	Ha.	100	0	0	70	2.22	6.529	10	0	0	756	0.59	1.3863	-90	0	480	-73.42
Silvipasture	Ha.	150	20	60.606	100	14.18	41.71	0	0	0	0	0	0	-100	-100	-100	-100
Mini Kits	No.	0	0	0	150	0.49	1.441	0	0	0	0	0	0	0	0	-100	-100
Mushroom Development	No.	0	0	0	0	0	0	20	0	0	1020	0.53	1.2453	2000	0	102000	53
Implements	No.	0	0	0	1	0.63	1.853	0	0	0	0	0	0	0	0	-100	-100
Chaff Cutter	No.	0	0	0	45	0.78	2.294	0	0	0	0	0	0	0	0	-100	-100
Hand Tokas	No.	0	0	0	110	0.05	0.147	0	0	0	0	0	0	0	0	-100	-100
Total	Ha.	788	33		810	34		1115	25.6		2763	42.56		41.5	-22.45	37.65	25.176

Source: Physical and Financial Achievements, 1999-2000 and 2001-2002, IWDP, Hills-II, J&K

Note: R. F. C. Demonstration: Rain fed Crop Demonstration
O. F. F. Production: On Farm Fodder Production

operations because of easy availability of wood from the local forests. The wooden agricultural tools and equipment are not as efficient as the factory made iron tools and equipment. At the same time, heavy dependence on forest resources for agricultural tools and equipment also add to the problem of deforestation. Thus, there is urgent need to propagate and popularize the use of factory made iron tools and implement in agricultural operations, which would help in restoring the degraded forests.

Over the period, the sustained efforts by the project have improved the performance of agriculture. As per the crop cut experiments conducted in different sub-watersheds on standing crops during 2001-2002 in Kharif season, the yield of maize has increased 2.5 times. The yield of maize and wheat shown an increase from 22 quintals per hectare (maize) and 19 quintals per hectare (wheat) in the base period to 32 quintals per hectare and 27 quintal per hectare respectively, after the project intervention. In percentage terms, the increase in yield of maize and wheat respectively was recorded at 45 and 42 per cent. The area covered under HYV crops has shown remarkable progress over the period. In 1999-2000, 4201 hectares of agricultural land has been covered under HYV crops. In subsequent years 2000-2001 and 2001-2002 (ending September 2001), 5779 hectares and 2338 hectares more arable land has been covered under HYV crops. Over the period, more than 12,000 small and marginal farmers have been benefited from arable land treatment (IWDP, Hills-II, 2001).

Other important sector treated under arable land component is the horticulture. The rain fed (R. F.) demonstration has been carried out over 278 hectares of arable land in Ramnagar sub-watershed, whereas, 197 hectares of arable land has been exposed to R. F. demonstration for horticulture in sub-watershed of Akhnoor in the year 1999-2000. In the year 2001-2002 too, greater attention has been paid on R. F. horticultural demonstration. The table 2 makes it evident that comparatively more attention has been given in Akhnoor sub-watershed on R. F. demonstration than Ramnagar. It has been observed during the field visits that the sub-watershed of Ramnagar, being predominantly forested has already better horticultural plantations. The horticultural potential in Akhnoor is also tremendous, which has picked up after the project interventions. The horticultural rejuvenation activities have been carried out in both the selected sub-watersheds. About 300 hectares and 244 hectares of arable land respectively have been covered under horticultural rejuvenation in Ramanagar and Akhnoor in 1999-2000. In the next year, a significant increase has taken place in hectare coverage under horticulture rejuvenation in both the sub-watersheds, which stood at 440 hectares in Ramnagar and 814 hectares in Akhnoor. In terms of percentage change, the figures stood at 46.67 in Ramnagar and 233.6 in Akhnoor. Thus, it is clear that R. F. demonstration and horticulture rejuvenation has received more attention in sub-watershed of Akhnoor than Ramnagar.

Table 2
Achievements of Watershed Development and Protection (Arable Land: Horticulture) in Sample Sub-Watersheds (Rs. in lacs)

Component	Unit	1999-2000 Ramnagar Phy.	1999-2000 Ramnagar Fin.	1999-2000 Ramnagar Fin. %	1999-2000 Akhnoor Phy.	1999-2000 Akhnoor Fin.	1999-2000 Akhnoor Fin. %	2001-2002 Ramnagar Phy.	2001-2002 Ramnagar Fin.	2001-2002 Ramnagar Fin. %	2001-2002 Akhnoor Phy.	2001-2002 Akhnoor Fin.	2001-2002 Akhnoor Fin. %	Ramnagar Change 1999-2000 to 2001-2002 Phy. %	Ramnagar Change 1999-2000 to 2001-2002 Fin.	Akhnoor Change 1999-2000 to 2001-2002 Phy. %	Akhnoor Change 1999-2000 to 2001-2002 Fin. %
Arable Land																	
Horticulture																	
R.F.Demonstration	Ha.	278	13.76	69.215	197	17	45.31	455	24.4	70.888	540	30.01	58.912	63.67	15.3	174.1	76.63
Hoticuture rejuvenations	Ha.	300	2.7	0.1988	244	1.51	0.375	440	3.92	11.412	814	5	9.8155	46.67	45.19	233.6	231.13
Tools	No.	0	0	0	0	0	0	0	0	0	0	0	0	0	0	0	0
Nursery Development	Ha.	0	4.5	22.636	0	19	50.67	0	6.08	17.7	0	15.93	31.272	0	35.11	0	-16.16
Total	Ha.	578	19.88	441	37.5	895	34.4	1354	50.94	54.84	72.79	72.79	35.84				

Source: Physical and Financial Achievements, 1999-2000 and 2001-2002, IWDP, Hills-II, J&K

The system of nursery development has also been implemented in both the sub-watersheds and location specific fruit plants were raised in these nurseries and supplied to the needy farmers on cost-sharing basis. In both the sub-watersheds, a few horticultural nurseries have been created and maintained by the project with active participation of local villagers. During the field visit, it has been observed that horticultural nurseries have been managed more effectively in the sub-watershed of Ramnagar than Akhnoor. The fruit plants of improved varieties such as mango, orange, guava, walnut, lemon, etc. conducive to local climatic conditions of the sub-watershed have been provided to the beneficiaries. The plants of wild pomegranate, mango, etc. have been found well distributed all over the selected sub-watersheds. The rejuvenation of wild pomegranate, mango etc. has taken place after conducting grafting and budding with improved variety. It has been noticed that despite severe drought over the period, the survival percentage has been very encouraging, ranging between 40-85, the lowest for papaya and highest for pomegranate.

Non-Arable Land Treatment

In the selected sub-watersheds, the non-arable lands include public lands, forestlands, community lands and private lands. The non-arable lands have been treated under various components depending upon suitability or topography of the areas. The main component is afforestation in contour trenches besides other treatments. The tables 3 to 5 gives the data on the performance of non-arable land treatment under the project in selected sub-watershed for the years1999-2000 and 2001-2002. Under the non-arable land treatment, afforestation activities have been carried out on 90 hectares of village common lands (VCLs) in the sub-watershed of Ramnagar, whereas, 347 hectares of VCLs have been covered in Akhnoor in 1999-2000. In the next year, the hectare coverage of VCLs increased significantly in Ramnagar and stood at 200 hectares. Similarly, 338 hectares of VCLs have been enclosed in the sub-watershed of Akhnoor. Thus, it is evident that degraded VCLs have received more attention in agriculturally dominated watershed than the forested watershed. However, in proportional sense, marginal decline has been recorded in hectare coverage between 1999-2000 and 2001-2002 in Akhnoor. It is to be noted that all the VCLs closures have been created with active participation of the people inhibiting the sub-watershed.

The silvi-pasture activities have also been under taken initially in both the sub-watersheds and 175 hectares of non-arable VCLs have been treated in Ramnagar, whereas, 95 hectares of degraded VCLs have been treated in Akhnoor. Later, the silvi-pasture treatment on VCLs has been abandoned in both the sub-watersheds. Besides, vegetative shrub barriers have also been created in project area to maintain and improve the soil moisture on the degraded VCLs. Over the period, project interventions on VCLs have

Table 3
Achievements of Watershed Development and Protection (Village Common Land) in Sample Sub-Watersheds (Rs. in lacs)

		1999-2000						*2001-2002*						*Ramnagar Change 1999-2000 to 2001-2002*		*Akhnoor Change 1999-2000 to 2001-2002*	
		Ramnagar			*Akhnoor*			*Ramnagar*			*Akhnoor*						
Component	*Unit*	*Phy.*	*Fin.*	*Fin. %*	*Phy.*	*Fin.*	*Fin. %*	*Phy.*	*Fin.*	*Fin. %*	*Phy.*	*Fin.*	*Fin. %*	*Phy. %*	*Fin.*	*Phy. %*	*Fin. %*
Village																	
Common Land																	
Afforestation	Ha.	90	12.630.732		347	45.5365.986		20027.71		100	338	42.03	100	122.22	119.9	-2.594	-7.69
Silvipasture	Ha.	175	20.5	50	95	13.319.275		0	0	0	0	0	0	-100	-100	-100	-100
Veg. Shrub Barriers	Ha.	103	7.919.268		213	10.1714.739		0	0	0	50	0	0	-100	-100	-76.53	-100
Total	Ha.	368	41		655	69		20027.71			388	42.03	-45.65		-32.4	-40.76	-39.1

Source: Physical and Financial Achievements, 1999-2000 and 2001-2002, IWDP, Hills-II, J&K

shown their impact and the degraded patches have been regenerated and helped improving the productivity of these lands. With the regeneration of VCLs, system of rotational grazing and equal usufruct sharing has been created with active community participation of the beneficiaries. The fodder scarcity has declined considerably and fodder demand is met from the protected VCLs. The idea of fodder bank is propagated to accumulate the surpluses for selling to the scarce households/villages.

In addition to afforestation, silvi-pasture and vegetative shrub barriers, pasture development and forest rejuvenation activities have also been carried out on non-arable forestlands in the selected sub-watersheds. The afforestation activities have been implemented on degraded forests in both the sub-watersheds. In 1999-2000, 340 hectares of degraded forests have been covered for afforestation in Ramnagar, whereas, 246 hectares of degraded forests have been included under afforestation programme in Akhnoor. In 2001-2002, the area covered under afforestation activities has increased significantly, and stood at 573 hectares in Ramnagar and 618 hectares in Ahknoor. In proportional terms, the increase was recorded at 68.53 per cent and 151.2 per cent respectively in Ramnagar and Akhnoor. In the sub-watersheds of Ramnagar and Akhnoor, pasture development have also been implemented on 110 hectares and 145 hectares of degraded pastures respectively initially, however, later the interventions related to pasture development has been abandoned.

To begin with, the forest rejuvenation activities have been implemented in Ramnagar and laterally it has been extended to Akhnoor also. In 2001-2002, 409 hectares (Ramnagar) and 570 hectares (Akhnoor) of degraded forests have been enclosed for rejuvenation for which local population was actively engaged. The vegetative shrub barriers have also been created in the degraded forestlands in the project area. For instance, to begin with 203 hectares (Ramnagar) and 345 hectares (Akhnoor) of degraded forestlands have been treated with vegetative shrub barriers. In the year 2001-2002, additional 165 hectares and 345 hectares of degraded forestlands have been treated with vegetative shrub barriers respectively in Ramnagar and Akhnoor. Besides above, the silvi-pasture activities have also been carried out in both the sub-watersheds initially, but laterally it has been abandoned (see table 4).

Over the period, the interventions on non-arable VCLs and forests have resulted in increased availability of fodder, forage/grasses, fuel wood and small timber, which is reflected in increased milk production, wool production, meat production, soil moisture regime, bio-diversity and reduction in soil run-off and soil erosion. For instance, significant increase has been recorded in forage/green grass production in the project area from a yield level of 3.40 ton per hectare (base line yield) to 10.60 ton per hectare as per crop-cut experiment carried out by the field functionaries

Table 4
Achievements of Watershed Development and Protection (Forests) in Sample Sub-Watersheds (Rs. in lacs)

Component	Unit	1999-2000 Ramnagar Phy.	Fin.	Fin. %	1999-2000 Akhnoor Phy.	Fin.	Fin. %	2001-2002 Ramnagar Phy.	Fin.	Fin. %	2001-2002 Akhnoor Phy.	Fin.	Fin. %	Ramnagar Change 1999-2000 to 2001-2002 Phy. %	Fin.	Akhnoor Change 1999-2000 to 2001-2002 Phy. %	Fin. %
Forests																	
Afforestation	Ha.	340	51	39.844	246	43.24	34.454	573	122.74	82.944	618	105	67.753	68.53	141	151.2	142.9
Pasture Development	Ha.	110	9.19	7.1797	145	20.31	6.175	0	0	0	0	0	0	-100	-100	-100	-100
Forest Rejuvenation	Ha.	241	13.18	10.297	0	0	0	409	21.08	14.245	570	33	21.288	69.71	59.9	57000	3300
Veg. Shrub Barriers	Ha.	203	15.11	11.797	345	33.73	26.876	165	4.16	2.8112	345	17	10.96	-18.72	-72.5	0	-49.63
Silvipasture	Ha.	333	39.53	30.883	175	28.23	22.494	0	0	0	0	0	0	-100	-100	-100	-100
Total	Ha.	1227	128		911	125.5		1147	147.98		1533	155		-6.52	15.6	68.28	23.52

Source: Physical and Financial Achievements, 1999-2000 and 2001-2002, IWDP, Hills-II, J&K

of the project. Overall, the increase in yield of forage/grasses production has been recorded at 211 per cent in the project area after the IWDP (Hills-II) intervention (IWDP, Hills-II, 2001).

The drainage line treatment interventions have also been implemented in both the sub-watersheds, which include crate-wire structures, landslide treatment, earthen bag check dam, and vegetative gully control and road drainage line. Besides, repair of traditional water points such as bowallies/ tanks, ponds, and water harvesting structures have also been implemented to improve the water regime in the project area. More than 150 traditional water points (bowallies) have been repaired in water scarce-forested watershed of Ramnagar over the period. Similarly, 75 ponds and 60 ponds have been repaired respectively in Ramnagar and Akhnoor. A few water-harvesting structures have also been repaired in both the sub-watersheds (see table 5). There are significant variations across the selected sub-watersheds in terms of activities implemented for drainage line treatment. The comparatively more crate-wire structures have been created in Akhnoor, whereas, traditional water points have been repaired comparatively more in Ramnagar. Actually, the predominantly forested watersheds have more problem of water scarcity, especially drinking water, which have been attended on priority in Ramnagar by rejuvenating the traditional water points and switching to gravity based water supply for drinking purposes. In Akhnoor also, the problem of water scarcity is acute, but being predominantly agricultural watershed with plain topography, it is easier to meet drinking water demand from piped water supply.

The agricultural land along the nalahs (rainfed streams) is heavily subjected soil erosion particularly during rainy season, which have been treated with crate wire structures on priority basis in Akhnoor sub-watershed. The road drainage line has also been attended in both the sun-watersheds. It is important to note that each of the interventions under drainage line treatment have been implemented in the selected sub-watersheds on the basis of need analysis carried out by the field functionaries of the project through participatory rural appraisals (PRAs) excercises. The drainage line interventions have significantly reduced the water run-off and soil erosion and rejuvenated the traditional water harvesting structures. For instance, in the year 2001 for the months of June, July, August and September, rainfall (in mm) has been recorded respectively at 228.1, 475.9, 285.1 and 201.7, whereas, the water run-off (in mm) has been recorded at 106.40, 223.8, 120.9 and 87.4. The percentage water run-off (4/3 x 100) was estimated at 46.65, 47.03, 42.41 and 43.33 per cent respectively for the months of June, July, August and September 2001. On the whole, the soil loss, which was 23.70 tons per hectare per annum in the base line period, has declined to 9.65 tons per hectare per annum after project intervention in the *Shivalik* hills (IWDP, Hills-II, 2001).

Table 5
Achievements of Watershed Development and Protection (Drainage Line) in Sample Sub-Watersheds (Rs. in lacs)

		1999-2000						2001-2002						Ramnagar Change 1999-2000 to 2001-2002		Akhnoor Change 1999-2000 to 2001-2002	
		Ramnagar			Akhnoor			Ramnagar			Akhnoor						
Component	*Unit*	*Phy.*	*Fin.*	*Fin. %*	*Phy.*	*Fin.*	*Fin. %*	*Phy.*	*Fin.*	*Fin. %*	*Phy.*	*Fin.*	*Fin. %*	*Phy. %*	*Fin.*	*Phy. %*	*Fin. %*
Drainage Line																	
Cratewire Str.	Cum	9531	40.62	16.82	11583	30.6627.722		313320.04		20.216	1680	57.82	39.15	-67.13	-50.66	-85.49	88.58
DRSM	Cum	128725	171.89	71.18	49244	59.6753.9514057044.7745.16371617						68.19	46.17	-68.48	-73.95	45.43	14.27
Bowallies/tanks	No.	79	8	3.313	0	0	0	8510.72		10.814	0	0.49	0.332	7.59	34	0	49
Village Ponds	No.	45	5	100	60	21.8083		3010.54		10.633	0	11.96	8.098	-33.33	110.8	-100	498
Water Harvesting Str.	No.	3	10	4.141	4	109.0416		0	55.0439		0	0.23	0.156	-100	-50	-100	-97.7
Landslide Treatment	Rmt.	50805	5.49	2.273	73402	4.273.86081310708.068.130769645						6.5	4.401	157.99	46.812	-5.11	52.22
Veg. Spurs	Rmt.	0	0	0	8540	0	0	0	0	049000		0	0	0	0	473.77	0
Earthen Bag Check Dam	Cum	0	0	0	0	0	0	0	0	0	2000	0	0	0	0	200000	0
Veg. Gully Control	Rmt.	4000	0.5	0.207	44760	2.522.2785		0	0107700		2.5	1.693	-100	-100	140.62	-0.79	
Road Drainage Line	Rmt.	10000	0	0	44760	1.481.338214020			0	0106600		0	0	40.2	0	138.16	-100
Total		203188	241.5		232353	110.6		18879399.13			408242147.69			-7.085	-58.95	75.69	33.53

Source: Physical and Financial Achievements, 1999-2000 and 2001-2002, IWDP, Hills-II, J&K

Livestock Institutional Support

The cattle population in the State of Jammu and Kashmir mainly consists of cows, buffaloes, sheep, goat, horses and camels. The camel and horses are few in numbers in the project area and are used for transport purposes. The cows and buffaloes are reared for dairy, whereas, sheep and goat are reared for meat and wool. The cattle population in the project area is 2.38 lacs, spread over an area of 1.11 lacs hectares, out of which the cows are 1.23 lacs, buffaloes 0.35 lacs, and sheep and goats are 0.80 lacs. The nomadic cattle population increases with their arrival during summer in *Karewas* and winter in *Shivalik* hills. The livestock predominantly consists of local breeds with low yield of milk and wool. PRA exercises were carried out by the field functionaries to ascertain the problems in livestock development in the project area, which identified the problems like (a) absence of effective health care system; (b) low availability of fodder; and (c) inadequate artificial insemination facility for breed improvement.

IWDP (Hills-II), Jammu and Kashmir has established 3 veterinary hospitals and 7 veterinary dispensaries in the project area. It is significant to note that all the veterinary hospitals have been established in the sub-watershed of Ramnagar. The veterinary facilities created across the project area are catering to the requirement of the cattle population. More specifically, attention have been given on (i) mass vaccination against foot-mouth diseases, haemoharrahic speticimia etc.; (ii) dosing against intestinal worms and liver fluke; (iii) artificial insemination; (iv) natural services by way of providing breeding cow bulls and rams; (v) castration of scrub bulls; (vi) dipping of sheep against parasite; (vii) distribution of cross-bred ram bouliet/marino rams; (viii) health cover for different diseases; and (ix) feed for calf rearing, pregnant eves and lambs born.

The data presented in table 6 makes it clear that under the component of genetic improvement of livestock, artificial insemination centres have been established in both the sub-watersheds, which have revolutionized the dairy farming elsewhere. However, the impact of artificial insemination programme is reportedly modest in project area, which needs to be propagated by the field functionaries. In order to improve the livestock health and productivity, veterinary hospitals and dispensary have been started with project interventions. The dairy and poultry units have also been given assistance, however, very marginally. The problem of poverty, unemployment and underemployment is very acute in the project area and dairy and poultry rearing could be alternative viable income generating activities (IGAs) for the people inhibiting the sub-watersheds, if undertaken on group basis. The products of these micro-enterprises have ready market in the local villages and nearby townships. It is ironical to note that livestock development infrastructure has been created, but the maintenance of the infrastructure has not been given adequate attention. As already noted, a significant proportion of the population in the project

Table 6
Achievements of Watershed Development and Protection (Livestock: Genetic Improvement) in Sample Sub-Watersheds (Rs. in lacs)

		1999-2000						2001-2002						Ramnagar Change 1999-2000 to 2001-2002		Akhnoor Change 1999-2000 to 2001-2002	
		Ramnagar			Akhnoor			Ramnagar			Akhnoor						
Component	*Unit*	*Phy.*	*Fin.*	*Fin. %*	*Phy.*	*Fin.*	*Fin. %*	*Phy.*	*Fin.*	*Fin. %*	*Phy.*	*Fin.*	*Fin. %*	*Phy. %*	*Fin.*	*Phy. %*	*Fin. %*
Livestock: Genetic Improvement																	
Cryocans	No.	0	0	0	0	0.42	3.09	0	0	0	0	0	0	0	0	0	-100
A.I Centre	Rs.	0	0.3	2.19	0	0.27	1.98	0	0.3	3.958	0	0.19	1.482	0	0	0	-29.63
Dairy Unit	No.	0	1	7.3	10	2.04	15	10	3.38	44.59	25	5	39	1000	238	150	145.1
Veternary Hospital	No.	1	6.45	47.1	0	2.37	17.4	0	2	26.39	0	1.93	15.05	0	-68.99	0	-18.57
Veternary Dispensary	No.	2	2.25	16.4	0	6.94	51	0	1.9	25.07	0	1.95	15.21	0	-15.55	0	-71.9
Equipments	Rs.	0	1.3	9.49	0	0.71	5.22	0	0	0	0	0	0	0	-100	0	-100
Poultry Unit	No.	0	0	0	0	0	0	0	0	0	0	2.49	19.42	0	0	0	249
Preg. Cows and Calves Feed	No.	0	0	0	0	0	0	0	0	0	0	1	7.8	0	0	0	100
Sterlity Control	Rs.	0	0.6	4.38	0	0.29	2.13	0	0	0	0	0	0	0	-100	0	-100
Castrators	No.	0	0.6	4.38	0	0.57	4.19	145	0	0	4	0.26	2.028	0	-100	400	-54.39
Feeding Stall	No.	0	0.61	4.45	0	0	0	0	0	0	0	0	0	0	-100	0	0
Bull/Buffalo	No.	0	0.3	2.19	0	0	0	0	0	0	0	0	0	0	-100	0	0
Maintenance of Livestock	Rs.	0	0.29	2.12	0	0	0	0	0	0	0	0	0	0	-100	0	0
Total		3	13.7		10	13.6		155	7.58		29	12.8	0		-44.67	190	-15.81

Source: Physical and Financial Achievements, 1999-2000 and 2001-2002, IWDP, Hills-II, J&K

area is rearing of sheep and goat as a source of livelihood. The transhumance pastoral and semi-pastoral communities depend heavily on sheep and goat rearing, for which provision has been made for migratory health-cover facilities (see table 7). New modified techniques have also been propagated under sheep husbandry to improve the quality of sheep and goats reared by the transhumance pastoral and semi-pastoral communities in the selected sub-watersheds. However, the sheep husbandry has not been promoted in a desired way, keeping in view the problems being faced by transhumance and local residents.

Besides above, the project has also been providing facilities in the villages, where veterinary hospital or dispensary is not available. The veterinary camps have also been organized in each village of the selected sub-watersheds. On the whole, 2251 artificial insemination has been done, 178 veterinary camps has been organized and 104, 877 and 182 of dairy, poultry and sheep units were set up across the project area over the period ending September, 2001. The concerted efforts of the project have resulted in a significant increase in the yield of milk and wool. For instance, the yield of cow milk has increased from 2.75 litre per day to 4.25 litre per day in case of local cow and from 6.75 litre per day to 8.60 litre per day in case of crossbred cow after project intervention. Similarly, over the project period, the yield of buffalo milk has increased from 3.50 litre per day to 5.25 litre per day in case of local buffalo and from 8.60 litre per day to 10.30 litre in case of cross-bred buffaloes. Likewise, the yield of sheep wool has increased significantly from a base line yield of 2.25 Kg. per annum to 3.40 Kg. per annum (IWDP, Hills-II, 2001).

In brief, watershed areas of the State of Jammu and Kashmir is facing multiple problems like poverty, backwardness, rapid environmental degradation, stagnant agriculture, lack of infrastructure and employment opportunities. Because of persisting poverty, unemployment and underemployment, the poor inhibiting the watersheds have over-utilized the resources causing severe damage to VCLs, forestlands, etc., which have resulted in rapid fall in productive potential of *Shivaliks* and further decline in livelihood base of the rural poor, requiring immediate attention. After the project intervention, an integrated approach has been adopted to mitigate the problem of environmental degradation and poverty alleviation, through evolving participatory approaches. The various project interventions, as discussed above, have improved significantly the environmental resources of the project area in the shape of: (i) significant improvement in vegetal cover and bio-mass thereof; (ii) multifold increase in the production of local as well as HYV of grasses; (iii) decrease in biotic pressure on the natural forests; (iv) revised the trend of deforestation; (v) slowing down of water run-off and increase in water regime of the catchments; and (vi) increase in yield of different crops, milk, wool and

Table 7
Achievements of Watershed Development and Protection (Livestock: Genetic Improvement) in Sample Sub-Watersheds (Rs. in lacs)

		1999-2000						*2001-2002*						*Ramnagar Change 1999-2000 to 2001-2002*		*Akhnoor Change 1999-2000 to 2001-2002*	
		Ramnagar			*Akhnoor*			*Ramnagar*			*Akhnoor*						
Component	*Unit*	*Phy.*	*Fin.*	*Fin. %*	*Phy.*	*Fin.*	*Fin. %*	*Phy.*	*Fin.*	*Fin. %*	*Phy.*	*Fin.*	*Fin. %*	*Phy. %*	*Fin.*	*Phy. %*	*Fin. %*
Sheep Husbandary																	
Rams	No.	0	1	20.6	0	0	0	359	0.14	3.867	0	0	0	359	-714.1	0	0
Sheep Units	No.	0	0	0	0	0	0	0	0	0	0	2	28.78	0	0	0	2
Health Cover (Migratory)	Ls.	0	2.55	52.6	0	0.74	25.2	15	2.5	69.06	0	1.99	28.63	15	-99.5	0	-35.2
Rams Feed	No.	0	0.5	10.3	0	0.4	13.6	0	0	0	0	0	0	0	-100	0	-100
Preganant Eves Feed	No.	0	0	0	0	0	0	0	0.98	27.07	0	1.5	21.58	0	0.98	0	1.5
DV/T																	
Dipping Vats/Tubs	No.	0	0.8	16.5	0	1.8	61.2	5360	0	0	4.4	1	14.39	5360	-100	4.4	-179
Dipping Troughs	No.	0	0	0	0	0	0	0	0	0	0	0.46	6.619	0	0	0	0.46
Total		0	4.85		0	2.94		5734	3.62		4.4	6.95		5734	-130.4	4.4	-35.35

Source: Physical and Financial Achievements, 1999-2000 and 2001-2002, IWDP, Hills-II, J&K

meat. It is significant to note that soil loss has decreased from 23.70 tons to 9.65 tons and on average grass biomass has recorded a growth of 6 quintals per hectares. Besides, the selected sub-watersheds have shown considerable improvement in bio-diversity and better aesthetic look. The treated areas under afforestation, rejuvenation and silvi-pastures interventions have also produced more biomass in the shape of large-scale production of grasses and bushes.

BIOPHYSICAL GAINS

Common property resources have been a major source of livelihood for the poor farmers in ecologically fragile areas. The degradation of CPRs has put the livelihood of the poor farmers at risk. Watershed development projects have a strong livelihood improvement dimension and aims at improving the welfare of the rural community through rehabilitation of degraded CPRs and conservation. In the project area, agriculture is the main source of livelihood of poor farmers, which is characterized by low and uncertain yields. The agriculture is unable to fully meet livelihood security. Keeping this in view, supplementary source of livelihood diversification strategies have assumed importance. In the context of IWDP (Hills-II), the project staff has provided training to the poor farmers in better cropping techniques, which have resulted in improved crop yields. The project has demonstrated the farmers regarding improved agricultural practices such as vegetative bunding, proper crop rotation, on-farm fodder production, restriction on growing water intensive crop, use of vegetative fencing, etc. The project has also propagated agro-forestry and rainfed horticultural demonstration. All these project interventions have resulted in motivating the individual farmers in switching to improved on-farm and off-farm activities, which has contributed significantly towards improving the soil moisture conservation, improved ground water regimes, increased availability of fuelwood and fodder on private lands, decline in soil erosion and reduced pressure on CPRs. The project interventions have facilitated the farmers in improving the productivity and thereby their livelihood conditions. The project has also promoted livestock development by introducing artificial insemination techniques, which has resulted in hybridization of livestock with improved yield and consequent reduction in number of animals. This has significant reduced the grazing pressure on watershed. The system of stall-feeding has also been propagated, as most of the CPRs are enclosed for rehabilitation and regeneration. The project has also popularized the use of improved cooking stoves and LPG gas by demonstrating the benefits to the farming households. With more use of these alternative sources of energy, the pressure for on CPRs has been reduced significantly. Besides above, some of the project interventions such as rehabilitation of gravity based irrigation system, water harvesting structures, enclosures on common property resources and village common lands, soil and water conservation measures required collective action through group formation, as the benefits from such interventions cannot be reaped singly but collectively, for which user groups have formulated suitable resource use rules to ensure equity in watershed development.

Reduction in Rainwater Loss and Sediment Yield

Under IWDP (Hills-II), Jammu and Kashmir, soil and water conservation measures have been promoted using evolving participatory approach. The micro-level watershed planning was carried out using the sweeping transect and emphasis has been given on soil erosion control on hill slopes and gullies, regulation of water flow system in the watershed drainage, and rearrangement of farmlands. In the entire cultivated area, wastelands and area other than hills and hillocks which generally come under grazing lands, were fully treated with appropriate soil and water conservation measures like contour bunding, field bunding, gully plugging, field to field drainage outlet structures, etc. In addition to this, the diversion channels were also made along all the hills and hillocks to guide the high velocity runoff from these sites on a safe and controlled way, so that the runoff with soil loads should not enter the arable land and cause further degradation. For the stabilization of bunds and gully plugs, the entire bund area was sown with pastures grass and legume. Similarly, in the hills and hillocks several of the soil conservation measures were adopted. The adverse climatic factors of the *Shivaliks* necessitated the adoption of micro catchment techniques for run-off harvesting and conservation practices as done in similar areas elsewhere. The usefulness and scope of rainwater harvesting and conservation practices in improving tree growth in arid zones have been amply demonstrated. The staggered trenches ensured higher survival rate of out planted seedlings in the experimental plots. The contour trenches also helped in moisture retention but the most of the sub-watershed of Akhnoor being an undulated terrain without sufficient slopes – the contour interval had to be very wide. The soil and moisture conservation measures in the demonstration plots were much effective as evidenced by the enhanced survival rate of out planted seedlings. The effect of regeneration of vegetation, along with soil and water conservation measures on hillslopes and wastelands was substantial. In *Shivaliks*, the run-off soil loss on barren hills was 23.0 70 tons in the baseline period, which fell to 9.65 tons per hectare after treatment. In *Karewas*, it declined to 4.80 tons per hectare from 8.24 tons per hectare in baseline period (see table 8). The progressive reduction in soil loss and sediment yield as a result of quick recovery of vegetation on hillslopes and lands adjoining the foothills has resulted in a greatly improving surface and ground water regime of the selected sub-watersheds.

Table 8
Run-off Soil Measures (April 2001 to Sept. 2001)

Month	*Rainfall (in mm)*	*Run-off (in mm)*	*Run-off %4/3 .100)*
		Shivaliks	
1-Apr	85.6	-	-
1-May	57.4	-	-
1-Jun	228.1	106.4	46.65
1-Jul	475.9	223.8	47.03
1-Aug	285.1	120.9	42.41
1-Sept	201.7	87.4	43.33
Soil loss (Tons per Ha/ per Year): Baseline: 23.70, Current: 9.65 Tons/Ha.			
		Karewas	
1-Apr	107.1	20.33	18.98
1-May	35.5	4.05	11.41
1-Jun	26	5.2	20
1-Jul	48.1	16.53	34.37
1-Aug	37.1	9.45	25.47
1-Sept	20.5	2.7	13.17
Soil loss (Tons per Ha./ Per Year): Baseline: 8.24, Current: 4.8 Ton/Ha.			

Source: Status Report for Supervision Mission of World Bank, IWDP (Hills-II), J&K, 2001

Status of Water Resources and Irrigation

The table 9 gives the data on surface and ground water rescurces across the sampled sub-watersheds. Both the average number of water points (*bowlies*) as well as the gravity based water points per village were comparatively more in project area with VDC than without VDC and non-project area. In the project area with VDC, there were 6.8 and 5 natural water points and gravity based water points respectively in forested watershed and 4.3 and 2 respectively in agricultural watershed. In non-project area, the average number of natural and gravity based water points were comparatively very low and stood respectively at 1.6 and 1 in forested watershed, and 2.6 natural water points in agricultural watershed, whereas there was no gravity based water points in agricultural watershed. This clearly reflects the benefits of participatory natural resource management through which the project has implemented the repairs of water resources points and installed new water points, mainly gravity based water resources. In the forested watershed, the difference between project area with VDC and non-project area in average number of natural water points and gravity based water points was as high as 3.2 and 4 respectively. The average number of hand pumps per village was also reportedly high in project area with VDC than project area without VDC and non-project area.

A mere perusal of the table 9 makes it clear that the number of water harvesting structures were more in project area with VDC than project area without VDC and non-project area. Similar is the case with the length of irrigation channel. However, significant differentials were noticed across project area with VDC and without VDC as well as non-project area (see table 10). The number of water harvesting structures as well as length of irrigation channel was significantly higher in project area than non-project area. In the project area with VDC, the average number of water harvesting structures per village was 3.5 and 2.9 respectively in forested and agricultural watersheds, whereas, it was 2 and 1.8 respectively in project area without VDC. In non-project area, the average number of water harvesting structures was 1.2 and 0.7 respectively in forested and agricultural watersheds. It is significant to note that the quality of water harvesting structures were also reportedly better in project area with VDC than non-project area, due to participatory repairs and renovation activities as well as maintenance initiated by the project. Similarly, the average length of irrigation channel in the project area with VDC was 2.8 km and 1.7 km per village respectively in forested and agricultural watershed, whereas, it was 1.8 km and 1.0 km respectively in project area without VDC. In the non-project area, the average length of irrigation channel was comparatively low and it was just 1.0 km and 0.8 km per village respectively in forested and agricultural watersheds. As a result of increased water potential in the selected sub-watersheds, the area under irrigated cropping increased significantly.

Table 9
Status of Water Resources and Irrigation (per village)

		Project area with VDC		*Project area without VDC*		*Non-project area*	
Item	*Unit*	*Forested WS*	*Agrl. WS*	*Forested WS*	*Agrl. WS*	*Forested WS*	*Agrl. WS*
Surface Water Supply							
Natural Water Points (Bowlies)	No.	6.8	4.3	5.2	3.4	3.6	2.6
Gravity based Water Points	"	5	2	2	1	1	0
Ground Water Supply							
Hand Pumps	"	2	3	0	1	0	1
Water Harvesting Structures	"	3.5	2.9	2	1.8	1.2	0.7
Irrigation							
Irrigation Channel	No.	2.8	1.7	1.8	1.2	1	0.8
Gross Irrrigated Land	Ha.	98	67	48	64	69	44
Net Irrigated Land	Ha.	68	36	34	37	27	18
Irrigation Intensity	%	144.11	186.11	141.17	172.97	255.56	244.44

Note: Irrigation Intensity = Gross Irrigated Area/Net irrigated Area × 100

Since changes in irrigated area depend on numerous factors, many of which could not be measured in the current study, the findings here must be treated cautiously. The data presented in table 9 reveals that the changes in gross and net irrigated area in project area with VDC and without VDC, and non-project area. A perusal of the table makes it clear that irrigation intensity in forested and agricultural sub-watersheds in project area with VDC was 144.11 per cent and 186.11 per cent respectively. In project area without VDC, it was 141.17 per cent and 172.97 per cent, whereas in non-project area, it was 255.66 per cent and 244.44 per cent respectively in forested and agricultural sub-watersheds. The data shows the marked differences in irrigation intensity across the sample area. The difference in irrigation intensity was 2.94 per cent and 13.14 per cent respectively in forested and agricultural sub-watersheds of project area with VDC and without VDC. It is significant to note that the differences in irrigation intensity in project area with VDC and non-project area (see table 10).

Table 10
Difference in Status of Water Resources and Irrigation in Sample Area (per village)

Item	Unit	*Project area with VDC - Project area without VDC*		*Project area without VDC - Non-project area*		*Project area without VDC - Non-project area*	
		Forested WS	*Agrl. WS*	*Forested WS*	*Agrl. WS*	*Forested WS*	*Agrl. WS*
Surface Water Supply							
Natural Water Points (Bowlies)	No.	1.6	0.9	3.2	1.7	1.6	0.8
Gravity based Water Points	"	3	1	4	2	1	1
Ground Water Supply							
Hand Pumps	No.	2	2	2	2	0	0
Water Harvesting Structures	"	1.5	1.1	2.3	2.2	0.8	1.1
Irrigation							
Irrigation Channel	No.	1	0.5	1.8	0.9	0.8	0.4
Gross Irrigated Land	Ha.	50	3	29	23	-21	-20
Net Irrigated Land	Ha.	34	-1	41	36	7	19
Irrigation Intensity	%	2.94	13.14	-111.45	-58.33	-114.39	-71.47

Change in Crop Intensity and Crop Productivity

As a result of the improved soil moisture regime, the increase in supplementally irrigation resources and increased use of fertilizer (including cow-dung), the cropping intensity has improved in the project area compared to non-project area. The data on grossed crop area, net area sown and crop intensity is presented in table 11, which clearly reveals significant variations across the selected sub-watersheds. There are considerable differences in cropping intensity in project area with VDC and non-project area, which can be attributed to the project interventions in the field of agriculture through rainfed crop demonstration, propagation of the use of modern inputs through extension agents of the project. It is to be noted that differences in cropping intensity was reportedly negative across the project area with VDC and without VDC as well as non-project area. For instance, it was – 58.26 per cent and – 46.27 per cent respectively between project area with VDC and non-project area (see table 12).

Table 11
Crop Intensity and Crop Yields

		Project area with VDC		*Project area without VDC*		*Non-project area*	
Item	*Unit*	*Forested WS*	*Agrl. WS*	*Forested WS*	*Agrl. WS*	*Forested WS*	*Agrl. WS*
Gross Cropped Area	Ha.	163	74	73	107	113	68
Net Sown Area	Ha.	158	68	57	93	70	42
Crop Intensity	%	103.16	115.63	128.07	146.58	161.42	161.9
Non-seed Inputs and Crop Yields							
Maize	Qntl./Ha	96	9.9	8.2	8.5	7.5	7.6
Paddy	"	5.2	6.2	4.5	8.5	4.4	4.5
Pulses	"	2.64	2.8	2.3	2.4	2	0.3
Wheat	"	8.96	9.9	8.2	8.4	7.6	7.8
Seed Inputs and Crop Yields							
Maize							
Local	Qntl./Ha	13.13	14.2	12.6	13.7	10.3	11.2
Hybrid	"	15.6	15.8	14.9	14.1	11.7	11.9
Difference	"	2.3	1.6	2.3	0.4	1.4	0.7
Paddy							
Local	"	10.8	11.2	9.4	10.3	9	9.2
Hybrid	"	12.3	13.1	11.4	11.7	10.1	10.7
Difference	"	1.5	1.9	2	1.4	1.1	1.5
Wheat							
Local	"	18.4	19.7	16.9	17.4	16.1	16.7
Hybrid	"	21.3	21.8	20.4	21.1	17.3	17.9
Difference	"	2.9	2.1	3.5	3.7	1.2	1.2

Note: Crop Intensity = Gross Cropped Area/Net Area Sown × 100

It is very difficult to isolate the impact of various inputs used on crop yields. Due to creation of VCLs and forest closures, an increase in the vegetation has taken place. The over all soil-moisture regimes have also improved after project intervention. The project has repaired the traditional irrigation water distribution channels (khuls), consequently the irrigation water availability and flow has improved. Keeping the influence of these factors aside, the data presented in table 12 shows differential levels of yields of various crops such as maize, paddy, pulses and wheat across the selected sub-watersheds, largely due to increased use of non-seed inputs. A cursory glance at table 12 makes it evident that average crop yield per hectare has been comparatively more in project area with VDC than project area without VDC, and further the yield per hectare has been comparatively higher in project area without VDC than non-project area. The productivity of selected crops is comparatively higher in agricultural sub-watersheds than forested sub-watersheds. However, a significant variation in crop yield is reported across the project and non-project area.

Table 12
Difference in Crop Intensity and Crop Yields in Sample Area

		Project area with VDC - Project area without VDC		*Project area without VDC - Non-project area*		*Project area without VDC - Non-project area*	
Item	*Unit*	*Forested WS*	*Agrl. WS*	*Forested WS*	*Agrl. WS*	*Forested WS*	*Agrl. WS*
Gross Cropped Area	Ha.	90	-33	50	6	-40	39
Net Sown Area	Ha.	101	-25	88	26	-20	51
Crop Intensity	%	-24.91	-30.95	-58.26	-46.27	-33.35	-15.32
Non-seed Inputs and Crop Yields							
Maize	Qntl./Ha	1.4	1.4	2.1	2.2	0.7	0.8
Paddy	"	0.7	1.6	0.8	1.7	0.1	0.1
Pulses	"	0.34	0.4	0.64	0.5	0.3	0.1
Wheat	"	0.76	1.5	1.36	2.1	0.6	0.6
Seed Inputs and Crop Yields							
Maize							
Local	Qntl./Ha	0.7	0.5	3	3	2.3	2.5
Hybrid	"	0.7	1.7	3.9	3.9	3.2	2.2
Difference	"	0	1.2	0.9	0.9	0.9	-0.3
Paddy							
Local	"	1.4	0.9	1.8	2	0.4	1.1
Hybrid	"	0.9	1.4	2.2	2.4	1.3	1
Difference	"	0.5	0.5	0.4	0.4	0.9	-0.1
Wheat							
Local	"	1.5	2.3	2.3	3	0.8	0.7
Hybrid	"	0.9	0.7	4	3.9	3.1	3.2
Difference	"	-0.6	1.6	1.7	0.9	2.3	2.5

The yield difference in the project area with VDC and without VDC was low compared with project area with VDC and non-project area in forested as well as agricultural sub-watersheds. The yield difference between project area with VDC and non-project area was 2.1 quintals, 0.8 quintal, 0.64 quintal and 1.36 quintal per hectare respectively for maize, paddy, pulses, and wheat in forested watershed. Whereas, in agricultural watershed, yield differences between project area with VDC and non-project area was markedly high and stood at 2.2 quintals, 1.7 quintals, 0.5 quintal, and 2.1 quintals respectively for maize, paddy, pulses, and wheat. However, such differentials were not so marked in project area without VDC and non-project area. The yield differences were 0.8 quintal, 0.1 quintal, 0.1 quintal and 0.6 quintal respectively for maize, paddy, pulses and wheat in agricultural watershed, and more or less the same is true for forested watershed (see table 12).

The data presented in table 12 further substantiate the yield differentials across the selected sub-watersheds. As can be noticed elsewhere too, there are significant differences in yield of local and hybrid varieties of crops in forested as well as agricultural sub-watersheds. For instance, yield difference in local and hybrid maize is 2.3 quintal per hectare in forested sub-watershed in project area with VDC and without VDC. Thus, one can infer that new institutional arrangements have no impact on the yield of local and hybrid variety. The yield difference is 1.2 quintal per hectare in agricultural sub-watershed in project area with VDC and without VDC. A significant differential in yield is noticed between project area with VDC and non-project area, as well as project area without VDC and non-project area. More or less the same applied to other crops with minor differences across the forested and agricultural sub-watersheds. On the whole, the average crop yield per hectare has significantly improved after the project area.

Afforestation, Horticulture Plantations and Fodder Production

Major efforts were made for revegetation of VCLs and grazing lands under hill and hillocks. Before project, these sites were totally devoid of vegetation and barren and cannot support and provide grazing resources to the animals of the villages of the watershed. The availability of the firewood was also negligible from these sites. The fast growing plant species like pasture grasses, legumes, shrubs and trees were planted. It is not easy to work and vegetate these highly degraded sites even without skeletal soils and also arrest runoff from rains in such steep slopes. Several of the soil conservation measures coupled with vegetational support helped in checking runoff soil loss and vegetate these sites. The trees and shrubs were planted in the pits putting earth and pebbles on the lower slopes so that these pits can hold rainwater. Small staggered trenches were also made for *in-situ* moisture conservation.

Table 13
Afforestation, Horticulture Plantations and Fodder Production

Item	Unit	Project area with VDC		Project area without VDC		Non-project area	
		Forested WS	Agrl. WS	Forested WS	Agrl. WS	Forested WS	Agrl. WS
Afforestation							
Community Land	No./per Ha.	84	73	64	58	42	24
Forest Land	"	68	57	54	47	38	19
Horticulture							
Plantation	"	338	217	224	177	83	52
Demonstration	"	87	78	72	68	52	38
Rejuvenation	"	76	68	63	51	43	28
Fodder Production							
Green	Qntl./per Ha.	410	360	385	330	318	115
Dry	"	48.4	37.6	29.3	24.7	22.4	14.2

The afforestation plantations on community land and forestland was comparatively high in project area with VDC than project area without VDC and non-project area. It is clear from table 13 that more afforestation activities have been implemented significantly in forested watershed than agricultural watershed across the selected sub-watersheds. For instance, afforestation plantations on community land were 84 and 73 per hectare in forested and agricultural watersheds in project area with VDC. It was 68 and 17 per hectare respectively in forested and agricultural watershed in project area without VDC. Similarly, in non-project area, afforestation plantations on community land in forested and agricultural watersheds was 42 and 24 per hectare respectively, whereas on forestland, it was 38 in forested watershed and none in agricultural watershed.

In the project area with VDC, horticultural plantations were reportedly higher than project area without VDC and non-project area. A perusal of the data presented in table 13 makes it clear that horticultural plantations per hectare were higher in forested sub-watershed than agricultural watershed. The table 3.13 also makes it evident that horticultural plantations were 338 and 227 per hectare in forested and agricultural watersheds respectively in project area with VDC, whereas in project area without VDC, it was 224 and 177 per hectare. In non-project area, horticultural plantations were 83 and 52 per hectare respectively in forested and agricultural watersheds, which were significantly lower than project area. Not only this, the survival rate of horticultural demonstration and rejuvenation were comparatively higher in forested watershed than agricultural watershed. For instance, survival rate of horticultural demonstration and rejuvenation was 87 per cent and 76 per cent respectively in forested watershed in project area with VDC and 52 per cent and 45 per cent respectively in non-project area (see table 14).

The average yield of fodder production across the selected sub-watersheds is shown in table 13. In this case also, significant differences can be noticed across project area with VDC and without VDC and non-project area. The project area with VDC has shown a remarkable performance in terms of both the green and dry fodder productions. The net difference in green fodder yield in project area with VDC and without VDC was 25 kg and 30 kg respectively in forested and agricultural sub-watersheds, and the net difference in yield of dry fodder was 19.1 kg and 12.9 kg respectively in forested and agricultural sub-watersheds. When we see the net differences in fodder yield in project area with VDC and non-project area, very remarkable differences can be noticed. The net difference in yield of green fodder was as high as 95 kg in forested watershed and 245 kg in agricultural watershed, whereas, the net difference in yield of dry fodder was 26 kg and 23.4 kg respectively (see table 14).

Table 14
Difference in Afforestation, Horticulture Plantations and Fodder Production in Sample Area

		Project area with VDC - Project area without VDC		*Project area without VDC - Non-project area*		*Project area without VDC - Non-project area*	
Item	*Unit*	*Forested WS*	*Agrl. WS*	*Forested WS*	*Agrl. WS*	*Forested WS*	*Agrl. WS*
Afforestation							
Community Land	No./per Ha.	20	15	49	49	22	34
Forest Land	"	14	10	38	38	16	28
Horticulture							
Plantation	"	114	40	255	165	141	65
Demonstration	"	15	10	35	40	20	30
Rejuvenation	"	13	17	33	40	20	23
Fodder Production							
Green	Qntl./per Ha.	25	30	95	245	67	215
Dry	"	19.1	12.9	2.6	23.4	6.9	10.5

Livestock Development

IWDP (Hills-II), Jammu and Kashmir have made significant attempt to organize and develop animal husbandry in the project area and the facilities were provided for artificial insemination. The programme has made a modest dent on crossbreeding. The semen bank and deep-freezing of bovine semen as a part of this project was an important landmark in the field of cattle breeding in rainfed area of Jammu and Kashmir. However, artificial insemination and frozen semen technology has been restricted to few villages, the cattle sheds for good quality bulls have not been properly managed and they are underfed. On the whole, the breed improvement programme has been only moderately successful. There has been significant increase in milk yield, which may be attributed to higher productivity of crossbred animals. The yield rates of local milch animals remained stagnant, whereas, the yield rates of crossbred animals continued to increase at a rapid rate initially. The initiation of protection of VCLs and forestlands by creating enclosures and the poor usufruct sharing arrangements there from has resulted in marginal decline in the yield of even crossbred animals lately.

The growing economic opportunities for undertaking dairy farming as a commercial proposition combined with the interaction among the number of factors in agrarian rainfed economy, households have been progressively reducing their holding of drought animals and increasing their stock of milch animals. It has been observed that the requirement of work animals in the project area has been declining rapidly. The factors underlying this process are decline in the average size of cultivated holding, shift in cropping pattern, increase in the cost of rearing work animals, less availability of VCLs, common grazing lands, protection and closures of VCLs and forests. On the whole, the decline in the requirement of work animals and increase in the requirement of milch animals has resulted in significant changes in the composition of cattle population. While the adult male cattle population showed a sharply falling trend, the adult female cattle population has shown an increasing trend. Consequently, sex ratio of adult cattle has shifted in favour of females; the population of buffaloes has shown a declining trend. At the same time the rise in the profitability of milk production has resulted in an intensive selection process in the rearing and maintenance of cows for milk. Thus, while attempts are being made to rear the best young female to adulthood, the unproductive and low productive animals are eliminated from the herd by disposing them off in the market.

The average number of milch animals per household and milk yield is given in table 15. The data presented in table makes it evident that average number of milch animals was reportedly more in forested watershed than in agricultural watershed across the sample area. In the non-project area, the average number of milch animals were reportedly higher than project area with VDC and without VDC, which clearly reflects that households in project area were rearing lesser number of milch animals than households in project area without VDC and non-project area. The differences in average number of milch animals in project area with VDC and without VDC were quite small than project area with VDC and non-project area. For instance, in the forested watershed, the difference in average number of milch animals in project area with VDC and without VDC was – 0.4, whereas in agricultural watershed, the difference in project area without VDC and non-project area was – 4. The table also reveals that average milk yield per animal across sample area. A mere perusal of the data given in table 4.18 makes it evident that milk yield was comparatively high in project area with VDC than project area without VDC and non-project area (see table 16).

Table 15
Livestock and Yield (per household)

		Project area with VDC		*Project area without VDC*		*Non-project area*	
Item	*Unit*	*Forested WS*	*Agrl. WS*	*Forested WS*	*Agrl. WS*	*Forested WS*	*Agrl. WS*
Milch Animals	No.	2.8	3	3.2	3.6	4.2	4
Milch Yield							
Cow	Litre/per Animal	4.25	4	3.75	3.5	3.25	2.5
Buffalo	"	6.5	5.25	4.5	4	4	3.5
Sheep	No.	4.8	3.6	3.2	2.8	2	1.2
Wool Yield	Kg./per Sheep	3.1	2.4	1.5	1.3	1	0.8

The average number of sheep per household was comparatively high in forested watershed across the sample area. It is significant to note that average number of sheep were more in project area with VDC than project area without VDC and non-project area. Similarly, wool yield was as high as 3.1 kg per sheep in forested watershed and 2.4 kg per sheep in agricultural watershed in project area with VDC. In project area without VDC, it was reportedly 1.5 kg and 1.3 kg respectively in forested and agricultural watershed, whereas, in non-project area, it was comparatively low and stood at 1 kg and 0.8 kg respectively (see table 16).

Table 16
Difference in Livestock and Yield in Sample Area (per household)

		Project area with VDC - Project area without VDC		*Project area without VDC - Non-project area*		*Project area without VDC - Non-project area*	
Item	*Unit*	*Forested WS*	*Agrl. WS*	*Forested WS*	*Agrl. WS*	*Forested WS*	*Agrl. WS*
Milch Animals	No.	-0.4	-0.6	-1.4	-1	-1	-4
Milch Yield							
Cow	Litre/per Animal	0.5	0.5	1	1.5	0.5	1
Buffalo	"	2	1.25	2.5	1.75	0.5	0.5
Sheep	No.	1.6	0.8	6.8	2.4	1.2	1.6
Wool Yield	Kg./per Sheep	1.6	1.1	2.1	1.6	0.5	0.5

The indigenous livestock breeds of the forested watersheds in *Shivaliks* have been exposed to natural selection for a very long time and are thus well adapted to harsh environmental conditions. IWDP (Hills-II) focus on new breeds under insemination programme at subsidized price. These changes weakened indigenous breeding system that could improve livestock hardiness. The crossbred livestock of better quality is replacing the local cattle, sheep and goat. Thus, after project intervention indigenous livestock diversity has declined precariously more in the forested sub-watershed of Ramnagar than Akhnoor and the decline is reportedly more in those of the villages where the VDCs are taking all types of decisions related to watershed development and protection. It is, thus, urgently needed to speed up the development of sustainable and long-term crossbreeding programmes.

The sampled population reared the milch animals to supplement household nutrition and earnings through the sale of milk and milk products in the village nearby townships. In the project area with VDC, a higher proportion of the milk is consumed within household and more than one-third and one-fourth of the milk is sold respectively in forested and agricultural sub-watersheds. The milk-based products such as cheese, ghee, kalari, etc. are sold in the market, but often at very low price due to undeveloped rural markets. The data given in table 17 also revels that sampled population in non-project area sell a small proportion of the milk and milk-based products in market, due to low milk yields of the local milch animals kept by them. It has been reported that the project has distributed improved grass varieties for plantation in fields and field bunds, which have resulted in improved fodder availability. The creation of user groups have facilitated in usufruct sharing on more or less equitable basis and ensured better availability of fodder. The non-project area is reportedly scarce in fodder and VCLs and forestlands are severely degraded, however, with project intervention regeneration of the enclosed VCLs and forestlands has taken place in project area, which has increased the availability of feed and fodder.

Fodder is a very crucial biomass needed for maintaining their livestock. The immediate livelihood impact of creation of enclosures on VCLs, forestlands and grazing lands has been felt with regard to the grazing practices and the availability of fodder from CPRs. The enclosed CPRs are forbidden for open grazing. The cost of restrictions imposed on open grazing in the enclosed area is to be compensated by increased grass production to some extent; the major concern remains one of ensuring fair or equal distribution. However, the usufruct sharing is reportedly inequitable. Most households in fact facing fodder scarcity and resort to one or the other mean to fill the deficit. Under IWDP (Hills-II), Jammu and Kashmir attempts has been made to protect the VCLs, forestlands and grazing lands so that CPRs should be conserved and rehabilitated.

Table 17
Livestock Production Objectives by Sample Population

	Project area with VDC				*Project area without VDC*				*Non-project area*			
Production Objectives	*Forested WS*		*Agrl. WS*		*Forested WS*		*Agrl. WS*		*Forested WS*		*Agrl. WS*	
Milk	No.	%	No.	%	No.	%	No	%	No.	%	No.	%
Self	147	67.74	69	72.63	34	56.66	84	70.59	103	88.79	91	92.8
Both (M & S)	70	32.25	26	27.37	26	43.33	35	29.41	13	11.20	7	7.14
Total	217		95		60		119		116		98	
Milk-Based Product												
Self	187	86.17	75	78.95	50	83.33	95	79.83	85	73.27	80	81.63
Both (M & S)	30	13.82	20	21.05	10	16.66	24	20.17	31	26.72	18	18.36
Total	217		95		60		119		116		98	
Agriculture												
Self	230	100	84	100	53	100	119	100	116	100	92	100
Both (M & S)	0	0	0	0	0	0	0	0	0	0	0	0
Total	230		84		53		119		116		92	

Note: M & S: Market and Self

Table 18
Animal Grazing on Common Property Resources by Sample Population

Grazing on CPRs	Project area with VDC				Project area without VDC				Non-project area			
	Forested WS		Agrl. WS		Forested WS		Agrl. WS		Forested WS		Agrl. WS	
Livestock	No.	% Change	No.	% Change	No.	% Change	No.	% Change	No.	% Change	No.	% Change
Before	7.65		6.0364		8.04167		4.1597		5.9297		5.605	
After	3.7	-51.63	3.7364	-38.10	3.77778	-53.02	3.1513	-24.24	3.3672	-43.21	3.6303	-35.23
Difference	-3.95		-2.3		-4.2639		-1.0084		-2.563		-1.975	
Grazing Days												
Before	365		365		365		365		365		365	
After	180	-50.68	190	-47.94	210	-42.46	220	-39.72	320	-12.32	245	-32.87
Difference	-185		-175		-155		-145		-45		-120	

Such change has been noticed in both the project area as well as non-project area (See table 18). However, the incidence of animal grazing on CPRs has declined comparatively more rapidly in project area than project area without VDC, and more decline in animal grazing has been reported in project area than non-project area. For instance, before project intervention, on average more than 7 animals per household were grazed on CPRs in forested watershed and about 6 animals per household in agricultural watershed. After project intervention, users groups have been created to protect and maintain the enclosed CPRs. Consequently, the average number of livestock grazed on CPRs declined to less than 4, and such a decline is in the range of 38 per cent to 52 per cent, the lowest in agricultural watershed and highest in forested watershed.

Not only the number of animals grazed on CPRs has declined, but the number of animal grazing days on CPRs has also declined significantly. For instance, before project intervention, due to open access to CPRs, the animals were grazed throughout the year. Whereas, after the project intervention, the decline in number of animal grazing has been reported in the range of 48 per cent to 51 days in project area with institutional arrangement and 40 days to 42 days in project area without institutional arrangement. Similarly, a decline in animal grazing in the range of 12 per cent to 33 per cent has been recorded in non-project area. It is to be noted that recently the forest department has started enclosing the forests for regeneration and animal grazing has been banned. The creation of closures has resulted in heavy grazing pressure on non-protected and non-enclosed CPRs, which are facing severe degradation. Thus, there is urgent need to introduce new high yielding varieties of grasses and fodder, which could be planted in enclosed CPRs as well as private lands. Besides, more and more user groups need to be created so that equitable usufruct sharing mechanism should be established. The user groups would also be helpful in introducing the system of stall-feeding on wider scale.

The creation of closures on VCLs and forestlands has resulted in significant decline in the availability of forest products, which were freely available before project interventions. After the project interventions, the regeneration of CPRs have taken place, which have improved the availability of forest products significantly in forested watersheds than the agricultural watershed. However, there are wide variations in availability of forest products across the selected watershed (See table 19). The villagers, who are unable to meet their fuel wood, fodder, small timber and minor-forest produce opt for a deficit strategy. The data makes it clear that rotational use of forest resources has been in use only in forested watershed, where institutional arrangements has been made through formation of user groups. In other cases, use of buffer stock, feeling of own trees as well as illicit felling of trees from the local forests are the dominant deficit strategy adopted. It is to be noted that even after the project intervention, illicit felling of trees is most common in agricultural

Table 19
Availability of Forest Products in Sample Sub-Watersheds

	Project area with VDC				*Project area without VDC*				*Non-project area*			
Small Timber	*Forested WS*		*Agrl. WS*		*Forested WS*		*Agrl. WS*		*Forested WS*		*Agrl. WS*	
Month (s)	No.	%	No.	%	No.	%	N0.	%	No.	%	No.	%
Below 2	119	51.07	15	13.63	48	66.66	69	57.98	61	47.66	43	36.44
Between 2-4	58	24.89	20	18.18	15	20.83	33	27.73	22	17.19	53	44.91
Between 4-6	30	12.87	74	67.27	2	2.77	7	5.88	17	13.28	15	12.71
Between 6-8	10	4.29	1	0.90	3	4.16	3	2.52	14	10.94	0	0
Between 8-10	4	1.716	0	0	1	1.38	0	0	0	0	0	0
Above 10	10	4.29	0	0	2	2.77	0	0	0	0	0	0
Fuel wood												
Below 2	52	22.31	15	13.63	27	37.5	21	17.64	36	13.79	31	26.27
Between 2-4	86	36.98	20	18.18	28	38.88	56	47.05	41	32.03	59	50
Between 4-6	40	17.16	5	4.54	11	15.27	30	25.21	25	19.53	19	16.10
Between 6-8	12	5.15	70	63.63	3	4.16	5	4.20	15	11.72	1	0.84
Between 8-10	5	2.14	0	0	1	1.38	1	0.84	1	0.78	0	0
Above 10	11	4.72	0	0	2	2.77	5	4.20	0	0	0	0
Fodder												
Below 2	70	30.04	15	13.63	16	22.22	30	25.21	29	22.66	46	38.98
Between 2-4	64	27.46	21	19.09	23	31.94	38	31.93	31	24.22	25	21.18
Between 4-6	58	24.89	73	66.36	8	11.11	44	36.97	33	25.78	10	8.47
Between 6-8	23	9.87	1	0.90	4	5.55	0	0	13	10.16	1	0.84
Between 8-10	7	3.00	0	0	2	2.77	0	0	1	0.78	0	0
Above 10	2	0.85	0	0	6	8.33	0	0	0	0	0	0

Table Contd...

Small Timber	Project area with VDC				Project area without VDC				Non-project area			
	Forested WS		Agrl. WS		Forested WS		Agrl. WS		Forested WS		Agrl. WS	
Minor Forest Produce												
Below 2	83	35.62	15	13.63	26	36.11	5	4.20	35	27.34	28	23.72
Between 2-4	58	24.89	20	18.18	19	26.38	7	5.88	27	21.09	37	31.35
Between 4-6	58	24.89	5	4.54	6	8.33	11	9.24	26	20.31	24	20.33
Between 6-8	25	10.73	1	0.90	14	19.44	36	30.25	26	20.31	9	7.62
Between 8-10	8	3.43	0	0	0	0	2	1.68	3	2.34	1	0.84
Above 10	0	0	0	0	2	2.77	0	0	0	0	0	0
Deficit Strategy												
Purchase	22	9.44	0	0	2	2.77	32	26.89	1	0.78	0	0
Buffer stock	141	60.51	110	100	72	100	119	100	127	99.22	112	94.91
Fell own Trees	168	72.10	110	100	72	100	119	100	127	99.22	67	56.78
Illicit Felling	137	58.79	110	100	5	6.94	119	100	127	99.22	112	94.91
Rotational Use	8	3.43	0	0	0	0	0	0	0	0	0	0

watersheds, where the villagers reportedly intrude the forests in the vicinity to fell the trees to meet the fuel wood and timber requirements.

Consumption of CPRs Products

The table 20 gives the data on average annual household consumption and the value of CPRs products collected by the sample population. A mere perusal of the data makes it evident that fuel wood collected from CPRs has declined very sharply in agricultural watershed in project area without VDC followed by agricultural and forested watershed in project area with VDC. The decline in fuel wood collected from CPRs in non-project area has also been reported, however, in fewer amounts than project area, which may be attributed to protection of forests by forest department, and/or decline in the potential of CPRs due to severe degradation. Similarly, decline has been noticed in fodder collection after project intervention. The decline in fodder collection is reportedly very sharp in project area with VDC. In non-project area too, a marginal decline has been reported in fodder collection. A decline in collection of timber and non-timber products and manure has been noticed. However, with project interventions, the value of horticultural product collected from CPRs has increased though modestly. Similar trend can also be noticed in non-project area.

In the project area with VDC, the decline in fuelwood collected from CPRs was to the tune of 51.95 per cent and 60.86 per cent respectively in forested and agricultural sub-watersheds. It was 20.86 per cent and 49.58 per cent respectively in project area without VDC. The decline in fuelwood collection from CPRs has taken place in non-project area also, but at a very slow pace, and it was reportedly 15.35 per cent and 7.9 per cent respectively in forested and agricultural sub-watersheds. More or less, a similar trend can be noticed for fodder and non-timber products collection from CPRs across the sample area. In case of horticultural products collected from CPRs, an increased was noticed in the selected sub-watersheds. The increase was significantly marked in non-project area and project area without VDC, but when we observe the data in rupees terms, the amount of horticultural products collected from CPRs was significantly higher in project area with VDC than project area without VDC and non-project area. It is to be noted that collection of food and fodder from CPRs has declined significantly over the period in non-project area, whereas a positive trend was visible for project area with VDC and without VDC. Over the period, the fuelwood, manure and timber collection from CPRs have declined across the sample area, however, the decline is markedly high in non-project area than project area with VDC and without VDC. Thus, it can be inferred that with project interventions, the productivity of CPRs has increased very rapidly in project area with VDC followed by project area without VDC compared to non-project area.

Table 20
Average Annual Household Consumption and CPRs Products Collected by Sample Population

CPRs Products Collected	Average Annual Household Consumption (Rs.)											
	Project area with VDC				*Project area without VDC*				*Non-project area*			
	Forested WS		*Agrl. WS*		*Forested WS*		*Agrl. WS*		*Forested WS*		*Agrl. WS*	
Fuelwood	Rs.	% Change	Rs.	% Change	Rs.	% Change	Rs.	% Change	Rs.	% Change	Rs.	% Change
Before	3424		3240		4324		4645		4566		4765	
After	1645	-51.95	1268	-60.86	3422	-20.86	2342	-49.58	3865	-15.35	4388	-7.91
Difference	-1779		-1972		-902		-2303		-701		-377	
Fodder												
Before	4685		4586		4690		4885		4656		4322	
After	2386	-50.92	3215	-29.89	3425	-26.97	3864	-20.90	4536	-2.57	4020	-6.98
Difference	-2299		-1371		-1265		-1021		-120		-302	
Horticultural Products												
Before	657		334		458		342		272		124	
After	865	31.65	486	45.50	643	40.39	536	56.72	514	88.97	271	118.54
Difference	208		152		185		194		242		147	
Non-timber Products												
Before	865		654		579		634		565		520	
After	453	-47.63	435	-33.48	455	-21.41	598	-5.67	534	-5.48	485	-6.73
Difference	-412		-219		-124		-36		-31		-35	

Table Contd...

	Average Annual CPRs Products Collected (Rs.)											
	Project area with VDC				*Project area without VDC*				*Non-project area*			
CPRs Products Collected	*Forested WS*		*Agrl. WS*		*Forested WS*		*Agrl. WS*		*Forested WS*		*Agrl. WS*	
Food												
Before	435		534		342		245		185		230	
After	546	25.51	864	61.79	428	25.14	337	37.55	56	-69.72	48	-79.13
Difference	111		330		86		92		-129		-182	
Fodder												
Before	5399		4685		5320		5643		5434		4558	
After	6656	23.28	7215	54.00	5960	12.03	6756	19.72	4536	-16.52	4020	-11.80
Difference	1257		2530		640		1113		-898		-538	
Fuelwood												
Before	4224		3865		4534		4985		4985		4890	
After	1645	-61.05	1268	-67.19	3422	-24.52	2342	-53.01	3865	-22.46	4388	-10.26
Difference	-2579		-2597		-1112		-2643		-1120		-502	
Manure												
Before	400		435		356		170		249		286	
After	254	-36.5	234	-46.20	143	-59.83	67	-60.58	56	-77.51	79	-72.37
Difference	-146		-201		-213		-103		-193		-207	
Timber												
Before	1865		2436		1563		2306		1486		2366	
After	258	-86.16	324	-86.69	354	-77.35	263	-88.59	85	-94.27	124	-1808.06
Difference	-1607		-2112		-1209		-2043		-1401		-2242	

Table 21
Average Annual Households' Fuelwood Consumption from CPRs by Sample Population

	Project area with VDC				*Project area without VDC*				*Non-project area*			
	Forested WS		*Agrl. WS*		*Forested WS*		*Agrl. WS*		*Forested WS*		*Agrl. WS*	
Fuel wood Consumption	Kg.	% Change	Kg.	% Change	Kg.	% Change	Kg.	% Change	Kg.	% Change	Kg.	% Change
Monsoon												
Before	422		398		458		406		356		308	
After	248	-41.23	204	-48.74	286	-37.55	184	-54.67	324	-8.98	234	-24.02
Difference	-174		-194		-172		-222		-32		-74	
Winter												
Before	548		522		512		447		389		364	
After	288	-47.44	296	-43.29	324	-36.71	311	-30.42	346	-11.05	294	-19.23
Difference	-260		-226		-188		-136		-43		-70	
Summer												
Before	346		335		423		349		314		265	
After	198	-42.77	176	-47.46	256	-39.47	224	-35.81	226	-28.02	182	-31.32
Difference	-148		-159		-167		-125		-88		-83	
Average												
Before	438.6		418.3		464.33		400.66		353		312.33	
After	244.6	-44.22	225.3	-46.13	288.66	-37.83	239.66	-40.18	298.66	-15.39	236.66	-24.22
Difference	-194		-193		-175.67		-161		-54.34		-75.67	

The annual fuel wood consumption from CPRs has declined after the project intervention. The decline in average fuel wood consumption was reportedly high in project area than non-project area. (See table 21). The average decline in fuel wood consumption from CPRs was high in project area with VDC followed by project area without VDC and non-project area. On average, the decline in food wood consumption was 44.22 per cent and 46.13 per cent respectively in forested and agricultural sub-watersheds in project area with VDC. In project area without VDC, the decline was to the tune of 37.83 per cent and 40.18 per cent respectively in forested and agricultural sub-watersheds. The decline was comparatively low in non-project area and stood at 15.39 per cent and 24.22 per cent respectively in forested and agricultural sub-watersheds. It is significant to note that decline in fuel wood consumption varies over the year. The highest decline is reported in winter followed by monsoon and summer seasons. The selected sub-watersheds are experiencing extreme weather conditions, where fuel wood consumption is comparatively high in winter, whereas, the highest decline is also recorded in the same season. They meet their fuel wood requirement in winter from the buffer stock maintained to meet the contingencies.

CONCLUSION

The agricultural intensification and diversification has occurred with project interventions consisting of technological transfers, access to agricultural extension services, improved access to local markets and subsidized agricultural input prices. After repairing irrigation drainage system, water harvesting structures and check dams, the conditions of the farmland have improved and the crop production has increased. The water resource was also enhanced by soil conservation, protection of green cover and reforestation, orchards, etc. The erosion control programme carried out by the project has reduced the erosion significantly. The vegetation has increased modestly and the forestland cover has also increased. Vegetation management practices included planting trees in the enclosed forestland, VCLs and private agricultural and community land. In most of the cases, only degraded natural forests have been handed over to the user groups, as there is an informal rule within the IWDP (Hills-II) not to hand over well-stocked forests. Consequently, access to many community forests has been restricted temporarily in order to allow these forests to recover. As a result, many user group members have to rely on unprotected forest to meet their basic needs.

In agricultural sub-watersheds in project area, there is severe fuel wood and fodder scarcity. Grazing cattle in community forest is either prohibited to defined periods during the year. As a result, people rely heavily on unprotected forests and VCLs for fodder, which further add to the forest degradation. Livestock composition has changed substantially, however, the number of crossbred cows increased, but other livestock fell in number.

An increase in milk production and productivity has been reported in project area. The decline in the sheep population was particularly steep, which was due to the fact that mostly landless and marginal farmers were keeping sheep and goats, which had free access to common grazing lands and wastelands. As a result of the project intervention, grazing was stopped on VCLs under regeneration, reducing the grazing area and depleting the forage potential. In spite of a reduction in total area available for grazing, dry forage production increased significantly compared to the pre project phase. The increase in dry forage production is mainly due to reseeding of field funds and hill slopes with grasses and legumes, and regenerating old rootstocks on hill slopes. The average dry forage productivity went up during the post project phase.

Degraded VCLs and forestland, which are temporarily closed and effectively protected, will yield more forest-produce in the future. In this sense, community forestry leads to a Pareto improvement for the members of a user group. In the short run, however, the poor have to suffer the most as the VCLs and forestland has been closed temporarily. Opposition to VCLs and forest closures has been noticed during the field visits. On the other hand, the better-off factions of a user group, e.g. land owners with tree on their own lands, do not oppose community forestry because they have alternative to forest use. At the user groups level, the heterogeneity and fragility of land resources along with the variable rainfall made it difficult to fully harness the potential of CPRs and adequately meet the environmental risks through private resources based on crop farming alone. The balancing of intensive (by cropping) and extensive (by pasture forest) use of land, as required by the resource characteristics became a part of collective strategies for risk management and production enhancement.

REFERENCES

Anderson, J. R. and Thamapillai, J. (1990), Soil Conservation in Developing Countries: Project and Policy Intervention, *Policy and Research Series Paper 8*, World Bank Publication, *World Bank*, Washington, D.C.

Blaikie, Piers (1983), The Political Economy of Soil Erosion, in T. O. Rioardan and R.K.Turner (eds.), *Progress in Resources Management and Environmental Planning*, Vol. 4, John Wiley and Sons Ltd.

Blaikie, Piers (1985), *The Political Economy of Soil Erosion in Developing Countries*, Longman: London.

Brooks, K. N., Tejwani K. J. Ffolliott, P. F. and Gregersen H. M. (1990), *Review of Role of Watershed Management in World Bank Forestry Projects*, World Bank: Washington, D.C.

Chopra, Kanchan; G. K. Kadakodi and M. N. Murty (1989), People's Participation and Common Property Resources, *Economic and Political Weekly*, Vol. 24.

Chopra, Kanchan; G. K. Kadekodi and M. N. Murty (1990), *Participatory Development: People and Common Property Resources*, Sage: New Delhi.

Conway, G. R. and J. A. McCracken (1990), Rapid Rural Appraisal and Agroecosystem Analysis, in M. A. Altieri and S. B. Hect (eds.), *Agroecology and Small Farm Development*, CRC Press: Florida.

Deshpande, R. S. and V. Ratna Reddy (1991), Differential Impact of Watershed Based Technology: Some Analytical Issues, *Indian Journal of Agricultural Economics*, Vol. 46, No. 3.

Easter, K. W., J. A. Dixon, and M. M. Hufschmidt (eds.), (1986), *Watershed Resources Management: An Integrated Framework with Studies from Asia and the Pacific*, Westview Press: Boulder, Colo.

Farrington, J. C. Turton and A. J. James (eds.) (1999), *Participatory Watershed Development*, Oxford University Press, New Delhi.

Government of India (1994), Guidelines for Watershed Development, Department of Wastelands Development, Ministry of Rural Areas and Employment, Government of India, New Delhi.

Hazra, C. R. (2000), Development and Strengthening of Village Level Institutions for Management of Common Property Resource in Vidyan Hills - A Case Study fo Kharaiys Nala Watershed, in Proceedings of *Regional Workshop on Watershed Development Management and Evaluation in the Western Ghats Region of India*, 28-29 Feburary, 2000, Centre for Water Resources Development and Management, Kozhikode.

Integrated Watershed Development Project (IWDP), Hills-II (2001), *Status Report for Supervision Mission of World Bank*, IWDP (Hills-II), Jammu and Kashmir.

Jain, T. C. (1997), Watershed Management Programmes: World Bank Experience-Achievements, Constraints and Future Strategy, in Katyal, J. C. and John Farrington (eds.), *Research for Rainfed Farming*, Proceedings of the ICAR-ODA Joint Workshop, September 11-14, 1995, National Academy of Agricultural research Management (NAARM), Hyderabad.

Kerr, J. M., Ganesh Pangare, Vasudha L., Pangare and P. L. George (1998), The Role of Watershed Projects in Developing Rainfed Agriculture in the Indian Semi-Arid Tropics, (Draft), *ICAR-World Bank Research Project*, World Bank, New Delhi.

Kolavalli, S. (1998), Scaling up Participatory Approaches, Paper prepared for National Workshop for Watershed Approaches to Wastelands Development, New Delhi, 27–29 April, 1998.

Pereira, H. C. (1989), *Policy and Practice in the Management of Tropical Watersheds*, Westview Press: Boulder, Colo.

Rajagopalan, V. (1991), Integrated Watershed Development in India: Some Problems and Perspective, *Indian Journal of Agricultural Economics*, Vol. 46, No. 3.

Rajendran, D. V. and V. Mashar (2000), Watershed Development Programme in the Region — NABARD Experience, in Proceedings of Regional

Workshop on *Watershed Development Management and Evaluation in the Western Ghats Region of India*, 28-29 Feburary, 2000, Centre for Water Resource Development and Management, Kozhikode.

Ramanna, R. (1991), Watershed Approach to Dryland Agricultural Development, *Indian Journal of Agricultural Economics*, Vol. 46, No. 3.

Sadhu, A. N. and Sudan, F. K. (1999), *Retrospective Study: IWDP (Hills-I)*, Integrated Watershed Development Project, Hills-II, Jammu and Kashmir, World Bank and Integrated Watershed Development Project, Hills-II, Jammu and Kashmir.

Shah, Amita (1997), Watershed Development Programmes in India: Emerging Issues for Environment-Development Perspectives, Paper presented at the Workshop on *Environment and Agriculture*, Delhi School of Economics, Delhi, December.

Shah, Anil (2000), *Eloquent Silent Revolution*, Development Support Centre, Ahmedabad.

Singh, A. J., Joshi, A. S., Singh, R. P. and Gupta, R. (1991), An Economic Appraisal of *Kand*i Watershed and Area Development Project in Punjab, *Indian Journal of Agricultural Economics*, July–Sept.

Singh, Katar (1989), Dryland Watershed Development and Management: A Case Study in Karnataka, Paper presented at the Golden Jubilee Conference of the Indian Society of Agricultural Economics, Mumbai.

Singh, Katar (1995), The Watershed Management Approach to Sustainability of Renewable Common Pool Natural Resources: Lessons of India's Experience, *Research Paper* 14, Institute of Rural Management, Anand.

Wankhede S. D. and J. P. Mahalle (2000), Integrated Watershed Development Programme in Western Ghats in Maharashtra, in Proceedings of Regional Workshop on *Watershed Development Management and Evaluation in the Western Ghats Region of India*, 28-29 Feburary, 2000, Centre for Water Resources Development and Management, Kozhikode.

White, Thomsan A. (1992), Peasant Initiatives for Soil Conservation: Case Study of Recent Technical and Social Innovations for Miassade, Hatti, EPAT/MUCIA, Working Paper No. 4.

White, Thomsan A., Ane Quinn, Robert M. (1992), An Economic Analysis of Miassade, Hatti, Integrated Management Project, EPAT/MUCIA, Working Paper No. 2.

World Bank (1992), *World Development Report, 1992*, The World Bank: Washington.

World Bank (1999), *Project Appraisal Document: Integrated Watershed Development (Hills-II) Project*, Rural Development Sector Unit, South Asia Region, New Delhi.

World Bank (1998), *Project Concept Document: Integrated Watershed Development (Hills-II) Project*, Rural Development Sector Unit, South Asia Region, New Delhi.

6

IMPROVING IRRIGATION EFFICIENCY IN KERALA THROUGH PARTICIPATORY IRRIGATION MANAGEMENT: AN ANALYSIS

George Chackacherry and K. Madhava Chandran

INTRODUCTION

The conventional understanding of the irrigation sector was primarily limited to design, construction, operation and maintenance of irrigation systems. However, the gap existing between irrigation potential created and utilized in different parts of the world forced governments to look for suitable alternatives. Accordingly, the concept of irrigation management was recognized. This perspective added several non-physical functions such as planning, monitoring, financing and evaluation to the realm of modernization of irrigation systems. Irrigation is not only a physical concept, but also one involving a much broader set of management issues (World Bank, 1998). The International meeting on modernization of irrigation system operations, held at Aurangabad in 1998, defined irrigation modernization as a process of technical and managerial upgrading of irrigation systems (as opposed to mere rehabilitation), combined with institutional reforms with the objective of improving resource mobilization and water delivery service to farms. Countries, which once promoted more government involvement in irrigation projects, are now adopting policies that create incentives for users to take over management of operation and maintenance of the projects, while government agencies focus on improvements at the main system level.

PARTICIPATORY IRRIGATION MANAGEMENT

Participatory Irrigation Management (PIM), also known as Irrigation Management Transfer (IMT), is an attempt to increase direct involvement of farmers in irrigation management. It implies expansion of the role of farmers and the private sector and reduction of the role of government in irrigation management through transfer of authority and responsibilities, either in part or full, to farmer organizations. Such reforms are intended to improve irrigation performance by involving farmers, who are the most important stakeholders. PIM would lead to some form of joint management of irrigation systems, with the State responsible for more tasks at the higher level of projects (main system) and farmer organizations for tasks at lower levels (branch canal/ distributory/ field channel). PIM is the keystone of water resources modernization reforms being implemented in about 60 countries around the world. Efforts to adopt PIM are often initiated by governments mainly due to shortage of funds to manage irrigation projects, which is also an outcome of factors like poor recovery of water charges, high overhead costs etc. It is expected to reduce operation and maintenance costs for governments, while enhancing the efficiency and sustainability of irrigation projects. In addition, the policy orientation of funding agencies like World Bank, and the shift towards decentralization/privatization also contribute towards adoption of measures such as PIM.

The Ninth Five Year Plan of the Government of India had laid thrust on promoting PIM (MOWR, 1996). At the national level, efforts to initiate farmer participation started during the 1980's with the establishment of Command Area Development Authority (CADA) under various irrigation projects. Improvement in irrigation management and crop productivity through promotion of participation through registered farmer organizations was the main objective of this programme. This participatory approach was envisaged as a forerunner to PIM, to be implemented at a later stage under the irrigation projects. CADA was established in Kerala during 1985. A three-tier system, with outlet- based farmer associations at the base level, canal committees at the middle level and project committees at the upper level were envisaged under CADA. According to the report of CADA (2004), 4011 farmer associations were formed in 16 irrigation project commands in the State. However, studies have shown that only about 30% of them are functional, while the extent of farmer participation in the functioning associations is also low (Chackacherry, 1995; CWRDM, 1999). However, CADA programme has been able to provide a platform for a change in the outlook of farmers towards a decentralized and democratic system of management.

Kerala has no rich tradition in the management of irrigation systems by farmers, though farmer managed irrigation systems exist in various parts of the country, for hundreds of years. This may be attributed to better availability of water resources in Kerala in the past, which might not have encouraged community action for irrigation management (Chackacherry, 1995). Serious initiatives had not been taken up in Kerala for irrigation management transfer to farmers through PIM, especially in the major/ medium irrigation sector. However, neighbouring States like Tamil Nadu and Andhra Pradesh are well ahead with regard to implementation of

PIM. On the one hand, under the present set up in our State, no part of the irrigation system can be handed over for management to farmer organizations, since they are not socially capable of taking up the tasks assigned to them; they should be either reorganized/restructured or adequately strengthened. On the other hand, the government agency concerned with these activities is neither physically nor socially equipped for taking up joint management with farmers. There are several problems that may hinder the implementation of PIM in Kerala such as less number of full-time farmers, small sized holdings, lack of sufficient interest among the farmers, reluctance of officials, etc. It was in this background that a study was undertaken jointly by the Centre for Water Resources Development & Management (CWRDM), Kerala and CADA of the State, with the main objectives of suggesting measures for reorganizing/restructuring the existing farmer associations for implementing PIM in Kerala, and to suggest improvements in the agency set up for supporting the process.

METHODOLOGY

The study was carried out in three irrigation projects in Kerala, namely, Neyyar, Chalakkudy and Malampuzha. Primary data was collected through Participatory Rural Appraisal (PRA), questionnaire and interview method from farmers and officials. Farmer associations were selected randomly from the list of live associations maintained by CADA. PRA was carried out with 41 farmer associations under the irrigation projects. The tools used for PRA include focus group discussion, semi-structured interview and pair-wise ranking. 650 farmers were consulted. Discussions were held with canal committee chairman, canal committee members and project committee members. 30 women randomly selected from ten farmers' association were consulted regarding their involvement in irrigated agriculture, following semi-structured interview method.

Besides farmers, 300 officials of the Irrigation department, CADA and Agriculture department from Lascar to Chief Engineer level were covered during group discussions and interviews. Their hopes and apprehensions about PIM were assessed, and opinions and suggestions were pooled. Members of Legislative assembly, leaders of political parties, Secretaries to Government etc. were also interviewed under the study. Secondary data from earlier studies carried out by the investigators, reports of Ministry of Water Resources etc. were also made use of in the study.

FINDINGS AND DISCUSSION

The responses obtained from farmers and officials have been analyzed and interpreted as factors to be considered/pre-requisites for implementing PIM in Kerala, as outlined below (CWRDM & CADA, 2001).

Favourable Orientation of Politicians and Administrators

At the outset, it is necessary for the political, administrative and irrigation agency leadership to take interest in adopting PIM. Andhra Pradesh pioneered in implementing PIM mainly because of the strong will and commitment of the political and administrative machinery there, including the Chief Minister, which resulted in the enactment of *The Andhra Pradesh Farmers Management of Irrigation System Act*, 1997. The present study has also shown a positive orientation on the part of the administrative and irrigation agency in Kerala towards PIM.

Changes in Organizational Set up

Irrigation agencies (CADA and Water Resources Department) and farmer associations constitute the present organizational structure under irrigation projects. This structure is primarily designed for full agency management, and there is no real opportunity for farmer or joint management. FAO (1995) mention that the major functions under irrigation projects such as governance, financial management, and execution mode and accountability pattern are all adapted to agency management only. According to CWC (1992), the task of organizational and procedural changes is to re-focus the emphasis of irrigation management from serving the macro level bureaucratic priorities to the dynamic field conditions and micro demand for services. Hence, changes are required in organizational structure with regard to clear water rights at the system and farmer levels, clearly designed line of authority between farmers and agency, and supportive accountability and incentive systems for management for implementing PIM in Kerala. Farmers should also be equipped with new managerial and technical skills related to irrigation management. Farmers, who were initially treated as beneficiaries under CADA, should become partners in planning, implementation and evaluation of irrigation development programme, while implementing PIM. The irrigation agency must be made multi-disciplinary, and the bureaucracy should back up efforts for management transfer and assign specific roles to officials to facilitate this. Farmers should have more opportunities for communication and interaction with officials. Sufficient number of awareness/ training programme should be organized for farmers on various aspects of PIM. Systems for regular monitoring/evaluation of PIM should also be instituted.

The process of management transfer to farmers from the irrigation agency requires setting up of farmer organizations, which are capable of assuming responsibilities. Reorganization of farmer associations existing under CADA, and providing them statutory powers would be very helpful in ensuring effectiveness under PIM. Community organizers will have to be employed to help farmers in establishing Water User Associations

(WUAs). NGO's can also offer valuable service on this aspect. For WUAs to be socially and economically viable, a minimum area of operation has to be decided, so that the cost of management does not become too high. However, the area should not be too big, in which case, the number of farmers will become unmanageable. One WUA can be made to serve an area of 200 to 300 ha, depending on hydraulic boundaries.

Water User associations (WUAs) will have to enter into partnership with the irrigation agency and perform functions such as procurement, allocation and distribution of water, generating contributions from members in cash and kind (labour) for maintenance works, checking indiscipline, settling disputes and overall supervision and control of the irrigation system handed over to them under PIM. At the same time, the irrigation agency should be in a position to deliver the agreed quantities of water at the right time and place. Then only, farmers will feel motivated to carry out the tasks assigned to them under PIM. A study carried out by CWRDM in Kerala has revealed that the cardinal factor influencing farmer participation under CAD programme is timely, adequate and equitable water availability (Madhava Chandran and Chackacherry, 2004)

Accountability of WUAs to all the members is an important principle for long-term viability of the organizations. Proper definition of membership and leadership are mechanisms for ensuring this accountability. Identification of good leadership is also important. This has to be ensured while handing over management under PIM to WUAs (CWRDM, 1998).

Legislative Backing

The provisions made in the CADA act of 1986 do not meet the requirements for implementing PIM in Kerala. Even though there exists provisions for formation of farmer associations, canal and project level committees under the act, no specific powers have been given for carrying out duties assigned to them. Also, there is no provision in the act for agreement regarding supply of water to farmer associations. The same is the case with regard to the Kerala Irrigation and Water Conservation Act 2003. The Chapter on PIM provided in the Act is incomplete. There is only one tier organization mentioned in the Act, namely, WUAs at the outlet level. Transfer of the irrigation system, agreement between government and WUAs, etc. are not mentioned in the Act. It is therefore essential to either amend the existing Irrigation and Water Conservation Act or enact a new PIM Act.

Clear Water Rights

Under the existing physical infrastructure of irrigation projects, water rights could be defined as volume of water at the main canal, share of flow at the head of tertiary units and time duration within the command area of the watercourse (Strosser, 1988). Water is treated as the property of the government and most of the irrigation acts in India proceed from the basic premise that water is not a right of farmers. However, under PIM, there has to be a clear concept regarding allocation of water to the WUAs (either as a defined volume or defined proportion), which they purchase for a fee, and distribute among farmers as per accepted rules of equity and sustainability. The water quota for an average year may be decided on the basis of provisions made in the concerned project, and as declared by the general policy guidelines issued by the State government. The document on amendment to irrigation act for PIM of the Government of India (MOWR, 1998) directs that it is necessary for irrigation acts to permit long-term allocation of water to WUAs and incorporate such allocations in the agreements with them.

Under PIM, WUAs should be given freedom in deciding their cropping pattern, within the quantum of water allocated under the operational schedule of the irrigation canal transferred to them. At the same time, they should also have some powers to restrict the type of crops grown by its members, taking into consideration the crop water requirements and need for equity in water distribution among members.

Rehabilitation of the Irrigation System

One of the major reasons for the low efficiency of irrigation projects is lack of proper maintenance of the physical system, primarily due to lack of sufficient government fund. Hence, proper rehabilitation of the system before irrigation management transfer to farmers is absolutely essential in order to ensure water availability, and accordingly, motivate farmers to implement PIM. It should also be ensured that rehabilitation involves farmers in its various stages, and that the work is compatible with farmers' needs.

Meaningful Transfer Agreement

The transfer of irrigation system to farmers under PIM has to be preceded by a number of activities, including preparation of the inventory of works, structures and equipment to be transferred, joint inspection of these by WUAs and irrigation agency, carrying out necessary testing and repairs at the cost of the irrigation agency, and finally, handing over the management of the system, along with all necessary documents to the WUAs. The transfer agreement should also include provisions stating that the system transferred is in good working condition.

The transfer agreement through a memorandum of understanding (MOU) should specify the structures transferred by the agency to WUAs, the area of operation, the rights and responsibilities of both the parties, the legal support available, the conditions regarding remitting of water charges by the WUAs and the agreement from WUAs to carry out the responsibilities assigned to them.

Co-ordination between Stakeholders

The main stakeholders of an irrigation project are farmers and officials. One of the important factors responsible for low performance of irrigated agriculture has been found to be the unwieldy and inefficient bureaucracy spread over various departments, operating within complex departmental procedures and maintaining very less interaction with farmers. Under CADA set up, activities are restricted to below the outlets. Hence, the officials are not in a position to influence water availability, which falls under the purview of the irrigation/water resources department. Hence, while implementing PIM, the concerned officials should be in apposition to co-ordinate their activities, in line with the requirements of farmers, leading to the delivery of adequate and timely water to the WUAs. Similarly, officials should develop a positive attitude towards better interaction/communication with farmers, and also ensure that this is achieved through conducting suitable training/awareness programmes.

Consultation and Involvement of Farmers

Inculcating a sense of ownership is an essential ingredient for effective people's participation. This requires sufficient consultation with farmers and enlisting their involvement in different stages of PIM such as organizing farmers, planning, operation and maintenance, monitoring etc. There are examples of successful farmer participation under minor irrigation in the Kerala Community Irrigation Project (KCIP) and the Kerala Minor Irrigation Project (KMIP) because there were provisions for prior consultation and involvement of farmers under these projects.

Motivation and Incentive Systems

Package of incentives for farmers as well as officials to make the management transfer programme effective can be expected to yield good results under PIM. Sufficient encouragement should be given to officials, who do good work though awards/honours. It is also required to eliminate apprehensions of officials regarding aspects such as job insecurity, professional status, new responsibilities etc. when PIM is implemented. Similarly, farmers should also be motivated to perform their duties properly through provision of relevant incentives. Further, disincentives for free rider farmers should also be incorporated under the programme.

Training/Awareness Programmes

It is required to bring in an attitude change among officials, orienting them towards a set up, wherein, farmers will play a major role in managing irrigation projects. They should also be properly exposed to aspects such as knowledge and skills related to improved irrigation management, PIM related procedural, administrative and organizational issues, available sources of incentives and funding, commercial matters and accounting procedures. Similarly, managerial and technical skills have to be developed among farmers and farmer leaders to take on their new roles under PIM. This requires training and awareness programmes, workshops involving farmers and officials, and exposure visits to projects where PIM has been successfully implemented.

Involvement of Women

Most often men are found to manage irrigation. Women's participation in this line has not been given due attention, even though they often provide considerable assistance in agriculture-related activities of farming households. Throughout most of the developing world, women lack either ownership or effective control of land, water and other natural resources, even though their labour is widely used. Women, actually play a very important role in irrigated farming, but it is mostly indirect and invisible. Many of the problems in irrigation management can be tackled if women are equipped properly. Studies (CWRDM & CADA, 2001) have found that women feel they can involve in various aspects of canal management such as maintenance of the system, water distribution and settlement of disputes. Male farmers would also take into consideration many of the suggestions offered by their wives. Hence, emphasis on involvement of women would add to the sustainability of WUAs under PIM. A gender assessment study conducted in Thrissur district of Kerala under the KCIP, the first irrigation project in Kerala to involve women farmers, has revealed that when only 25% of men depend on agriculture; about 46% women are involved in it. It has also been found that more than 75% of the men farmers are part-time agriculturists, who depend on other vocations for their livelihood, and hence, have limited interest in farming (Chackacherry, 1996). In the training programme organized under KCIP, the main participants have been women. They were also found to show interest in knowing about improved agronomic practices. Further, about 45% of members of WUAs under KCIP were women. Chackacherry (1996) reported that men do not have any problem in bringing women into irrigated agriculture. In short, efforts to orient women and equip them to take on responsibilities and tasks under PIM can be expected to offer considerable help in implementing irrigation management transfer in Kerala.

Mechanisms to Settle Disputes

Many of the problems in irrigation management arise due to lack of transparency in the functioning of the irrigation system as such. Of prime importance in this regard is the awareness on water account. Aspects such as the quantity of water available in the reservoir / main system, how much will be available to the users, the policy directives of government which influence water availability / use etc. are to be made aware to the farmers by the PIM implementing agency. Designated officers of the agency should be given powers to penalize offending irrigators. Most importantly, the WUA should be empowered to stop supply of water to a user, in case an offence is committed by him/her. PIM will involve some form of sharing arrangement for water supplies between members. This can be helpful in building up good social capital of trust and co-operation between farmers, a condition naturally favourable for settlement of disputes among them.

PILOT PROJECTS ON PIM

As mentioned earlier, Kerala lags behind many states in India in participating farmers in the management of irrigation, and implementing PIM. Based on the outcome of the study (CWRDM & CADA, 2001), two pilot projects on PIM were formulated. They are now being implemented in Neyyar (mostly catering garden land crops) and Malampuzha irrigation projects in the State (mostly catering rice), with the objective of evolving a strategy for implementation of PIM in all the irrigation projects of Kerala, by demonstrating its possibility and convincing officials, farmers and others concerned. The programme envisaged is to transfer the operation and maintenance of one branch canal each at Neyyar Irrigation Project (Olathanni branch canal – 6.41 km length, 501 ha of ayacut area with garden land crops) and at Malampuzha Irrigation Project (Kuthannur branch canal – 14.63 km length, 1664 ha of ayacut area with rice crop) to the farming community. The strategy and methodology adopted, and the phasing of the pilot projects are based on the suggestions of the study referred above. Since the existing farmer organization structure and government set up are not congenial for the implementation of PIM, changes are brought in on these aspects. A three-tier system with outlet based Water User Associations (WUAs), branch/distributory level WUAs, and project level Project Management Council is the structure tried in the farmer organization set up. Overseer, Assistant Engineer/Assistant Executive Engineer, and Executive Engineer respectively, will be attached as competent authority to these associations to help them in implementing their decisions. Works in the irrigation systems will be identified, prioritized and implemented by the WUAs. The competent authority will render technical advice and ensure that the works are carried out as per technical specifications. During turnover of operation and maintenance to WUAs, the WUA concerned and Water Resources Department will enter into a MOU, which clearly mentions about the commitments of both parties. Encouraging group farming, bringing women to the mainstream of irrigated agriculture, establishing relations with panchayaths, creating opportunities for coordination among the departments/agencies concerned, mobilizing tie-ups with marketing establishments, etc. are other allied activities carried out under the project. The programmes are carried out through five different phases, namely, Preparation, Organisation, Rehabilitation, Capacity Building and Turnover phases. Presently the pilot projects are in the last part of the Rehabilitation phase.

CONCLUSION

Till recently, irrigation management was considered as a technical or engineering subject, but it is now looked upon more and more as an integrated socio-technical enterprise, involving not only technical skills but also managerial tasks relating to the irrigation system. Irrigation management in almost all countries has been undergoing a great change during the last two decades. Management practices of the past era are no longer sufficiently cost effective, efficient in water use, sustainable, environmentally sound, or in line with trends for decentralization, privatization and accountability to users. Countries that once promoted more and more government involvement in irrigation management are adopting new policies that do just the opposite. Therefore, PIM is considered as a strategic intervention to improve the irrigation sector and mitigate the existing problem of farmers, water managers, and government. Interestingly, PIM is seen as a means to relieve budgetary pressures, especially when the cost of deferred maintenance cumulates year after year. Apart from the prerequisites mentioned in this article, and the strategy worked out from the study, what is more important is the fact that implementation of PIM needs to be carried out in a carefully staggered manner, which allows adequate time and effort in training and motivation, making use of the farmers' organizational skills and experience.

ACKNOWLEDGEMENT

We are extremely grateful to Dr E. J. James, Executive Director, Centre for Water Resources Development and Management, Kozhikode, Kerala for all the encouragement given for conducting the study and preparation of this research paper.

REFERENCES

Central Water Commission (1992), *An Approach to Organisational and Procedural Changes in Irrigation Sector,* Report of Central Water Commission, New Delhi.

Centre for Water Resources Development and Management (1998), *Evaluation of Community Irrigation Systems,* Final Report submitted to the State Committee on Science, Technology and Environment, Government of Kerala.

Centre for Water Resources Development and Management (1999), *Evaluation of Beneficiary Farmers' Associations under Command Area Development Programme in Kerala.* Final Report submitted to INCID, New Delhi.

Centre for Water Resources Development and Management and Command Area Development in Kerala (2001), *Implementation of Participatory Irrigation Management in* Kerala, Final Report of the Study on Re-organisation of Farmer Associations for Implementing PIM in Kerala, submitted to Government of Kerala.

Chackacherry, George (1995), *Trends in Decentralisation in Water Management in Kerala,* In Proceedings of the International Conference on Water Management, Chennai, India: Confederation of Indian Industry.

Chackacherry, George (1996), *Kitchen to Farm: Suggestions for an Action Oriented Plan and Working Paper on Women in Agro-Economic Activities*, Kerala Community Irrigation Project, Kerala.

Command Area Development Authority (2004), *Annual Report 2003-2004,* Kerala.

Food and Agriculture Organisation (1995), *Irrigation Management Transfer in Asia,* Papers from the Expert Consultation on Irrigation Management Transfer in Asia, RAP Publication.

Madhava Chandran, K. and Chackacherry, George (2004), Factors Influencing Farmer Participation in Irrigation Management, *Journal of Tropical Agriculture,* 42(1-2) : 77-79.

Ministry of Water Resources (1996), *Report of the Working Group on Participatory Irrigation, Management for Formulation of Plan Proposals for Ninth Five Year Plan* (Draft), New Delhi.

Ministry of Water Resources (1998), *Amendment to Irrigation Act for Participatory Irrigation Management,* New Delhi.

Strosser, Pierre (1998), Water Markets in Pakistan, *INPIM Newsletter*, No.8.

World Bank (1998), *Handbook on Participatory Irrigation Management*, Economic Development Institute, Washington.

7

PARTICIPATORY IRRIGATION WATER MANAGEMENT IN JAMMU AND KASHMIR

Falendra K. Sudan

INTRODUCTION

Over the past few decades, water scarcity has emerged as a global problem. In course of time, the level of human interventions has grown to such an extent quantitatively that the need for a more informed approach to this vital natural resource endowment is being recognized and articulated all over the world (Falkenmark *et al.*, 1980). As human increasingly started to intervene in the flow of rivers (Deursen, 2000), the gradual success of these interventions resulted in the upscaling of their operational dimensions that finally led to the emergence of large dams as instruments for river basin development (Bandyopadhyay *et al*, 2002). The large-scale supply-side augmentation has been prescribed by engineers as the only way of meeting the growing water requirements of developing countries. As a result, policy–makers in India have been adhering to the idea of solving the water problem by harnessing the huge monsoon run-off in the main rivers, primarily with the help of such large interventions (Bandyopadhyay and Perveen, 2004). The proposal for the interlinking of rivers offering a combination of storage dams and transfer canals has been offered as a long-term supply-side solution to water shortages (IWRS, 1996).

The idea of interlinking rivers is described as the perfect win-win solution that will address the problems of water scarcity and floods. In contrast with the official prescription for interlinking of rivers, there is a strong opinion that India's water crisis is caused by the mismanagement of water resources, and the solution do not lie merely in supply-side augmentations. There has been no serious attempt to work out a series of area specific answers by way of local conservation and augmentation to the maximum extent possible. Recently, the drought conditions were the result of bad water management in the past and the answer lay in better resource management in the future (Iyer, 2000). Thus, the proposed interlinking of river needs to be assessed in the light of the emerging paradigm changes in the subject of water management. What is needed is a fundamental shift away from present reductionist engineering paradigm to a holistic and trans-disciplinary one. The new ways of managing water has increasingly been identified as integrated water resource management (IWRM). A higher priority should be assigned to projects that meet basic and unmet needs of water (Gleick, 1996). Following these paradigmatic shifts in notions worldwide, various other means to conserve water in stream is emphasized. In the Indian context, the old paradigm is still strongly entrenched in the official water administration. The exclusively supply-side solutions offered by the present paradigm tend to either impinge on demands by others, or cause serious damage to nature. The demand-side management is therefore slowly becoming a new paradigm for water governance.

Nigam *et al.* (1997) makes it clear that if the precipitation available within the respective watersheds or sub-basins is harvested and conserved properly, the satisfaction of domestic water needs would not be a problem in most parts of the country. For promoting domestic water security, local level water conservation is a suitable option compared with developing large storage and long distance diversion facilities, as these often carry high financial, social and ecological costs (World Commission on Dams, 2000). The construction of dams on the Himalayan rivers cannot be undertaken by ignoring vital questions on the uncertainty associated with taking a mechanical and traditional view of development of the Himalayan rivers (Ives and Messerli, 1989). Due to the verticality and the consequent fragility of the Himalaya, large dam on Himalayan rivers will be subject to high levels of seismic hazards, since the potential for earth quakes at the plate boundary all along the Himalayan foothills is well known and widely accepted (Khattri, 1987). Keeping this in view, the newer interventions may be counter-productive.

Besides, there has been the inherent failure of policy-makers to perceive the dimensions of the upstream-downstream linkages. Such gaps have led to a major source of conflict over water use in river basins all over the world. The inequitable distribution of fresh water leads to violent inter- and intra-basin conflicts. The long-drawn Cauvery dispute is a case in point. The situation with respect to Satluz-Yamuna Link (SYL) canal is no different. Inevitably, interstate trans-boundary water-related conflicts may become inter-country. The proposal risks major confrontation with Bangladesh, which receives much of its water from the Ganga and the Brahmaputra, after they leave India. The government of Bangladesh protested to India but has so far not received any response (Vidal, 2003). Unfortunately, many plans for river basin development in developing countries are inflexible and the requisite spirit of cooperation is often lacking (Serageldin, 1995). The results are that damages to the ecosystem services or losses in downstream population have been ignored. In the case of India, the situation is more complex, with the downstream areas belonging to another country, as is evident from the recent controversies regarding construction of Baglihar dam in Jammu and Kashmir and the steps to abrogate 1960 Indus Water Treaty. Keeping the above in view, the only solution to the problem of domestic water scarcity in dry upland lies in demand management through local-level harvesting and conservation of rainfall in decentralized and sustainable manner.

India has embarked upon Participatory Irrigation Management (PIM), under which the management of some of the systems is being turned over to the Water Users' associations (WUAs). The conservation of land and groundwater resources through rainwater harvesting is another area where market failure as well as government failure has been glaring. As a solution to this, Participatory Watershed Development (PWD) is being evolved and practiced in many countries, including India, for the protection and sustainable use of land and water resources. The users' associations reflect definite institutional arrangements in which rewards and sanctions were clearly defined in order for the systems to work. Therefore, a solution that has been tried in various places is to transfer the responsibility for the operation and maintenance (O and M) of the canal systems below certain point to the local farmers. At present, most state governments in India encourage the formation of WUAs and eventually want to transfer irrigation management to irrigators (Parthasarthy, 2000). Integrated Watershed Development Project (IWDP), Hills-II, Jammu and Kashmir decision on PIM is a step in this direction. WUAs are expected to serve as channels for feedback and feed-forward information to improve the performance and services provided by O and M irrigation personnel. The initiatives of IWDP, Hills-II should be seen as the first steps towards attaining sustainable water resource management. With this backdrop, an attempt has been made to analyze the process of evolving participatory irrigation water management under IWDP, Hills-II; and to examine the impact of watershed interventions on reduction in rainwater loss and sediment yield; status of water resources and irrigation; and change in crop intensity and crop productivity.

STUDY AREA AND METHODOLOGY

Integrated Watershed Development Project (IWDP), Hills-II, Jammu and Kashmir covers two sub-watersheds in *Shivaliks*, viz. Ramnagar and Akhnoor and two sub-watersheds in *Karewas*, viz. Rajwar and Rambiyara. The present study was confined to two sub-watersheds of Akhnoor (Jammu district) and Ramnagar (Udhampur district). While Ramnagar is in the inner *Shivaliks*, Akhnoor is on its outer part. Ramnagar sub-watershed is the catchment area of Ramnagarwali *Khad* (ephermal) in the middle catchment of river Tawi. It has an area of 32,630 Ha. and is subdivided into 39 micro-watersheds. Akhnoor sub-watershed with an area of 42350 Ha. is subdivided into 37 micro-watersheds. It covers about a dozen rivulets (*nallahs*) and a large number of small nallahs originating from *Kalidhar* ridge and draining into Chenab river on the western portion of *Kalidhar* ridge, drains join Manawar Tawi which in turn also meets the Chenab river. In terms of physiographic features, there may not be 100 per cent similarity between the selected sub-watersheds but, certainly, in term of accessibility, natural resource endowment, infrastructure development and general index of socio-economic awareness, two sub-watersheds are comparable.

The primary data has been collected using well-structured and pre-tested questionnaires, participatory rural appraisal (PRA) techniques and group meetings. Whenever possible and required, secondary data sources have also been utilized. Three WUAs each have randomly been selected from the sub-watersheds of Akhnoor and Ramnagar. Therefore, a total of six WUAs have been selected with the consultation of project functionaries at sub-watershed level. The sample size has been restricted to the members of the WUAs selected under IWDP (Hills-II) in the selected sub-watersheds. A total of 15, 12 and 19 WUAs' members have been selected from village Sunetar, Johnu and Dehari respectively in Ramnagar sub-watershed and 21, 12 and 11 WUAs' members have been selected from village Chohara, Maira and Ambarain respectively in Akhnoor sub-watershed. For the collection of primary data and information on environmental and economic impact, three micro-watersheds/villages each from the two sub-watersheds of Akhnoor and Ramnagar, where a maximum number of people are known to have benefited from each project intervention have been selected. In the non-beneficiary category, three sample villages each have been selected on random basis, each 20-25 km away from the sampled micro-watersheds/villages (*i.e.*, project area). Therefore, a total of six WUAs (three each from selected sub-watersheds) and twelve villages (six each from 'project' and 'non-project' area) have been selected for collection of primary data and information. The stratified sampling technique has been used to select villages (where WUAs have been created, as well as 'project' and 'non-project' villages). During the stratification, care has been taken to include both "forested watersheds' villages and agricultural watersheds' villages" in order to makes a comparative study. For the collection of primary data related to environmental and economic variables, a purposive sample of approximately 20 per cent of household level respondents in 'project' and 'non-project' area. The team strategy has been used to collect data and information. The content analysis technique has been used to analyze the data and information qualitatively and quantitatively (using descriptive statistics).

PIM UNDER IWDP (HILLS-II)

In the watershed areas of Jammu and Kashmir, the irrigation system consists of local systems such as medium and small water harvesting structures and distribution channels constructed along the slopes of hills (locally known as *khuls*). Over the period, due to neglect and mismanagement, their utilities in irrigating crops have declined remarkably. IWDP (Hills-II) has started the process of modernization and repair of the existing water harvesting structures and gravity based irrigation water distribution channels. Besides, treatment of the catchment areas has also been initiated to check erosion, restore the vegetative cover in the upper reaches, and remove impediments to the free flow of water into the water harvesting structures. Together, in course of time, these measures are also expected to increase the proportion of rainfall that is harnessed for local irrigation. User groups have been involved in repair, maintenance and improvement of the physical structures, as well as water management on cost-sharing basis mainly in the form of voluntary free labour. The project has introduced significant institutional reforms to revive and strengthens community institutions to make them more effective and flexible. It has played a vital role in clearly defining the relative rights and obligation for users communities, which are part of a larger integrated system as well as the rights of user communities with those outside. The attention has also been given to the basis on which entitlements can be changed, and putting in place an institutional mechanism for enforcing regulations. The project has recognized that better use of available infrastructure with user participation is probably more cost effective than creating new systems. As a result, improvement of the existing physical structures received high priority than the construction of new structures. For this, the project has allocated substantial resources.

Formation of WUA

WUA is the basic foundation of the irrigation reform process under the IWDP (Hills-II). The process of formation of a WUA was relatively simple. The election process to form the WUAs was highly democratic and fair and was conducted in a participatory manner. Initially, WUAs have been formed in those areas where physical infrastructure for irrigation already existed in the form of water harvesting structures and small canal outlets in upper hill slopes fed by rains, which were in need of repairs and maintenance. The participatory development staff of the project along with village *panchayat* identified the user-farmers. The elections were held for the formation of WUAs, with active support of the *panchayat* and project functionaries. On the specified date, elections were conducted. Each voter casts his vote. The candidates securing the maximum numbers of votes were declared elected. In some cases, the elections may be unanimous.

The consensus candidate was then declared elected. The tenure of the WUA is for a period of one year. When a vacancy arises due to death or resignation of a WUA president or a member, the vacancy is filled up following the similar process. Each WUA has a separate bank account. The president and a member are the signatories to the bank account.

Operation of Irrigation System

To begin with, the entire reform process has been repair and renovation oriented. The majority of the works have been done through WUAs along with *panchayats*. WUAs have executed the works in a fair and impartial manner, with proper records of all financial transactions maintained, and contracts given in a fair and just procedure. In general, the cost of the works done by WUAs is 25 per cent lower than those carried out by the contractors. Most of the works are done at the estimated rate. A part of the sanctioned amount was paid in advance and remaining on the completion of the works. But in majority of the cases, even after the completion of the works, the balance amount was not paid. Whatever funds were released go directly to the WUA account. The project officials have no role in handling the funds. Proximity to the process of technical and financial estimates, funds disbursements, works execution by the local farmers, have all brought more transparency in WUA functioning. Contractors were almost out of the picture. The project authorities have given clear instructions that there will be no new structures, or alteration of original designs. So, the major emphasis is on restoring the original designs. The repairs and strengthening of structures, adding of missing structures, desilting the structures, installation of water regulation structures wherever damaged or not installed earlier, and strengthening the distribution channels and bunds are some of the major works carried out by the WUAs. In the process of users identifying and executing the works in case of drainage system, a lot of attention has been paid on users' needs like cattle pathways, bathing steps for women, removal of encroachments and clearing shrubs and silt for smooth flow of water till the tail-reach. Likewise in the case of water-harvesting structures, in addition to the above-mentioned works, the major focus was on repairing and strengthening the existing infrastructures, raising and strengthening bunds to original designs, and improving the water-harvesting structures through voluntary labour contribution. In some cases, the president of WUAs have been trapped in difficulties owing to lack of proper manpower and material planning, which resulted in severe losses to them.

Allocation Rules

The operation aspects are necessary to ensure that physical facilities are maintained in good working condition and regulate access to facilities provided by it. One of the crucial factors on which the authority and credibility of the WUA depends is its ability to ensure that equitable distribution of the benefits takes place and any free rider is penalized. Towards this end several rules have been formulated. One's share of water is determined by the land acreage. This ensures an equitable distribution of the common resources and the prevention of the free riding. In addition, members also patrol together in order to avoid any breaching of the channel. The enforcement of the rules and making sure that whatever penalties are imposed are paid is an arena for the WUA to prove its authority and impartiality. In the project area, irrigated water is used essentially for paddy cultivation. The extent of area to be cultivated as well as the timing of the start of irrigation in a particular season is generally decided by the users' association as a whole in the light of available water supply and rainfall at the beginning of the season. As the entire service area cannot be irrigated, the system is confronted with the problem of rationing supplies. The mechanisms have been evolved to ensure equitable sharing of the shortfall in supplies among all users. Acreage rationing is but one element of the water rationing system. In times of shortage, besides rationing acreage, the water allocation is regulated even more strictly. It is ensured that all users share the reduction in both area and water supply in more or less the same proportion. As paddy is relatively water intensive crop grown in the project area, a system of rotational use of water is adopted even in a normal year to permit all users to grow the crop. The collective decisions are being taken at the beginning of the season regarding which land and how much is to be cultivated, which implies certain coordination in the timing of operations.

Maintenance of Irrigation System

The purpose of 'maintenance' is to ensure that physical facilities function smoothly and at the level of performance for which they were designed. If the maintenance is inefficient, the volume of water made available to the fields get reduced, which hampers the level of output in relation to the potential of the system. The quality of the maintenance affects the interests of both the organizations and the users. The strength of this common interest is stronger in PIM, as the users have made a substantial contribution to the cost of developing it. The common interest of user organizations is also strong, where the neglect of maintenance leads to heavy loss in productivity. WUAs have established conventions regarding the timing of repairs, division of work, responsibility of members and the obligations of users. The custom and social pressure facilitates the smooth working of the arrangements. However, they do not work entirely on the strength of mutual interest reinforce by custom. There is provision of sanction (ranging from fines to loss of water rights) against non-compliance, for which WUA has the authority to enforce the sanction to ensure compliance. However, not even a single case has been reported of sanction against non-compliance. The large landowners have dominated the WUA functioning, as they cultivate a large proportion of land and benefits relatively more from the irrigation system. Thus, the land tenure system has very closely influenced the management of local water control systems.

Benefits and Costs

The potential beneficiaries were mobilized on issue related to fair sharing of the benefits and costs. It was emphasized that users' participation will result in cost-effectiveness in repairs and maintenance and efficient management once it is commissioned. Since a part of the cost of resource development and operation mainly in the form of labour contribution is to be borne by the potential beneficiaries, they have keen interest in minimizing the costs, ensuring a fair balancing of competitive interests and providing economical and efficient services. There was provision of equitable distribution of water based on the size of land holdings in the area of their operation and this has been adhered to in actual practice also. However, there was an unscrupulous distribution here and there, and conflicts occur among the users and members of users' association over the way allocations are managed and over attempts to violate the allocations. In such situations, the executive committees of the WUA mediate disputes among its members. Disputes that cannot be resolved at WUA level are referred to village *panchayat*. The conflicts, which cannot be resolved by the WUA as well as *panchayat*, are referred to project officials. It does not however follow that the latter will always intervene decisively.

Transparency and Accountability

Transparency is one of the key principles of PIM. In order to strengthen confidence over the functioning of the users' organizations, the activities undertaken by them have to be informed to all the water users in the area of their operation. PRA exercises reveal that farmers have been informed in advance by the WUAs about the various activities to be undertaken. The management committees of the WUAs were disseminating information on various activities undertaken for efficient management of the irrigation systems in the area of their operation. The dissemination of information was through general body meetings on regular intervals. The steps have been undertaken to prioritize the repair works on the system so that the available funds should be utilized precisely. Keeping in view the limited amount of funds available at their disposal, the prioritization of the work was also considered necessary by the majority of the users. The quality of the works undertaken by the WUAs were comparatively better than the quality of the works undertaken by the contractors prior to the system of PIM. The conveyance losses in the distribution channels have been minimized, which facilitated in distribution of water up to the tail end area. With PIM, the project functionaries are playing the role of facilitators and made accountable and responsible to the farmers' organizations. On the other hand, the executives of the WUAs are also accountable to their members. The provision of re-election of the WUAs at regular interval also ensured the downward accountability in the reform process.

IMPACT OF PIM

There are perceptible economic gains to the farmers after the initiation of PIM. The land values have increased in both forested as well as agricultural sub-watersheds in the range of 10-15 per cent. The irrigated land value is marginally more in agricultural sub-watersheds due to the field drainage provided by the WUAs, which also arrested the waterlogged conditions in the area of operation of the WUAs. The formation and operationalization of the WUAs have resulted in an increase in area under cultivation and ensure timely sowing of crops and receipt of water at the tail end. New acreage has been added to irrigate lands due to better coverage and maintenance. As a result, irrigated area has increased mainly in tail-reaches. With better maintenance of the irrigation system, the availability of water for various crops has increased, which resulted in increased productivity in terms of rice, pulses, vegetables and orchards. The level of income derived from various crops has increased significantly as compared to the income levels prior to the formation of WUAs. The rate of increase in income has been reported to be comparatively higher in rice and vegetables than other crops. The area of operation of WUAs was under the rice-growing belt, and the farmers reported that the cropping pattern may conveniently be shifted to some extent to commercial crops like vegetables and pulses, which may assure higher returns in future.

Reduction in Rainwater Loss and Sediment Yield

The micro-level watershed planning was carried out using the sweeping transect and emphasis has been given on soil erosion control on hill slopes and gullies, regulation of water flow system in the watershed drainage, and rearrangement of farmlands. The diversion channels were made along all the hills and hillocks to guide the high velocity runoff from these sites on a safe and controlled way, so that the runoff with soil loads should not enter the arable land and cause further degradation. The adverse climatic factors of the *Shivaliks* necessitated the adoption of micro catchment techniques for run-off harvesting and conservation practices as done in similar areas elsewhere. The usefulness and scope of rainwater harvesting and conservation practices in improving tree growth in arid zones have been amply demonstrated. The staggered trenches ensured higher survival rate of out planted seedlings in the experimental plots. The contour trenches also helped in moisture retention but the most of the sub-watershed of

Akhnoor being an undulated terrain without sufficient slopes – the contour interval had to be very wide. The effect of regeneration of vegetation, along with soil and water conservation measures on hillslopes and wastelands was substantial. In *Shivaliks,* the run-off soil loss on barren hills was 23. 70 tons in the baseline period, which fell to 9.65 tons per hectare after treatment. In *Karewas,* it declined to 4.80 tons per hectare from 8.24 tons per hectare in baseline period (IWDP, 2001). The progressive reduction in soil loss and sediment yield as a result of quick recovery of vegetation on hillslopes and lands adjoining the foothills has resulted in a greatly improving surface and ground water regime of the selected sub-watersheds.

Status of Water Resources and Irrigation

The table 1 gives the data on surface and ground water resources across the sampled sub-watersheds. Both the average number of water points (*bowlies*) as well as the gravity based water points per village were comparatively more in project area with WUA than without WUA and non-project area. In the project area with WUA, there were 6.8 and 5 natural water points and gravity based water points respectively in forested watershed and 4.3 and 2 respectively in agricultural watershed. In non-project area, the average number of natural and gravity based water points were comparatively very low and stood respectively at 1.6 and 1 in forested watershed, and 2.6 natural water points in agricultural watershed, whereas there was no gravity based water points in agricultural watershed. This clearly reflects the benefits of PIM through which the project has implemented the repairs of water resources points and installed new water points, mainly gravity based water resources. In the forested watershed, the difference between project area with WUA and non-project area in average number of natural water points and gravity based water points was as high as 3.2 and 4 respectively. The average number of hand pumps per village was also reportedly high in project area with WUA than project area without WUA and non-project area.

A mere perusal of the table 1 makes it clear that the number of water harvesting structures were more in project area with WUA than project area without WUA and non-project area. Similar is the case with the length of irrigation channel. However, significant differentials were noticed across project area with WUA and without WUA as well as non-project area (see table 2). The number of water harvesting structures as well as length of irrigation channel was significantly higher in project area than non-project area. In the project area,with WUA, the average number of water harvesting structures per village was 3.5 and 2.9 respectively in forested and agricultural watersheds, whereas, it was 2 and 1.8 respectively in project area without WUA. In non-project area, the average number of water harvesting structures was 1.2 and 0.7 respectively in forested and

Table 1
Status of Water Resources and Irrigation (per village)

Item	Unit	*Project area with WUA*		*Project area without WUA*		*Non-project area*	
		Forested WS	*Agrl. WS*	*Forested WS*	*Agrl. WS*	*Forested WS*	*Agrl. WS*
Surface Water Supply							
Natural Water Points (Bowlies)	No.	6.8	4.3	5.2	3.4	3.6	2.6
Gravity based Water Points	"	5	2	2	1	1	0
Ground Water Supply							
Hand Pumps	"	2	3	0	1	0	1
Water Harvesting Structures	"	3.5	2.9	2	1.8	1.2	0.7
Irrigation							
Irrigation Channel	No.	2.8	1.7	1.8	1.2	1	0.8
Gross Irrigated Land	Ha.	98	67	48	64	69	44
Net Irrigated Land	Ha.	68	36	34	37	27	18
Irrigation Intensity	%	144.11	186.11	141.17	172.97	255.56	244.44

Note: Irrigation Intensity = Gross Irrigated Area/Net irrigated Area × 100

Table 2
Difference in Status of Water Resources and Irrigation in Sample Area (per village)

Item	Unit	*Project area with WUA - Project area without WUA*		*Project area without WUA - Non-project area*		*Project area without WUA - Non-project area*	
		Forested WS	*Agrl. WS*	*Forested WS*	*Agrl. WS*	*Forested WS*	*Agrl. WS*
Surface Water Supply							
Natural Water Points (Bowlies)	No.	1.6	0.9	3.2	1.7	1.6	0.8
Gravity based Water Points	"	3	1	4	2	1	1
Ground Water Supply							
Hand Pumps	No.	2	2	2	2	0	0
Water Harvesting Structures	"	1.5	1.1	2.3	2.2	0.8	1.1
Irrigation							
Irrigation Channel	No.	1	0.5	1.8	0.9	0.8	0.4
Gross Irrigated Land	Ha.	50	3	29	23	-21	-20
Net Irrigated Land	Ha.	34	-1	41	36	7	19
Irrigation Intensity	%	2.94	13.14	-111.45	-58.33	-114.39	-71.47

agricultural watersheds. It is significant to note that the quality of water harvesting structures were also reportedly better in project area with WUA than non-project area, due to participatory repairs and renovation activities as well as maintenance initiated by the project. Similarly, the average length of irrigation channel in the project area with WUA was 2.8 km and 1.7 km per village respectively in forested and agricultural watershed, whereas, it was 1.8 km and 1.0 km respectively in project area without WUA. In the non-project area, the average length of irrigation channel was comparatively low and it was just 1.0 km and 0.8 km per village respectively in forested and agricultural watersheds. As a result of increased water potential in the selected sub-watersheds, the area under irrigated cropping increased significantly.

Since changes in irrigated area depend on numerous factors, many of which could not be measured in the current study, the findings here must be treated cautiously. The data presented in table 1 reveals that the changes in gross and net irrigated area in project area with WUA and without WUA, and non-project area. A perusal of the table makes it clear that irrigation intensity in forested and agricultural sub-watersheds in project area with WUA was 144.11 per cent and 186.11 per cent respectively. In project area without WUA, it was 141.17 per cent and 172.97 per cent, whereas in non-project area, it was 255.66 per cent and 244.44 per cent respectively in forested and agricultural sub-watersheds. The data shows the marked differences in irrigation intensity across the sample area. The difference in irrigation intensity was 2.94 per cent and 13.14 per cent respectively in forested and agricultural sub-watersheds of project area with WUA and without WUA. It is significant to note that the differences in irrigation intensity in project area with WUA and non-project area (see table 2).

Change in Crop Intensity and Crop Productivity

As a result of the improved soil moisture regime, the increase in supplementally irrigation resources and increased use of fertilizer (including cow-dung), the cropping intensity has improved in the project area compared to non-project area. The data on grossed crop area, net area sown and crop intensity is presented in table 3, which clearly reveals significant variations across the selected sub-watersheds. There are considerable differences in cropping intensity in project area with WUA and non-project area, which can be attributed to the project interventions in the field of agriculture through rainfed crop demonstration, propagation of the use of modern inputs through extension agents of the project. It is to be noted that differences in cropping intensity was reportedly negative across the project area with WUA and without WUA as well as non-project area. For instance, it was – 58.26 per cent and – 46.27 per cent respectively between project area with WUA and non-project area (see table 4).

Table 3
Crop Intensity and Crop Yields

Item	Unit	Project area with WUA		Project area without WUA		Non-project area	
		Forested WS	Agrl. WS	Forested WS	Agrl. WS	Forested WS	Agrl. WS
Gross Cropped Area	Ha.	163	74	73	107	113	68
Net Sown Area	Ha.	158	68	57	93	70	42
Crop Intensity	%	103.16	115.63	128.07	146.58	161.42	161.9
Non-seed Inputs and Crop Yields							
Maize	Qntl.7Ha	96	9.9	8.2	8.5	7.5	7.6
Paddy	"	5.2	6.2	4.5	8.5	4.4	4.5
Pulses	"	2.64	2.8	2.3	2.4	2	0.3
Wheat	"	8.96	9.9	8.2	8.4	7.6	7.8
Seed Inputs and Crop Yields							
Maize							
Local	Qntl./Ha	13.13	14.2	12.6	13.7	10.3	11.2
Hybrid	"	15.6	15.8	14.9	14.1	11.7	11.9
Difference	"	2.3	1.6	2.3	0.4	1.4	0.7
Paddy							
Local	"	10.8	11.2	9.4	10.3	9	9.2
Hybrid	"	12.3	13.1	11.4	11.7	10.1	10.7
Difference	"	1.5	1.9	2	1.4	1.1	1.5
Wheat							
Local	"	18.4	19.7	16.9	17.4	16.1	16.7
Hybrid	"	21.3	21.8	20.4	21.1	17.3	17.9
Difference	"	2.9	2.1	3.5	3.7	1.2	1.2

Note: Crop Intensity = Gross Cropped Area/Net Area Sown × 100

It is very difficult to isolate the impact of various inputs used on crop yields. Due to creation of VCLs and forest closures, an increase in the vegetation has taken place. The over all soil-moisture regimes have also improved after project intervention. The project has repaired the traditional irrigation water distribution channels (khuls), consequently the irrigation water availability and flow has improved. Keeping the influence of these factors aside, the data presented in table 4 shows differential levels of yields of various crops such as maize, paddy, pulses and wheat across the selected sub-watersheds, largely due to increased use of non-seed inputs. A cursory glance at table 4 makes it evident that average crop yield per hectare has been comparatively more in project area with WUA than project area without WUA, and further the yield per hectare has been comparatively higher in project area without WUA than non-project area. The productivity of selected crops is comparatively higher in agricultural sub-watersheds than forested sub-watersheds. However, a significant variation in crop yield is reported across the project and non-project area.

The yield difference in the project area with WUA and without WUA was low compared with project area with WUA and non-project area in forested as well as agricultural sub-watersheds. The yield difference between project area with WUA and non-project area was 2.1 quintals, 0.8 quintal, 0.64 quintal and 1.36 quintal per hectare respectively for maize, paddy, pulses, and wheat in forested watershed. Whereas, in agricultural watershed, yield differences between project area with WUA and non-project area was markedly high and stood at 2.2 quintals, 1.7 quintals, 0.5 quintal, and 2.1 quintals respectively for maize, paddy, pulses, and wheat. However, such differentials were not so marked in project area without WUA and non-project area. The yield differences were 0.3 quintal, 0.1 quintal, 0.1 quintal and 0.6 quintal respectively for maize, paddy, pulses and wheat in agricultural watershed, and more or less the same is true for forested watershed (see table 4).

Table 4
Difference in Crop Intensity and Crop Yields in Sample Area

		Project area with WUA - Project area without WUA		*Project area without WUA - Non-project area*		*Prcject area without WUA - Non-project area*	
Item	*Unit*	*Forested WS*	*Agrl. WS*	*Forested WS*	*Agrl. WS*	*Forested WS*	*Agrl. WS*
Gross Cropped Area	Ha.	90	-33	50	6	-40	39
Net Sown Area	Ha.	101	-25	88	26	-20	51
Crop Intensity	%	-24.91	-30.95	-58.26	-46.27	-33.35	-15.32
Non-seed Inputs and Crop Yields							
Maize	Qntl./Ha	1.4	1.4	2.1	2.2	0.7	0.8
Paddy	"	0.7	1.6	0.8	1.7	0.1	0.1
Pulses	"	0.34	0.4	0.64	0.5	0.3	0.1
Wheat	"	0.76	1.5	1.36	2.1	0.6	0.6
Seed Inputs and Crop Yields							
Maize							
Local	Qntl./Ha	0.7	0.5	3	3	2.3	2.5
Hybrid	"	0.7	1.7	3.9	3.9	3.2	2.2
Difference	"	0	1.2	0.9	0.9	0.9	-0.3
Paddy							
Local	"	1.4	0.9	1.8	2	0.4	1.1
Hybrid	"	0.9	1.4	2.2	2.4	1.3	1
Difference	"	0.5	0.5	0.4	0.4	0.9	-0.1
Wheat							
Local	"	1.5	2.3	2.3	3	0.8	0.7
Hybrid	"	0.9	0.7	4	3.9	3.1	3.2
Difference	"	-0.6	1.6	1.7	0.9	2.3	2.5

The data presented in table 4 further substantiate the yield differentials across the selected sub-watersheds. As can be noticed elsewhere too, there are significant differences in yield of local and hybrid varieties of crops in forested as well as agricultural sub-watersheds. For instance, yield difference in local and hybrid maize is 2.3 quintal per hectare in forested sub-watershed in project area with WUA and without WUA. Thus, one can infer that new institutional arrangements have no impact on the yield of local and hybrid variety. The yield difference is 1.2 quintal per hectare in agricultural sub-watershed in project area with WUA and without WUA. A significant differential in yield is noticed between project area with WUA and non-project area, as well as project area without WUA and non-project area. More or less the same applied to other crops with minor differences across the forested and agricultural sub-watersheds. On the whole, the average crop yield per hectare has significantly improved after the project area.

Besides above, the repairs and maintenance works have generated much needed wage employment for the poor labourers and small and marginal farmers. The undertaking of the construction works by the members of WUAs along with *panchayats* has resulted into cost-effectiveness to the tune of about 10-15 per cent. PIM have also empowered the farming communities in self-organization and equipped them with the supervision and management skills. The farmers' participation in various stages of the reform process has created a sense of ownership of the assets created. They have played a greater role in the choice of prioritizing and executing the work themselves through the own organizations. Above all, the institutional changes facilitated the closer interaction of the project functionaries and local stakeholders and unified the diverse groups around the water.

CONCLUSION

The institutional reforms in irrigation water management initiated under IWDP (Hills-II) are both bold and innovative. The WUAs have taken up the maintenance activity with the financial support provided by the project. Due to physical improvements in the irrigation system, the water availability has improved. It addresses the key issues pertaining to irrigation management: institutional structure, incentives, accountability, transparency, and sustainability. The institutional changes are still in progress. In the future, linkages need to be established which will make WUAs independent of project support. There is need to redefine the role of irrigation agency, which calls for suitable institutional restructuring. Ultimately, WUAs will need to be financially self-sustainable. The sustainability of the WUAs essentially depends on their capacity to operate and manage the system on their own. The financial sustainability has an important role in the sustainability of the WUAs. At present, WUA is completely dependent on the project functionaries for funds. The physical works will continue for some time, which would provide some finances for the day-to-day functioning of the WUA. But what happens when the funds from the donor stop. In the past, in other parts of the country where WUAs were formed, many became defunct due to financial bankruptcy. WUA will have to be financially self-sufficient to perform their duties on sustainable basis and to make the reform process a success. In the near future, it is envisaged that the WUAs will collect water charges from the users. This will be the most logical step in the reform process. However, the WUA executives may not be able to perform their duties in this regard due to inadequate manpower at their disposal. In order to operationalize the task of water tax collection, concrete steps need to be taken. The steps need to be initiated to strengthen the capacity building of WUAs to ensure financial sustainability, after the withdrawal of the project.

Another weak link has been the issue of accountability. Mechanisms to ensure accountability must continue to be evolved. The roles and responsibilities of all the agencies concerned must be further defined. There is need to transfer all the operation and maintenance function to the WUAs in near future. Likewise, transparency is critical if the organizations are to progress further. To a large extent, transactions are captured in the accounting system. There is need to put in place simple and standardized procedures for accounting and finance. There is need to ensure replicability of the successes achieved. In the tasks of operation and maintenance of the PIM, emphasis should be paid on getting rid of the middlemen and contractors. Users' participation should be ensured for identifying, and executing the essential works for repair and maintenance. The role of the project functionaries or line department needs to be curtailed. At the same time, there is need to form federations of WUAs, for which continued support and training are required. The sustainability of the farmers' organization also calls for involvement of NGOs and training institutes in upgrading the skill and capacity building. There is urgent need to forge suitable linkages with reputed local NGOs, training institutes and line departments to equip the farmers' organization in various aspects of PIM. No doubt, adequate groundwork and the required environment are created, there is need to give a closer look to the weaknesses and remedy them. More importantly, a regular monitoring mechanism has to be put in operation to initiate corrective measures as and when needed.

REFERENCES

Bandyopadhyay, J. and S. Perveen (2004), Interlinking of Rivers in India: Assessing the Justifications, *Economic and Political Weekly*. 39(50): 5307-16.

Bandyopadhyay, J., B. Malik, M. Mandal and S. Perveen (2002), Dams and Development Report on a Policy Dialogue, *Economic and Political Weekly*, 37 (4): 4108-12.

Brewer, J., S. Kolavalli *et al.* (1999), *Irrigation Management Transfer in India: Policies, Processes and Performance*, IIM, Ahmedabad and IIMI, Colombo, Oxford and IBH Publishers.

Deursen, W. P. A. (2000), *Humans, Water and Climate Change*, The Dutch National Research Programme on Global Air Pollution and Climate Change, Carthago Consultancy, Rotterdam.

Falkenmark, M. *et al.* (1980), *Water and Society: Conflicts in Development*, Part 2, Pergamon Press, Oxford.

Gleick, P. H. (1996), Basic Water Requirement for Human Activities: Meeting Basic Needs, *Water International*, 21 : 83-92.

IWDP (2001), *Status Report for Supervision Mission of World Bank*, IWDP (Hills-II), Jammu and Kashmir.

Ives, J. D. and B. Messerli (1989), *The Himalayan Dilemma*, Routledge, London.

IWRS (1996), Interbasin Transfers of Water for National Development: Problems and Perspectives, Indian Water Resources Society.

Iyer, R. R. (2000), *Water: Charting a Course for the Future*, Centre for Policy Research, New Delhi.

Khattri, K. N. (1987), Great Earthquake Seismicity Gaps and Potential for Earthquake Disaster Along the Himalayan Plate Boundary, Techno-physics, Vol. 138, pp. 79-82.

Nigam, A., B. Gujja, J. Bandyopadhyay, and R. Talbot (1997), *Freshwater for India's Children and Nature*, WWF and UNICEF, New Delhi.

Parthasarathy, R. (2000), Participatory Irrigation Management Programme in Gujarat: Institutional and Financial Issues, *Economic and Political Weekly*, 25 (35 and 36): 3147-54.

Serageldin, I. (1995), *Towards Sustainable Management of Water Resources*, World Bank, Washington, D. C.

Vidal, J. (2003), India's Dream Bangladesh's Disaster, *The Guardian*, July 24.

World Commission on Dams (20000, *Dams and Development — A New Framework for Decision-Making*, Earthscan Publishing, London.

8

BENEFIT-COST ANALYSIS OF SEMI-INTENSIVE PRAWN FARMS IN WEST BENGAL: AN EMPIRICAL STUDY

Debabrata Lahiri

INTRODUCTION

Environmental impact prawn farms have been well known all over the world. But as the intensity of prawn faming increases negative externalities to environment. Mostly, externalities to environment have been related to loss of mangroves, increased salinity to adjoining areas to prawn farms. On the other hand, negative environmental impact due to prawn farming has been due to its different technologies adopted in terms of stocking density, application of feed and medicine etc. Among the various forms of technology available for prawn farming the least harmful has been the paddy-cum-prawn cultivation. Whereas, the intensive form, has been the most harmful to the environment. In between, along intensity scale in mangrove areas (The Sundarbans) the semi-intensive form has been harmful to environment in form putting additional areas to salinity. As a result, production of crops decreases. In other words by pursuing semi-intensive prawn farming the social costs has been more than the benefits.

REVIEW OF LITERATURE

It has been evident that environmental degradation starts with clearance of mangroves and digging rice fields in the coastal areas. As coastal areas are dominated with acid clays or cat-clays (Pillay, 1990), its reclamation, for prawn ponds, results in exposure oxidation of pryrite (FeS_2) (Flaherty et al, 1995). The aerators oxidizes the existing hydrogen sulphide and convert it to sulphuric acid. This black affluent discharged into the neighbouring land makes them totally unfit for cultivation. Dumping of unhygienic waste in the open, results in pollution of air and water series for already unemployed farmers. The villagers have to walk 1.5 Kms to fetch potable drinking water (Naganathan, M. et. al., 1995). It has been also reported that the salinity in wells has decreased with the increase in distance from the cultured prawn ponds (Ramachandra Bhatta and Mahadev Bhat, 1998).

With the destruction of mangroves consumptive and non-consumptive benefits such as, sediment stabilisation, wildlife habitat and a renewable supply of forest products, namely wood for boat construction, bark for tanning, fuel wood, fodder, fertiliser and honey are lost (Mercer and Hamilton L.S, 1984; Parks and Bonfiaz, 1994).

The problem of villagers has been analysed by segregating the villages into fishing and farming villages by problems encountered such as, well water salinity, access to beach creek blocked, agricultural land salinity, under employment, poor health and availability fodder and fuel wood. Further, in order to assess gravity of the problem, correlation matrix has been formed with empirical data (Pawan G. Patil and M, Krishnan, 1997).

Attitudes of villagers towards access and use of mangrove forest, benefits from mangrove forests, causes for decline in the yield of fishery product, attitude towards local community participation and attitude of the villagers towards protection of forests has been empirically assessed in Thailand (Sathirathai, 1998).

As prawn culture mostly results in mangrove destruction, so for economic valuation Total Economic Value (TEV) method has been used which consists of Use Value (UV) and Non-Use Value (NUV). Use Values have been disaggregated into Direct Use Value (DUV), Indirect Use Value (IUV) and Option Value (OV). Non-Use Value more difficult to assess and could be segregated into Existence Value (EV), Bequest Value (BV). For total economic valuation, the components, functions/services and diversity/attributes could be cross tabulated by on-site quadrant and off-site quadrant (Sathirathai, 1998). Also the forgone benefits of the resources could be assessed by comparing them with economic benefits conservations of the area into alternative uses such as irrigated paddy, sugarcane and prawn farming (Dixon and Lal, 1994).

Cost-benefit analysis covering financial and economic analysis has been done for alternative land-use options e.g. private point of view (financial and economic analysis) covering mangrove forest, commercial shrimp farms, society's point of view (economic analysis) covering mangrove forest, commercial shrimp farms and converting mangrove forests into commercial shrimp farms (Sathirathai, 1997).

The Supreme Court of India in 1996 in its landmark judgment to prevent degradation of environment ordered banning of shrimp farms within 500 meters of high tide line from the sea. It also ordered closure of all such prawn by March 1997. But yielding to the pressure of prawn industry the Indian Parliament in 1997 passed the Aquaculture Authority

Bill 1997 overruling Supreme Court decision. After legislation of the bill the general flip-flop between Supreme Court and Parliament is bound to continue due to the significant loopholes in Coastal Zone Management (CZM) policy and the relative lack of objective analysis concerning this issue (Pawan G.Patil and M. Krishnan, 1997).

OBJECTIVES AND METHODOLOGY

The following are the objectives of the paper:

i) To analyse the effect of salinity on production and income of crops due to semi-intensive prawn farming;

ii) To estimate the financial and social cost benefit of semi-intensive prawn farms and also that of saline farms.

The semi-intensive prawn farms had been situated in Canning, Mathurapur, Joynagar, Patharpratima, Barasat, Kultali, Gosaba and Kakdwip Community Development Blocks (C.D Blocks). Of the 8 C.D. Blocks, Patharpratima had been selected purposively to study the externalities of semi-intensive prawn farms. The Durbachati (J.L.No.164), Kamdeb Nagar (J.L.No.160), Chintamanipur (J.L.No.161), Krishnapur (J.L.No.163) and Paschim Surendranagar (J.L.No.166) villages had been selected purposively for studying the semi-intensive prawn farms, non-saline farms and salinity-affected farm.

After house listing of all the households, 100 households had been selected randomly under each category viz., landless, marginal, small, medium and large for semi-intensive and saline farms. The total number of households categorized into landless, small, marginal, medium and large farms is given in Table 1.

Table: 2 gives the detail of sample households in extensive, semi-intensive and saline farms.

The sample farms have been chosen from the semi-intensive and sample farms each of 100 each categorised into marginal, small, medium and large farms respectively. For financial benefit cost analysis the benefit-cost ratio, net present value and has been worked out. Social benefits and cost has also been worked out.

Private v/s Social Cost Benefit

To study the externalities of various types of prawn farming private (financial) social cost benefit analysis has been used on the premise that if private cost benefit is more than social cost benefit the market forces would automatically allow the prawn farming to continue in spite of externalities. Alternatively, if social cost benefit is more than private cost benefit, the prawn farming should not be allowed to continue due to externalities.

Table 1
Categorization of Household as per House Listing Schedule

S.N.	*Name of the C.D. Block/villages with J.L.no.*	*Population (as per 1991 Census)*	*No. of households (as per 1991 Census)*	*No. of households as per house listing*	*Total number of households by types as per house listing schedule*				
					Landless	*Marginal farmers*	*Small farmers*	*Medium farmers*	*Large farmers*
1.	Patharpratima C.D. Block								
2.	Durbachati (J.L.No 164)	2623	412	420	172	239	5	4	-
3.	Kamdebnagar (J.L.No. 160)	1786	295	310	124	155	15	16	-
4.	Chintamanipur (J.L.No.161)	1868	256	270	109	150	4	7	-
5.	Krishnapur (J.L.No.163)	796	136	140	57	78	2	3	-
6.	Paschim Surendranagar (J.L.No.166)	4820	747	750	307	420	12	11	-

Table 2
Details of Sample Household

S.N.	*Village type/ Prawn farming type*	*Type of families*					
		Landless	*Marginal*	*Small*	*Medium*	*Large*	*Total*
1.	Semi-intensive prawn farms	4	101	2	-	-	107
2.	Saline farms	-	92	4	1	-	97
	Total	20	261	22	1	-	304

Private Cost Benefit

The private cost benefit analysis has been done by working out $B - C$ ratio internal rate of return (IRR) and net present value (NPV). The formulae for B - C ratio, IRR, NPV are as follows:

(i) Benefit-Cost ratio = $$\frac{\sum_{t=1}^{t=n} \frac{B_t}{(1+i)^t}}{\sum_{t=1}^{t=n} \frac{C_t}{(1+i)^t}}$$

(ii) Internal Rate of Return (IRR): The discount rate i such that

$$\sum_{t=1}^{t=n} \frac{B_t - C_t}{(1+i)^t} = 0$$

(iii) Net Present Value (NPV) = $$\sum_{t=1}^{t=n} \frac{B_t - C_t}{(1+i)^t}$$

For private cost benefit analysis, discount rate or internal rate has been taken as 6%. The duration of the project as per financial institutions i.e., commercial banks, central co-operative banks has been 5 years. But actually it has been observed that the project life of extensive and semi-intensive prawn farms is more than 5 years. It generally extends up to 15 years. For analysis the private cost benefit ratio, IRR and NPV has been calculated for 5, 10, 15 years.

In social cost benefit analysis, the shadow prices of the following

1. Land
2. Labour
3. Inputs

Are taken as follows:

Prawn farming is always beneficial from private (financial) point of view at any level of technology adoption i.e., extensive and semi-intensive but when compared to social benefit it is not more. Apart from private (financial) analysis the study also intends to analyse the social cost-benefits as per following:

Social Cost-Benefit Analysis

The evaluation of net aggregate consumption benefits from extensive and semi-intensive prawn farms has been carried out in successive stages of approximation. The first, and straight forward step is to assess the benefits and costs under the assumption that market prices adequately represent social opportunity cost and, hence, the ultimate consumption benefits and costs involved. On this basis consumption benefits of extensive and semi-intensive prawn farms include items (1) and (2) and consumption cost includes items (3), (4) and (5). Details of which has been given in Table 13 and 16.

Items (1) and (2) corresponds to real gains from extensive and semi-intensive prawn farms as evolve, which would not have come about in the absence of cultivation of prawn under different levels of technology. Analogously items (3), (4) and (5) represent resource payments which could have been used elsewhere, were it not for the extensive and semi-intensive prawn farms, and therefore measure the sacrifice of consumption possibilities that the economy sustains because of the project. The item (5) measures directly the consumption sacrificed in adopting new methods of cultivating the previously farm land.

The market value of net aggregate consumption in any given year of the project is thus defined as follows:

$$MC = (1) + (2) - (3) - (4) - (5)$$

MC represents the first approximation to the net aggregate consumption benefits of the extensive and semi-intensive prawn farms. The second approximation involved adjustment of the market prices of specific resources, wherever the price do not reflect the real contribution of resources to the aggregate consumption objectives, i.e., social opportunity cost. Such discrepancies can occur for a variety of resources and in variety of ways. In extensive and semi-intensive prawn farms two major resources have been singled out for price adjustment: foreign exchange and unskilled labour.

As far as foreign exchange is concerned a good quantity of prawns are exported outside the countries. Here it is assumed that out of total production of prawns in extensive and semi-intensive situation 25 per cent of the total production is consumed locally and the rest 75 per cent of the production is exported. As a result the foreign exchange earning to the nation would be on 75 per cent of the production. If foreign exchange is expressed in U.S. dollar terms, then a dollar of foreign exchange is worth Rupees forty seven (Rs 47.00) – its official exchange rate. In the analysis to follow opportunity cost of foreign exchange relative to its official market price would be denoted by $(1 + \Phi)$; Φ represent foreign exchange premium in the Indian economy.

The situation of unskilled labour in the Indian economy has been opposite to that of foreign exchange; while foreign exchange is relatively rationed at a price well below its opportunity cost, unskilled labour is found in surplus, and the market wage exceeds the opportunity cost of employing additional labour. The opportunity cost of unskilled labour - relative to the going market rate has been denoted by $(1 + \lambda)$; where λ represents unskilled labour premium, which is 0.70 in the Indian economy.

The use of skilled workers is still require in extensive and semi-intensive prawn farming @ 20 per cent of the total labour used. It is assumed that these labourers are fully employed in the economy at wages reflect their social opportunity cost.

It is now possible to express net aggregate consumption benefits from extensive and semi-intensive prawn farms, after incorporating the opportunity cost premium, as

$$SC = (1) + (2) + \Phi\,(2 - F) - (3) - \lambda^{E}(3) - (4) - (5) - \lambda(5.A - 2)$$

Rearranging the terms in the equation we get-

$$SC = MC + \Phi\,FE + \lambda^{E}\,E + \lambda L$$

The second approximation, *SC*, is obtained by adding the terms to the first approximation, *MC*. The first term corrects *MC* for the opportunity cost of foreign exchange by multiplying net foreign exchange component of net benefit and costs, *FE* by the (positive) foreign exchange premium Φ. The second term corrects the *MC* for the opportunity cost of land by multiplying the land component, *E*, by the (negative) land premium λ^{E}. The third term corrects the MC for the opportunity cost of unskilled labour by multiplying the net unskilled labour component, L, by the (negative) labour premium λ.

RESULTS

Results of the study has been presented as operational holdings, tenurial conditions, cropping pattern, fallow land due to salinity, causes of salinity, changes in cropping pattern due to salinity, changes in productivity level of crops, private (financial) cost-benefit and social cost-benefit.

Operational Holding

Semi-intensive Farms

Operational holding of marginal farmers had been 0.39 ha., comprising of 0.32 ha., (82.05%) as own land and an insignificant area of 0.07 ha., (17.95%) as leased-in land. For small farmers, the operational holding had been 1.31 ha., comprising 0.98 ha., (74.81%) as own land and 0.33 ha., (25.19) as leased-in land. For all farms the operational holding had been 0.40 hectares comprising of 0.33 ha., (82.50%) as own land and 0.07 ha., (17.50%) as leased-in land. Here also own land played a significant role in determining the size of operational holding. Details have been given in Table 3.

Table 3
Operational Holding of Crop Farms in Semi-intensive Farms

(ha./household)

S.N.	Type of farms	Type of land: Own	Leased-in	Leased-out	Operational holding
1.	Landless Agril. Labour	-	-	-	-
2.	Marginal farmers	0.32 (82.05)	0.07 (17.95)	-	0.39 (100.00)
3.	Small farmers	0.98 (74.81)	0.33 (25.19)	-	1.31 (100.00)
4.	Medium farmers	-	-	-	-
5.	Large farmers	-	-	-	-
	All farms	0.33 (82.50)	0.07 (17.50)	-	0.40 (100.00)

(Figures in parentheses indicate percentage to operational holding)

Saline Farms

In salinity affected farms of the total number of 100 sample farms only marginal, small and medium farmers had been found. For salinity affected farms there has been no incidence of leased-in and leased-out land. Thus the entire operational holding comprise of own land only. The operational holding for marginal, small, medium farmers and all farms had been 0.31, 1.55, 2.56 and 0.39 ha., respectively. Details have been given in Table 4.

Table 4
Operational Holding of Crop Farms in Saline Farms

(ha./household)

S.N.	Type of farms	Type of land: Own	Leased-in	Leased-out	Operational holding
1.	Landless Agricultural labour	-	-	-	-
2.	Marginal farmers	0.31 (100.00)	-	-	0.31 (100.00)
3.	Small farmers	1.55 (100.00)	-	-	1.55 (100.00)
4.	Medium farmers	2.56 (100.00)	-	-	2.56 (100.00)
5.	Large farmers	-	-	-	-
	All farms	0.39 (100.00)	-	-	0.39 (100.00)

(Figures in parentheses indicate percentage of operational holding)

Tenurial Conditions

Semi-intensive Prawn Farms

Under semi-intensive farming system- the landless agricultural labourer has 0.17 ha. of land of which the entire area has been leased-in. Marginal farmers had operated 0.19 ha. of which 0.18ha.(94.74%) as own land, 0.15 ha. (78.95%) as leased-in land and 0.14 ha. (73.69%) as leased-out land. Small farmers operated 0.34 ha. of land of which the entire area has been own land. For all farms the total area operated is 0.20 ha. of land of which 0.18 ha.(90.00%) as own land, 0.15 ha. (75.00%) as leased-in land and 0.13 ha. (65.00%) as leased-out land. Details have been given in Table 5.

Table 5
Operational Holding in Semi-intensive Prawn Farms

(ha./household)

S.N.	Type of farms	Type of land			
		Own land	*Leased-in land*	*Leased-out land*	*Total*
1.	Landless agricultural labour	-	0.17 (100.00)	-	0.17 (100.00)
2.	Marginal farmers	0.18 (94.74)	0.15 (78.95)	0.14 (73.69)	0.19 (100.00)
3.	Small farmers	0.34 (100.00)	-	-	0.34 (100.00)
4.	Medium farmers	-	-	-	-
5.	Large farmers	-	-	-	-
	All farms	0.18 (90.00)	0.15 (75.00)	0.13 (65.00)	0.20 (100.00)

(Figures in parentheses indicates percentage to total)

Cropping Pattern

Semi-intensive Farms

In semi-intensive form of prawn cultivation no part of cropland is either leased-in or leased-out for prawn cultivation. The entire land is cultivated if not salinised due to ingress of saline water through breach of embankment etc. For marginal farmers of the total gross cropped area, 79.60 per cent has been put to Kharif paddy. Rabi/Summer paddy and vegetable constituted only 10.20 per cent each. For small farmers the cropping intensity has been less at 106.11 per cent. Of the total gross cropped area, 94.24 per cent has been put to Kharif paddy. Rabi/Summer paddy and vegetable constituted 5.04 and 0.72 per cent respectively. For all farms of the total gross cropped area, 79.60 per cent has been put to Kharif paddy. Rabi/Summer paddy and vegetable constitutes only 10.20 per cent each. Details have been given in Table 6.

Table 6
Cropping Pattern of Semi-intensive Farm

(ha./household)

S.N.	Types of farms	Season				Net sown area	Gross cropped area	Cropping intensity (%)
		Kharif	Rabi					
		Paddy	Paddy	Mustard	Vegetables			
1.	Landless Agril. Labourer	-	-	-	-	-	-	-
2.	Marginal farmers	0.39 (79.60)	0.05 (10.20)	-	0.05 (10.20)	0.39	0.49 (100.00)	125.64
3.	Small farmers	1.31 (94.24)	0.07 (5.04)	-	0.01 (0.72)	1.31	1.39 (100.00)	106.11
4.	Medium farmers	-	-	-	-	-	-	-
5.	Large farmers	-	-	-	-	-	-	-
	All farms	0.39 (79.60)	0.05 (10.20)	-	0.05 (10.20)	0.39	0.49 (100.00)	125.64

(Figures in parentheses indicates percentage to gross cropped area)

Saline Farms

For marginal, small, medium farmers and all farms the cropping intensity had been 108.01, 100.13, 142.19 and 108.60 per cent respectively. For marginal farms, Kharif paddy occupied 91.99 per cent. In Rabi season, Rabi/Summer paddy, chilies, mustard and vegetables occupied 2.97, 2.37, 1.49 and 0.59 per cent respectively. For small farms Kharif paddy occupied 99.87 per cent. In Rabi/Summer season vegetable occupied 0.13 per cent of the gross- cropped area. For medium farmers of the total gross cropped area Kharif paddy occupied 70.33 per cent, 29.67 per cent of the gross cropped area had been occupied by Rabi/Summer paddy. For all farms Kharif paddy occupied 91.55 per cent of the gross cropped area. In Rabi/ Summer season paddy, chilies, mustard vegetables occupied 4.69, 1.64, 1.18 and 0.47 per cent respectively of the gross cropped area. Details have been given in Table 7.

From the above analyses it appears that the salinity has increased the cropping intensity of sample farms. The reason for such increase in cropping intensity is due to the fact that fallow land cannot be taken into account while calculating the cropping intensity. Because among sample farms there has been a considerable drop in net sown area due to salinity and it is the basis of calculating gross cropped area and cropping intensity. The Table indicates that there has been a reduction in net sown area for marginal, small, medium farmers and all farms to the extent of 45.16, 61.94, 89.45 and 51.28 per cent respectively. This has been shown in Table 8.

Table 7
Cropping Pattern of Saline Farms

(ha./household)

S.N.	Type of farms	Betel-leaf	Kharif	Rabi/Summer				Net sown area	Gross cropped area	Cropping intensity (%)
			Paddy	Chilli	Paddy	Mustard	Vegetable			
1.	Landless Agril. Labour	-	-	-	-	-	-	-	-	-
2.	Marginal farmers	0.002 (0.59)	0.31 (91.99)	0.008 (2.37)	0.01 (2.97)	0.005 (1.49)	0.002 (0.59)	0.312	0.337 (100.00)	108.01
3.	Small farmers	-	1.55 (99.87)	-	-	-	0.002 (0.13)	1.55	1.552 (100.00)	100.13
4.	Medium farmers	-	2.56 (70.33)	-	1.08 (29.67)	-	-	2.56	3.64 (100.00)	142.19
5.	Large farmers	-	-	-	-	-	-	-	-	
	All farms	0.002 (0.47)	0.39 (91.55)	0.007 (1.64)	0.02 (4.69)	0.005 (1.18)	0.002 (0.47)	0.392	0.426 (100.00)	108.67

(Figures in parentheses indicates percentage to gross cropped area)

Table 8
Extent of Fallow Land due to Salinity

(ha./household)

S.N.	Types of farms	Net sown area (NSA) before salinity	Net sown area (NSA), after salinity	Fallow land
1.	Landless agricultural labour	-	-	-
2.	Marginal farmers	0.31 (100.00)	0.17	0.14 (45.16)
3.	Small farmers	1.55 (100.00)	0.59	0.96 (61.94)
4.	Medium farmers	2.56 (100.00)	0.27	2.29 (89.45)
5.	Large farmers	-	-	-
	All farms	0.39 (100.00)	0.19	0.20 (51.28)

(Figures in parentheses indicates percentage of fallow land to net sown area before salinity)

Causes of Salinity

The causes of salinity have been high tide, breach of embankment, overflow of prawn farms and seepage. The embankments along rivers and estuaries are being maintained by the Irrigation and Waterway department of

Government of West Bengal. The embankments are poorly maintained by the department. Causes of breach of embankments are mainly due to damage of old asbestos water inlet and outlet pipes and high tide during full moon and no moon. Though there is no hard and fast rule of breach of embankments at a particular point but breach occurs 4-5 times in a year.

Salinity among sample forms has been caused by various reasons such as high tide, breach of embankments, overflow of prawn farms and seepage from prawn farms to adjoining paddy fields. Of the total area affected by salinity, breach of embankment for marginal farmers (35.48 %) and small farmers (61.94%) has been the significant reason, whereas, for medium farmers, seepage from prawn farms to adjoining paddy fields has been the major reason. For all farms, breach of embankment has been the significant reason (35.90%) followed by seepage, overflow of prawn forms and high tide. Details of which have been given in Table 9.

Table 9
Causes of Salinity

(ha./household)

			Causes of salinity				
S.N	*Types of farms*	*Operational area*	*High tide*	*Breach of embankment*	*Overflow of prawn farms*	*Seepage*	*Total*
1.	Landless agricultural labour	-	-	-	-	-	-
2.	Marginal farmers	0.31 (100.00)	0.001 (0.32)	0.11 (35.48)	0.003 (0.97)	0.03 (9.68)	0.144 (45.16)
3.	Small farmers	1.55 (100.00)	- (61.94)	0.96	-	-	0.96 (61.94)
4.	Medium farmers	2.56 (100.00)	-	-	-	2.29 (89.45)	2.29 (89.45)
5.	Large farmers	-	-	-	-	-	-
	All farms	0.39 (100.00)	0.001 (0.26)	0.14 (35.90)	0.003 (0.77)	0.06 (15.38)	0.20 (51.28)

(Figures in parentheses indicates percentage to operational area)

Changes in Cropping Pattern due to Salinity

The changes in cropping pattern due to salinity by breach of embankment or high tide or overflow of prawn farm or seepage is reflected at all farms level by decrease in area under Kharif paddy from 0.39 to 0.18 ha. (53.85%). Similarly the summer paddy has dropped from 0.02 to 0.007 ha (65.00%). Mustard and summer vegetables had not been grown due to salinity. For marginal farmers the reduction in Kharif paddy and summer paddy had been from 0.31 to 0.16 ha (48.39%) and 0.01 to 0.008 ha (20.00%). Details of which have been given in Table: 10.

Table 10
Changes in Cropping Pattern due to Salinity

(ha./household)

S.N.	Type of farms	Kharif paddy Before	Kharif paddy After	Chillies Before	Chillies After	Summer paddy Before	Summer paddy After	Mustard Before	Mustard After	Vegetables Before	Vegetables After
1	Landless	-	-	-	-	-	-	-	-	-	-
2	Marginal	0.31	0.16 (-48.39)	0.008	0.01 (+25.00)	0.01	0.008 (-20.00)	0.005	-	0.002	-
3	Small	1.55	0.49 (-68.39)	-	0.10	-	-	-	-	0.002	-
4	Medium	2.56	0.27 (-89.45)	-	-	1.08	-	-	-	-	-
5	Large	-	-	-	-	-	-	-	-	-	-
	Total	0.39	0.18 (-53.85)	0.007	0.01 (+42.86)	0.02	0.007 (-65.00)	0.005	-	0.002	-

(Figures in parentheses indicates percentage change in cropping pattern over before salinity conditions)

Changes in Cropping Intensity

Due to salinity, the cropping intensity has increased to 111.11 and 120.41 per cent for marginal and small farms respectively while for medium farms it has decreased to 100.00 per cent. While for all farms there has been an insignificant increase in cropping intensity. It appears that due to salinity, cropping intensity has increased because calculations for the same have been made on the basis of net sown area and gross cropped area. But after salinity the reduction in net sown area due to salinity as discussed in the previous section could not be acquainted. Details of changes in cropping pattern due to salinity have been given to Table: 11.

Table 11
Changes in Cropping Intensity

S.N.	Types of farms	Before salinity	After salinity
1.	Landless	-	-
2.	Marginal	108.01	111.11
3.	Small	100.13	120.41
4.	Medium	142.19	100.00
5.	Large	-	-
	All farms	108.67	109.34

Changes in Production of Crops due to Salinity

For all farms, except for chillies, being a salt resistant crop, production of all other crops e.g., Kharif paddy, summer paddy has decreased. The increase in production of chillies has been from 44.46 to 47.09 qtls per ha. (5.92 %). For marginal farmers, production of chillies has increased from 44.46 to 47.40 qtls per ha. (6.61 %). For marginal farmers, Kharif paddy and summer paddy production has decreased from 27.79 to 24.70 qtls per ha. (11.12 %) and 59.49 to 38.14 qtls per ha. (35.89 %) respectively. For small and medium farmers, production of Kharif pady has decreased from 28.77 to 24.70 qtls per ha (14.15 %) and 27.71 to 24.70 qtls per ha (10.86 %) respectively. Changes in production level of crops before and salinity has been given in Table 12.

Table 12
Changes in Production of Crops due to Salinity

(Qtls/ha.)

S.N.	Type of farms	Kharif paddy		Chillies		Summer paddy		Mustard		Vegetables	
		Before	After	Before	After	Before	After	Before	After	Before	After
1	Landless	-	-	-	-	-	-	-	-	-	-
2	Marginal	27.79	24.70 (-11.12)	44.46	47.40 (+6.61)	59.49	38.14 (-35.89)	5.56	-	9.04	-
3	Small	28.77	24.70 (-14.15)	-	46.39	-	-	-	-	9.06	-
4	Medium	27.71	24.70 (-10.86)	-	-	59.26	-	-	-	-	-
5	Large	-	-	-	-	-	-	-	-	-	-
	Total	27.94	24.70 (-11.12)	44.46	47.09 (+5.92)	59.38	38.14 (-35.77)	5.56	-	9.04	-

(Figures in parentheses indicates percentage change in production over before salinity conditions)

Benefit –Cost Ratio, Net Present Value and Internal Rate of Return

In semi-intensive prawn farms also the commercial bank finances the farmer for a period of 5 years, but the activity goes on continuing for more than 15 years. So, like extensive prawn farms the calculation for benefit-cost ratio, net present value and internal rate of return had made for 5, 10 and 15 year period.

For landless agricultural labourer, benefit-cost ratio has increased with the increase in project period from 3.09 to 4.20 and to 4.77 for 5, 10 and 15 years respectively.

Internal rate of return for 5, 10 and 15 years respectively has increased from 4.33 to 5.37 and 5.85 for 5, 10 and 15 years.

Per hectare net present value has also increased with the life of the project period from Rs. 6765.26 to Rs. 21392.21 and to 39097.89 for 5, 10 and 15 years period respectively.

For marginal, small and all farms the benefit-cost ratio, net present value and internal rate of return followed the same trend with the increase of the life of the project period.

With the increase in size of farms from landless agricultural labourers to small farmers the net present value for 5, 10 and 15 year project has increased. Details of benefit-cost ratio, internal rate of return and net present value has been given in Table 13.

Before Salinity

To make the results comparable with the extensive and semi-intensive farms of prawn cultivation the benefit-cost ratio, net present value and internal rate of return has been calculated for 5, 10 and 15 year period.

For marginal farms, small farms, medium farms and all farms the benefit-cost ratio had been 0.40, 0.27, 0.81 and 0.41 respectively. With the increase in project period the benefit-cost ratio did not change at all. The internal rate of return had been similar to benefit-cost ratio for all the project period.

For marginal farmers, net present value has increased from 62.78 to 191.02 to 347.62 respectively for 5, 10 and 15 year period respectively. For small farms the net present value has also increased from 35.45 to 107.86 and 196.28 for 5, 10 and 15 years respectively. For medium farmers, the net present value has increased from 137.51 to 418.38 and 761.38 for 5, 10 and 15 years respectively. Similarly, for all farms net present value has increased from 63.36 to 166.57 to 324.62 respectively for 5, 10 and 15 years period.

With the increase in size of farms from marginal farmers to medium farmers though the net present value has increased from 62.78 to 137.51 for 5 years period, no definite trend in increase of the value has been found. This is also true for 10 and 15-year period. Details of benefit-cost ratio, internal rate of return and net present value has been given in Table 14.

After Salinity

As discussed earlier, a major part of the net sown area had been affected by salinity. The production of different crops except chillies has declined a great extent. The benefit-cost ratio of marginal, small, medium and all farms had been 0.24, 0.04, 0.70 and 0.26 at the end of 5th year. The benefit-cost ratio remained more or less same for different category of farms at the end of 10th and 15th year. Except small farms, the net present value had been negative for all different categories of farms. As a whole for all farms, the net present value has been negative. Details of benefit-cost ratio, internal rate of return and net present value has been given in Table: 13.

Table 13
B/C Ratio, NPV and IRR for Semi-Intensive Prawn Farms

		B/C ratio, NPV and IRR after 5th 10th and 15th year								
		5th year			*10th year*			*15th year*		
S.N.	*Type of farmers*	*B/C ratio*	*NPV ('Rs 000/ha.)*	*IRR*	*B/C ratio*	*NPV ('Rs 000/ha.)*	*IRR*	*B/C ratio*	*NPV ('Rs 000/ha.)*	*IRR*
1.	Landless agricultural labourers	3.09	6765.26	4.33	4.20	21392.21	5.37	4.77	39097.89	5.85
2.	Marginal farmers	3.17	7266.00	4.36	4.24	22928.24	5.35	4.78	41901.03	5.79
3.	Small farmers	3.02	7326.32	4.12	4.02	23142.84	5.02	4.50	42317.05	5.42
4.	Medium farmers	-	-	-	-	-	-	-	-	-
5.	Large farmers	-	-	-	-	-	-	-	-	-
	All farms	6.38	78420.75	11.95	11.39	24646.15	23.55	15.81	45411.59	34.12

(NPV and IRR calculated at 6% discount rate)

Table 14
B/C Ratio, NPV and IRR before Saline Condition

		B/C ratio, NPV and IRR after 5th 10th and 15th year								
		5th year			*10th year*			*15th year*		
S.N.	*Type of farmers*	*B/C ratio*	*NPV ('Rs 000/ha.)*	*IRR*	*B/C ratio*	*NPV ('Rs 000/ha.)*	*IRR*	*B/C ratio*	*NPV ('Rs 000/ha.)*	*IRR*
1.	Landless agricultural labourers	-	-	-	-	-	-	-	-	-
2.	Marginal farmers	0.40	62.78	0.40	0.40	191.02	0.40	0.40	347.62	0.40
3.	Small farmers	0.27	35.45	0.27	0.27	107.86	0.27	0.27	196.28	0.27
4.	Medium farmers	0.81	137.51	0.81	0.81	418.38	0.81	0.81	761.38	0.81
5.	Large farmers	-	-	-	-	-	-	-	-	-
	All farms	0.41	63.36	0.41	0.35	166.57	0.41	0.38	324.62	0.41
	B/C Ratio, NPV and IRR after Saline Condition									
1.	Landless agricultural labourer	-	-	-	-	-	-	-	-	-
2.	Marginal farmers	0.24	-87.44	0.24	0.23	-266.04	0.23	0.23	-484.16	0.23
3.	Small farmers	0.04	23.42	0.04	0.04	71.24	0.04	0.04	129.65	0.04
4.	Medium farmers	0.70	-347.94	0.70	0.70	-1058.64	0.70	0.70	-1926.57	0.70
5.	Large farmers	-	-	-	-	-	-	-	-	-
	All farms	0.26	-112.26	0.26	0.26	-341.55	0.26	0.26	-621.58	0.26

(NPV and IRR calculated at 6% discount rate)

Table: 15
Data for Calculation of MC and SC in Semi-intensive Farms

S. N.	Items	0th year	1st year	2nd year	3rd year	4th year
	Benefits					
1	Agricultural output	834575.00	834575.00	834575.00	834575.00	834575.00
2	Fish output	25456340.23	22698025.00	22698025.00	22698025.00	22698025.00
2-D	Domestic currency	6364085.06	5674506.25	5674506.25	5674506.25	5674506.25
2-F	Foreign exchange	19092255.17	17023518.75	17023518.75	17023518.75	17023518.75
	Costs					
3	Land cost	6546000.00	-	-	-	-
4	Fixed cost	3010292.52	-	-	-	-
5	Variable cost	88478.00	27039.00	64589.32	31783.00	32717.00
5.A	Labour	38628.86	6425.00	18439.00	8645.00	9455.20
5.A-1	Skilled	7725.78	1285.00	3687.30	1729.00	1891.04
5.A-2	Unskilled	30903.08	5140.00	14751.20	6916.00	7564.16
5.B	Inputs	49849.14	20614.00	46150.32	23138.00	23261.80
6.	Net aggregate consumption (MC)	1646144.71	23505661.00	23468010.68	23500817.00	23499883.00
7.	Per ha. MC	471161.75	665317.32	664251 64	665180.21	665153.78
8.	Rs in '000'	471.16	665.32	664 25	665.18	665.15
9.	SC	23961372.65	31503152.10	31458773.94	31497064.90	31495677.20
10.	Per ha. SC	678216.04	891682.77	890426.66	891510.47	891471.19
11.	Rs in '000'	678.22	891.68	890.42	891.51	891.47

Comparison of Financial and Social Cost:

Similar to extensive farms, in semi-intensive prawn farms also the social cost has been more than the financial cost. As the financial cost from 0th year to 4th year has increased irrespective of the size of farms, the social cost has also increased. Details have been given in Table 16.

From the above analyses, as social cost has always been more than financial cost it indicates that there are externalities due to prawn farming. These externalities are borne by adjoining farms by way of increased in salinity and reduction in yield of crops keeping fallow land etc.

Table 16
Comparison of Financial and Social Cost in Semi-intensive Prawn Farms

(Rs.'000/ha.)

S.N.	Type of farms	0th year		1st year		2nd year		3rd year		4th year	
		Financial cost	Social cost	Financial cost	Social cost	Financial cost	Social cost	Financial cost	Social cost	Financial cost	Social cost
1.	Landless agricultural iabourers	345.50	678.22	397.36	891.68	443.73	890.42	482.54	891.51	515.07	891.47
2.	Marginal farmers	355.60	678.22	414.01	891.68	465.93	890.42	509.91	891.51	546.98	891.47
3.	Small farmers	368.65	678.22	433.65	891.68	491.44	890.42	542.81	891.51	584.13	891.47
4.	Medium farmers	-	-	-	-	-	-	-	-	-	-
5.	Large farmers	-	-		-	-	-	-	-	-	-
	All farms	272.99	678.22	258.15	891.68	245.27	890.42	232.25	891.51	219.71	891.47

CONCLUSION

The above analyses indicate that due to semi-intensive prawn farms salinity in adjoining farms have not only increased but also production of various crops has decreased. Due to the nature of technology adopted for prawn cultivation apart from salinity other forms environmental pollution has not been there. The social cost has been also more than financial cost which indicates that due to semi-intensive prawn farming the society has to pay much more than the individual prawn farms. But the answer to the question as to how the prawn farms have been continuing for years even when there has been a Government regulation. The Government regulations instead of being more strict it has been only restricted to imposition of licensing fees whenever a farm applies for institutional finance from the commercial banks/cooperative societies.

REFERENCE

Pillay, T. (1990), "Aquaculture: Principles and Practices", Fishing News Book, Oxford;

Flaherty, M amd C. Karnjanakesorn (1995), "Marine Shrimp Aquaculture and Natural Resources Degradation in Thailand" *Environmental Management*, Vol. 19, No. 1, pp. 27-37;

Naganthan, M and Sivagnanam, .J & Rajendran, C (1995), "Blue Revolution in Green Belt", *Economic and Political Weekly*, 30(12), pp. 607-8:

Bhatta, Ramachandra and Mahadev. Bhat (1998), "Impacts of Aquaculture on the Management of Estuaries in India" *Environmental Conservation*, 25(2); pp. 109-121;

Mercer, D.E & Hamilton, L.S. (1984), "Mangrove Ecosystems: Some Economic and Natural Benefits" *Nature and Resources*, Vol. 20, pp. 14-19;

Parks, J and Bonfiaz, M. (1994), "Non-sustainable Use of Renewable Resources: Mangroves Deforestation and Mari culture in Ecuador", *Marine Resources Economics*, Vol. 9, pp. 1-18;

Patil, Pawan G and M. Krishnan (1997), "The Kandaleru Shrimp Farming Industry and Its Impact on the Rural Economy", *Agricultural Economics Research Review*, Vol. 10, No. 2, pp. 293-308;

Sathirathai, S. (1998), "Economic Valuation of Mangroves and the Role of Local Communities in the Conservation of Natural resources: case Study of Surat Thani, South Thailand, EEPSEA Research Report Series;

Dixon, J.A and P.N. Lal (1994), "The Management of Coastal Wetlands, Economic Analysis of Combined Ecological economic systems", In P. Dasgupta, and K.G. Maler eds, *The Environment and Emerging Development Issues*, Oxford, Claredon Press, pp. 149-56.

9

AGRICULTURAL LAND TO SHRIMP POND IN ORISSA: PROBLEMS AND PROSPECTS

Dr. Tushar Kanti Das and Dr. Lopamudra Mishra

INTRODUCTION

The potential of aquaculture to meet the challenges of food security and to generate employment and foreign exchange has led to rapid expansion of this sector. Particularly shrimp culture has been the driving force behind the strong increase in the shrimp trade during the late 1980s and early 1990s making it the most important seafood product traded internationally. These are the products, which fetch huge returns on investment. However, this rapid development in the shrimp culture has been accompanied by increasingly controversial debates over the environmental, social and economic impacts of shrimp culture. The shrimp industries assume no responsibility for the negative externalities they generate from their activities.

In India shrimp farming in brackish water along the coastline has been a lucrative enterprise for quite some time. Shrimp farms have been constructed in low-lying coastal areas, depriving impoverished farms of agricultural land causing salinisation of ground water in coastal villages and polluting water ways with excess nutrients (ESCAP and ADB, 2000). With the advent of new technology and rising global demand for shrimp, the basic character of shrimp aquaculture in coastal communities is changing dramatically. Traditional, low production forms of aquaculture that have contributed to local food production is being supplanted by modern, resource intensive, high production systems that cater to international markets.

Chilika is a unique lagoon situated along the east coast of India and a Ramsar site. The highly productive lagoon is a hotspot of biodiversity and with its rich fishery resources sustains the livelihood of more than 0.2 million-fisher folk. The lagoon has been facing multidimensional ecological and anthropogenic pressure leading to siltation, choking of the seawater inlet channel, fall in salinity level, proliferation of invasive species, and decrease in fish yield, leading to an overall loss of biodiversity and productivity adversely affecting the livelihood of the local community who depend on the lagoon. Chilika is highly productive due to high nutrient concentration as a result of riverine and marine nutrient inputs as well as effective nutrient recycling between the sediments and the water column.

The present paper makes a modest attempt to raise one important issue that is the conversion of agricultural land to shrimp ponds along the coastline of Orissa. In the present analysis section-I introduces the problem of the study. Section-II illustrates some stylized facts about shrimp culture. Section-III describes the benefit and cost of shrimp culture and section-IV represents the implications of shrimp culture. Government policies regarding the conversion of agricultural land are described in section-V and section-VI draws the broad conclusion of the study.

STYLIZED FACTS ABOUT SHRIMP CULTURE

In the exploitation of natural resources, appropriators often fail to perceive the ecosystem dynamics in which the resource in question is embedded (Berkes and Folke, 1998). In the development of monoculture, a system's diversity is reduced to one that is more easily 'controlled'. This is usually achieved by increases in external inputs rather than relying on services of the original system (Holling and Meffe, 1996). Whether, a pattern of exploitation is ecologically sustainable or not, is seldom part of the management agenda or the institutional framework. Also it is not reflected in the economic measures of progress and growth.

Shrimp culture is the world's most rapidly expanding aquaculture sector (Phillips *et al.*, 1993). Shrimp biomass harvested from oceans has remained constant for the period 1990 to 1995, while production of farmed shrimp supplies has increased from 10% in 1985 to 22% in 1997 and may reach 57% by the year 2005 (Rosenberry, 1997). Moreover, no significant decline in the price of shrimp has occurred, indicating continued high demand. Rapid development of shrimp farming in many countries is a phenomenon of the last 25 years. Thailand is the leading producer with an industry centered on the production of the giant tiger shrimp followed by Indonesia, Ecuador, China, Vietnam and India (FAO, 2001).

In India from 1950-51 to 1994-95 marine fish production has increased from 5.34 to 26.92 lakh tons indicating 404.12 per cent increase (Indian Agriculture in Brief, 26th Edition). Marine exports have been the second largest foreign exchange earner in 1994-95 primarily because of high value shrimp exports to Japan, Europe and the United States. Both captured and cultured shrimp constituted more than 72 per cent of the export in 1994-95 and it has provided over 1500 crores of rupees to the economy (MPEDA, 1996). Whatever be the consequences on the environment and economy it is significant to mention that both the central and state governments have been so far not only supporting but also promoting it as, it has been earning huge foreign exchange (Naganthan, *et al.*, 1995). The economic benefits of shrimp aquaculture, in particular foreign exchange earning and provision of employment, are highly important to the Indian economy, but there is a need to minimise further its social and environmental costs, in particular through more enforcement of current regulation.

Export oriented shrimp industry in India developed when large-scale aquaculture in higher income East Asian countries such as Thailand, Indonesia, China, Philippines and Taiwan began to suffer from ecological, environmental and social damage. In the states like West Bengal, Orissa, Karnataka, Andhra Pradesh and Tamil Nadu, shrimp are grown in both traditional as well as through commercial cultures. Overexploitations of fish stocks and poor aquaculture practices are of concern in India. Overexploitation of shrimp resources in coastal waters has reduced exports from capture fisheries and encouraged the growth of aquaculture. Shrimp News International reports that in 1996, there were 10,000 shrimp farms in India with approximately 200,000 hectares in production: production totaled roughly 70,000 metric tons with the firms reflecting 60 per cent extensive and 35 per cent semi-intensive aquaculture.

The shrimp farming is practiced in coastal areas also under varying land-use options e.g. traditionally as paddy-cum-prawn culture, in leased-in land for natural shrimp production, leasing out of land for shrimp production, owner operated shrimp production (Bhatta and Bhat, 1998). Due to high profitability of the enterprise, various lands are converted into shrimp farms (Patil and Krishnan, 1998). In addition to the small traditional farms there are also farms owned by corporate houses (Naganathan *et al.*, 1995), which cultures mainly black tiger shrimp.

Environmental impacts of shrimp farming have been viewed in many ways in different studies and the extent of environmental degradation is also linked with the adoption of technology either traditional or non-traditional. Land put to only traditional shrimp culture for years cannot be brought back to normal paddy cultivation as the salinity increases considerably. Under this land use option nutrient uptake by paddy plants has been reported to be poor due to the presence of high salinity (Csavas, 1995). In places after many years of shrimp culture the yield of paddy has reported to drop from 3 tons/hectare to 0.5 tons/hectare (Gadgil *et al.*, 1990). Thus conversion of lands to shrimp farming could be regarded as non-trivial irreversible (Krutilla and Fisher, 1975). Due to congestion of commercial firms along the coastal line availability of even clean brackish water for daily exchange becomes limited due to discharge of affluent from shrimp ponds (Flaherty *et al.*, 1995).

It has been evident that environmental degradation starts with clearance of mangroves and digging of rice fields in the coastal areas. As the coastal areas are dominated with acid clays (Pillay, 1990), its reclamation for shrimp ponds results in exposure oxidation of pyrites

(Flaherty and Karnjanakesorn, 1995). The aerators oxidises the existing hydrogen sulphide and converts it into sulphuric acid. This blackens the soil making it infertile. Further, the seepage of brackish water with chemical affluent into the neighboring land makes them totally unfit for cultivation. The dumping of unhygienic waste in the open results in pollution of air water and series of chronic health problem arises. Loss of grazing land for cattle also aggravates the situation for already unemployed farmers. It has been reported that the salinity in wells has decreased with the increase in distance from the cultured shrimp ponds (Bhatta and Bhat, 1998).

Kadekodi and Gulati (1999) have mentioned that the rapid expansion of commercial aquaculture, particularly shrimp culture, in Chilika Lake is contributing directly and indirectly to the decline of the lake's fisheries and world-renowned bird population. The findings of the study leads to the conclusion that the root causes for biodiversity loss in the lake lie in regional socio-economic factors. Samal (2002) and Mohanty (2000) in their study indicated the resistance by people mostly fishermen against intensive shrimp cultivation and export – a joint venture agreed upon by Tata and the Government of Orissa. They pointed out that inequalities in the society provide context for collective action.

BENEFITS AND COSTS OF SHRIMP AQUACULTURE

The concept of 'externalities' is a useful tool to look into the problems associated with shrimp culture. An externality exists when an economic activity causes an impact on third party, which is not accounted for in the private decisions of the producers and consumers. Externality may be positive i.e. beneficial to the third party or it may be negative. Negative externality implies a cost to the third party. Where externalities exist, it implies that the costs and benefits faced by the shrimp farmer are different to those faced by the society.

There are 132 fisher folk villages and 900 other villages in the catchment with a population of more than 1 million, who depend on the lagoon and its catchments. The specific methods of fishing are a reflection of the particular caste characteristics of the community. The complex mix of resources in and around the lagoon (water, fish, land, forests and fauna) has an interrelated effect on community life. The shrimp culture industry has expanded is the last two decades under the rationale of providing economic benefits at national, regional, community and household levels. At the national level, the earning of foreign exchange from the export of shrimp is the most important. Community level benefits are seen as rural job creation and making (apparently) unproductive and marginal land productive. Also, the main beneficiary of shrimp culture is the diverse group of people of the state that control the production process and which may include large rural landowners, urban entrepreneurs, government officials, bankers and other socio-political elite. Usually, these groups of beneficiaries are seen in a pejorative sense. However, from a different perspective, they can be viewed as rational economic actors responding to prevailing opportunities and constraints and they are gaining appropriate returns to entrepreneurship and risk taking.

As far as the cost of shrimp pond is concerned, the most important is the increase in the salt loading in the shrimp ponds and it is one largest threat to the soils along the coastline of the state. The effects of salinity on rice are well established. Despite the common perception that rice is tolerant to salt, almost all rice varieties are sensitive to salinity (Greenland, 1997), Salinity reduces the growth of seedlings and seed yield of plants, even at low '*Nacl* ' concentrations (Khatun and Flouers, 1995).

The geographical area, within the catchment limit constitutes the boundary for the resource management plan for the lagoon. More specifically, the land use practices in the catchment affects the lagoon ecosystem. The farmland, mostly paddy fields, is spread all around the lagoon and the run off from the crop field laden with pesticides and chemicals from fertilizers finds its way into the lagoon. While types of land use change are associated with shrimp culture, one of the most concerned issues is the conversion of agricultural land in Orissa into shrimp ponds.

The pattern of land use of a country or a State at any particular time is determined by its physical, economic and institutional frameworks taken together. The idea of becoming quickly rich as it happened in shrimp culture has attracted the common agriculturists to shift from rice cultivation to shrimp culture in ponds[1]. Also as mentioned above in some cases the salinisation of agricultural lands nearer to the shrimp ponds by seepage has led to the failure of crops especially rice and rendered the land unsuitable for agriculture. As a result there is the decline in the yield rate, which led to the conversion of agricultural land to shrimp ponds. It is also important to note that in Orissa there is an increasing trend in converting the agricultural lands to shrimp ponds. In the year 1983 only 23.5 hectare of land were used for shrimp culture, which increased to 7075 hectare in the year 1990 and that has further increased to 11332 hectare in the year 1999. Thus in this process an established livelihood system is broken down and the land with multiple agricultural uses is turned over to mono-crop production.

Table 1
Areas under Shrimp Cultivation in Orissa (in hectares)

Year	*Area*
1983	23.5
1990	7075
1994	8500
1999	11332

Source: Compiled from data collected from MPEDA and Indian Hatcheries organization.

Secondly, most of the farmers remove sediments from the pond bottoms after each harvest and they typically dump them in the nearby land. These piles of sediment contain large amounts of salt and during the rainy season, with the onset of the rain, these are mixed with other fresh water streams, thereby contaminating both surface and subsurface bodies of fresh water. Without quick and effective intervention by the government, a major environmental problem is in the offing.

Thus, the benefits seem to be accruing to consumers of developed countries where the shrimp is exported and producers and elites of the State, while the costs are born by the rural poor.

IMPLICATIONS OF LAND CONVERSION

Shrimp culture ensures high return on investment, however, rapid expansion and intensification of shrimp farming has caused increased concern regarding potential environmental impacts. The fast return on the investment is often due to lack of appropriate institutions (norms and rules) to internalise the cost of extraction. The result is that the long term benefits are traded for short term benefits that often carry with them long term degradation of environmental and social conditions.

The behaviour of shrimp cultivators in ponds and agricultural farmers can be understood from their strategic behaviour. The agricultural farmers have two strategies i.e., the cultivation of rice or conversion of agricultural land to shrimp ponds. The shrimp culturists have the two strategies i.e., to do moderate shrimp culture or to follow intense culture. Each of them knows what the pay off of each strategy will be depending on what policy the other follows. So the game is:

Table 2
Game in Terms of Profitability from the Land

	Agricultural Farmers	
Pond Shrimp Culturist	*Cultivation of rice*	*Conversion of agricultural land to shrimp ponds*
Moderate Shrimp Culture	(3,3)	(3,10)
Intense Shrimp Culture	(10,2)	(10,10)

The above payoffs of the game represent the short term behaviour of both the players. Initially when the agricultural farmer is cultivating rice and the shrimp culturist follows a moderate policy, the pay off is (3, 3). When shrimp culturist is shifting from moderate to intense culture, his pay off is rising to 10 in terms of profitability, but that of the agricultural farmers is declining to 2 from 3. This is due to fact that the productivity of agricultural land nearby to the shrimp ponds will decline due to seepage of saline water from these ponds. Now-a-days the agricultural farmers nearer to the shrimp ponds are gradually converting their agricultural lands to shrimp ponds to earn more profits. The conversion of rice paddies to shrimp ponds is being viewed as an example of the restructuring and intensification of agriculture, as farmers switch to higher value crops. Unfortunately the underlying conditions of low income for rice farming households, indebtedness limited off-farm employment opportunities and the high profit potential associated with shrimp farming, intensify the pressure to choose short-term exploitation of the resource over the long term benefits. So the pay off of both the players becomes (3,10) if the pond culturists is adopting moderate shrimp culture and agricultural farmer is shifting to shrimp culture. Similarly, if shrimp culturist is following intense culture and the agricultural farmer are converting the land to shrimp ponds, the pay offs become (10, 10) in terms of profitability. However this pay off structure is not sustainable if we analyse the situation for a long period as after about twenty or twenty five years the shrimp ponds will be fit for nothing and that will be abandoned.

It is a well known fact that sustainable development is critically linked with land degradation especially in agrarian economy. When a long term perspective is taken and the same situation is viewed from the point of view of the land quality, the game becomes just the reverse. Just like the previous game, here also the agricultural farmers have the two strategies, i.e., the conversion of agricultural land to shrimp ponds and the shrimp culturist adopts either moderate shrimp culture or intense shrimp culture.

Table 3
Game in terms of the soil quality of the land

	Agricultural Farmers	
Pond Shrimp Culturist	*Cultivation of rice*	*Conversion of agricultural land to shrimp ponds*
Moderate Shrimp Culture	(12,12)	(12,5)
Intense Shrimp Culture	(5,10)	(3,3)

When the agricultural farmer is cultivating rice and the shrimp culturist is doing moderate shrimp culture, let the pay off be (12,12). Taking into consideration the profitability of the shrimp business, the agriculturist is converting his land into shrimp ponds. However after a long period, say, 10 years to 15 years, gradually the soil quality of the land will decline and salt loading of that particular land will increase. As a result the pay off will so in terms of quality of the soil will decline to 5 from 12. So when the agricultural farmer is converting his land to shrimp pond and the shrimp culturist is going for moderate shrimp culture, their pay off is (12,5).

In a similar manner, let us suppose the moderate shrimp culturist shifts from moderate to intense culture. Similar is the situation for him i.e. in the long run, the shrimp pond might be unsuitable for even shrimp culture. It is a well known fact that shrimp is not an efficient converter of feed (Macintosh and Philips, 1992) and therefore in commercial shrimp farms as much as 77.5% of Nitrogen and 86% of phosphorus is wasted. This either accumulates within the pond or discharged in the environment. Thus, over time land put to shrimp farming could become unproductive for crops. So the shrimp culturist's pay off will decline to 5 from 12. As far as agricultural farmers are concerned, the soil quality of their land will decline due to the negative externality of the intense shrimp culture. This is due to the fact that a large portion of nutrients in shrimp feed becomes waste which is directly discharged into the neighboring land making them unfit for cultivation. So in this above game, the pay off of agricultural farmers will decline to 10 from 12. If the agricultural farmers will shift to shrimp culture and shrimp culturist will follow the intense culture of shrimp, the pay off becomes (3,3) where the sustainability of the land is questionable.

In this process of land conversion from rice to shrimp usually, small landowners have been willing to sell their land and reinvest in other productive activities, though this change of livelihood can also lead to social dislocation and reduced livelihood support. So there is the transfer of land from the former small-scale farmers to larger private shrimp culturists. The means of transfer may be legal sale of deed, forced sale through harassment or sale following degradation of the land by pollution from the shrimp ponds. Again, shrimp culture in the State has increased the land price. As a result the landowners have started selling their lands, which are previously leased out to small farmers. Thus, these farmers now have less land to lease in.

Secondly, we know that many shrimp ponds are constructed on ex-agricultural land especially rice paddy. As a result, rural peasants become detached from reliance on the land and are effectively made available to the wage labour market. Food production for own and local markets declines. Food security worsens as livelihoods become more dependent on both the market for shrimps and the market for food stuffs (the actions of both of which are outside the control of rural population). Earlier, peasants worked on the land to which they had use rights to produce food and a livelihood for their family. Thus, in the process of incorporation of the rural economy into the national and international market economy, the peasants are now denied use of land and have to enter the national market economy as wage labourers in order to buy food and other livelihood requirements in the market.

So apart from environmental degradation, the conversion of agricultural land to shrimp pond has resulted in the increasing landlessness among the rural poor, breakdown of the traditional livelihood support systems, increasing poverty, diminishing food security and the transfer of land to local and State level elites.

GOVERNMENT POLICIES AND ITS EFFECTIVENESS

Improvement in the policy of the government and the operation of the legal framework to maintain the natural resources of a state in a sustainable way is very important. In the state of Orissa, until now, the role of the state in controlling the conversion of the agricultural land to shrimp ponds is very negligible. Rather the policy of the government is encouraging the people to do so. After privatization, with the assistance from the World Bank, the State government has planned to set up ice plants, shrimp processing plants and shrimp feed plants in addition to hatcheries which are encouraging shrimp culture. However it was investigated by the National Environmental Engineering Research Institute (NEERI) that the social and economic costs are far more than the benefits of aquaculture farms.

In February 1991, the Ministry of Environment and Forests, Government of India issued a notification under the Environment (Protection) Act, 1986 which declared all Coastal stretches of seas, bays and estuaries up to 500 meter from the high tide line on the landward side as the Coastal Regulation Zone (CRZ). The notification placed a number of restrictions on industrial activities in the CRZ and requested coastal States to prepare coastal zone management plans. In response to the various social and environmental conflicts that arose from the fast development of the shrimp agriculture sector, social workers and environmentalists filed a petition with the Indian Supreme Court in 1994. The petition sought a ban on non-traditional aquaculture farms in the CRZ through the enforcement of the 1991 CRZ Notification. In December 1996, the Supreme Court of India responded by placing a number of stringent restrictions on shrimp farming in the coastal zone. Some of the recommendations are

- No shrimp culture ponds were to be constructed within 500 meter of the high tide mark applicable for all seas, estuaries, creeks, rivers and black waters and farms constructed on this zone were to be demolished before 31st March 1997.
- No aquaculture ponds were to be constructed within 1000 meter of Chilika and Pulicat lakes.
- Agricultural lands, salt pan lands, mangroves, wetlands and land for village common purposes were not to been converted into shrimp ponds.
- The Government of India shall constitute an authority under the Environment (Protection) Act 1986 to handle all appeals and cases related to shrimp farms.

Full enforcement of the Supreme Court judgment would have had major implications for the shrimp aquaculture sector. The aquaculture lobby accordingly filed a petition in 1997 seeking a review of the judgment. The court responded by extending the implementation date of the order and provided a new interim order on 19 August 1997. The new interim order stated that farms that were to be demolished as per 1996 judgment would not be demolished until further order but that no fresh seeds could be put in these farms.

Following the Supreme Court Judgment, the Ministry of Agriculture formulated the Aquaculture Bill, which was enacted in1997. The bill specified the organization and the mandate of Agriculture Authority and provided a number of directives for shrimp aquaculture in the coastal zone. The major recommendations were:

- All aquaculture farms were to obtain a license within six months of notification of the bill.
- No license was to be granted for aquaculture farming proposed within 200 meter from high tide line.

However, this restriction on eligible location did not apply to aquaculture farms on the day of establishment of the Aquaculture Authority and thus providing a deviation from the initial Supreme Court judgment.

The environmental degradation resulting from shrimp culture is not controlled by the present coastal regulation. The prevalent government policies have failed to tackle the environmental degradation. So there lies the need to develop institutional mechanism for the better monitoring and enforcement of different environmental laws and regulations to manage the coastal land area.

CONCLUSION

Inland shrimp farming is now a highly controversial natural resource management issue. Innovations in the culturing of black tiger shrimp (*Penaeus Monodon*) have opened door to expansion of shrimp farming away from the coast into the irrigated river deltas of Orissa. Shrimp culture development has taken place in a vacuum of planning and regulation in Orissa. In reality shrimp culture is affected by a large number of laws and regulation including land laws, water laws, environment laws, and fish laws etc., which are yet to be fully integrated.

The lack of emphasis on enforcement of different legislation, particularly environmental legislation, has allowed the shrimp farms to operate along the coast using culture practices that degrade the environment. Thus, the problem with the State is not a lack of legislation, especially environmental legislation, but an unwillingness or inability to enforce it. As a result of this an established livelihood system is broken down as land with multiple agricultural uses is turned over to mono-crop production and that mono-crop production will be over after few years leaving behind the land which is of no use. Environmental activists must find some way to influence the state government to resist the temptation to support shrimp farming.

It is a well known fact that once shrimp farm is being established in an area, it is very difficult to dislodge. This is due to the fact that the shrimp culture creates well-funded investment groups that will fight vigorously for its continuation. So it is also important that the people who are adversely affected by shrimp culture development become organized by forming collectives, so as to initiate public debate and gain strength to compete with the powerful interests of the shrimp producers.

The hydrological interventions undertaken to restore the lagoon have resulted in considerable improvement of its fishery resources, water quality and a positive impact on the biodiversity of the lagoon. This has significantly contributed towards the increase in the per capita income of the community who depend on the lagoon for their livelihood. The increase in the productivity level both in the wetland as well as the watershed due to the good environmental practices facilitated the poverty alleviation of the community. The community participation, linked with the various national and international institutions, intensive monitoring and assessment system are some of the uniqueness of the management practices adopted by CDA for restoration of this unique wetland. At the same time, it is opportune to start a process to establish a participatory and a self-sustainable mechanism for conservation and wise-use of the resources of the Chilika lagoon; develop a formal mechanism for the dialogue between the community and the stakeholders; complete understanding of the ecosystem through intensive studies and develop a long-term management plan based on ecological principles.

NOTE

1. This is because "farmers can earn up to 25 times more by cultivating shrimp instead of rice".

REFERENCES

Berkes, F. and Folke, C. (eds.) (1998), *Linking Social and Ecological Systems: Management Practices and Social Mechanisms for Building Resilience*, Cambridge University Press, UK, p. 459.

Bhatta, Ramchandra and Bhat, Mahadev (1998), 'Impacts of aquaculture on the management of estuaries in India', *Environmental Conservation*, Vol. 25, No. 2, pp. 109-121.

Csavas, I. (1995), 'Developing of shrimp farming with special reference to South East Asia', *Fishing Chimes*, Vol. 15, No. 1, pp. 33-34.

ESCAP and ADB (2000), *State of the Environment in Asia and Pacific 2000*, Economic and Social Commission for Asia and the Pacific and Asian Development Bank, New York, United Nations.

Flaherty, M. and Karnjanakesorn, C. (1995), 'Marine shrimp aquaculture and natural resources degradation in Thailand', *Environmental Management*, Vol. 19, No. 1, pp. 27-37.

Food and Agricultural Organisation (2001), *FISHSTAT*, United Nations, FAO, Fishery Department, Data and Statistics Unit, June.

Gadgil, M., Chandran, M. Hegde, K.M., Hegde, N.S., Naik, P.K., Bhat, P.K. (1990), 'Report on the management of ecosystem for development of Karnataka's coastal region', *Times Research Foundation*, Calcutta.

Greenland, D. J. (1997), 'The Sustainability of rice farming', *CAB international*, New York.

Holling, C.S. and Meffe, G.K. (1996), 'Command and control and the pathology of natural resource management', *Conservation Biology*, Vol. 10, pp. 328-337.

Kadekodi, G. K. and S. C. Gulati (1999), 'Root causes of biodiversity losses in Chilika lake: reflections on socio-economic magnitudes', Institute of Economic Growth, Delhi and Center for Multi-disciplinary Development Research, Darwad.

Khatun, S. and Flowers, T. (1995), 'Effects of salinity on seed set in rice, *Plant Cell and Environment*, Vol. 18 (1), pp. 61-67.

Macintosh, J.V. and Phillips, M. J. (1992), 'Environmental issues in shrimp farming', in Shrimp 92, Hong Kong, *Proceedings of the 3rd Global Conference on Shrimp Industry*, Kualalampur, Malaysia, INFOFISH.

Krutilla, J. V. and Fisher, A. C. (1975), *The Economic Impacts of Natural Environment*, John Hopkins Press, Baltimore.

Marine Products Exports Development Authority (MPEDA), (1996), *Export Statistics*, Cochin.

Mohanty, R. (2000), 'Save the Chilika movement: a case study of people's resistance to the integrated shrimp farm project in Chilika, Orissa', Civil Society and Governance Programme, IDS.

Naganthan, M., Sivagnanam, J. and Rajendran, C. (1995), 'Blue revolution in green belt', *Economic and Political Weekly*, Vol. 30, No. 12, pp. 607-608.

Patil, P.G. and Krishnan, M. (1998), 'The Kandaleru shrimp farming industry and its impact on the rural economy', *Agricultural Economics Research Review*, Vol. 10, No. 2, pp. 293-308.

Pillay, T. (1990), 'Aquaculture: Principles and Practices', *Fishing News Book*, Oxford.

Phillips, M. J., Kwei Lin C. and Beveridge, M. C. M. (1993), 'Shrimp culture and the environment: lessons from the world's most rapidly expanding warm water aquaculture sector', in Pullin, R. S. V., Rosenthal H. and Maclean, J. L. (eds.), *Proceedings of the International Conference on Environment and Aquaculture in Developing Countries*, International Center for Living Aquatic Resources Management, Manila, pp. 171-198.

Rosenberry, B. (1997), 'World shrimp farming, 1997", *Shrimp News International*, San Diego, California, p. 284.

Samal, Kishor C. (2002), 'Shrimp culture in Chilika lake: case of occupational displacement of fishermen', *Economic and Political Weekly*, May 4.

10

MODELLING COASTAL DEGRADATION IN A SMALL ISLAND DEVELOPING STATE (SIDS): THE CASE OF FISHERIES RESOURCES IN MAURITIUS

Sanjeev K. Sobhee

INTRODUCTION

Human interaction with nature, in general, and developing economies, in particular, has long been a subject of keen interest among researchers (see for instance, WCED, 1987; Mellor, 1988; Cline-Cole *et al.*, 1990; Dasgupta, 1993; and Dasgupta and Mäler, 1996 for useful insights on how policies could be designed to address environmental degradation and the depletion of natural resources. The focus on developing countries is understood, given that the majority of poor people live in rural areas and derive a large part of their living from soil and forestry resources (WDR, 1990; Dasgupta, 1993). However, in recent years, attention has been diverged towards, yet a subset of the developing world, small island developing states (SIDS). Studies on SIDS are rather sparse and have become increasingly important because of their vulnerability, explained by their limited and fragile natural resource base, which is threatened by the constant stress of tourism. SIDS constitute wonderful tourist resorts because of their comparative advantage in terms of natural assets. As such they become highly exposed basically due to government failures, tourism market imperfections and absence of proper resource conservation programmes. As Dasgupta (1993) has observed, environmental degradation would raise fertility and reduce literacy. The implications of this statement would be disastrous when considering the future of agro-based economies or those who depend on natural resources. However, while Dasgupta (1993) looks at the environment as one cause for an increase in fertility and illiteracy, one should not rule out the possibility of reverse causality, that is, fertility and illiteracy leading to environmental degradation. Actually, the linkages are interactive and the causation could be bi-directional.

In this paper, we consider the latter linkage, that is, how illiteracy and the need to have a large family, by-products of poverty and characteristics of the coastal community, may lead to environmental degradation of the fisheries resources in Mauritius (see Sobhee, 2003) for an empirical analysis of fisheries degradation in Mauritius). The potential channels through which human interaction with nature may explain marine degradation are addressed. We organise the rest of the paper as follows: Section II deals with a theoretical framework that postulates theorems and proofs and in Section III we conclude.

A THEORETICAL FRAMEWORK

Definitions and Theorems

To capture precisely the relationship between socio-economic conditions and environmental degradation, we assume the following conditions to hold:

i. Complete public education (combined primary and secondary) constitutes the minimum qualification, which an individual must acquire to secure a job in any institution.

ii. Natural and non-natural capital (an institution) are the employers of labour.

iii. Each fisherman's family constitutes a business entity in itself where all members, parents and children, are indulged in the fisheries.

In addition, we introduce the following definitions:

Definition 1: Given assumption (i) above, we define A as a partitioned set such that $A = \bigcup_{i=1}^{2} A_i$ where $i = 1$ represents students from the high-income class and $i = 2$ represents low-income class individuals, basically fishermen's children. Obviously, $A_1 \cap A_2 = \Phi$, i.e, these two subsets are disjoint.

Definition 2: Let F be defined as a one-one function that maps A onto E where E is the set representing "Attending School", i.e. F: $A \rightarrow E$. Moreover, let E_1 represent the set "Regularity in Attending School" and E_2 represent the set "Non – Regularity in Attending School". Also, $E_1 \cup E_2 = E$ and obviously $E_1 \cap E_2 = \Phi$.

Definition 3: Let G be a one-one function which maps E onto T, the set which represents "Educational Attainment". Now $T = T_1 \cup T_2$ where T_1 is the subset of "Complete Educational Attainment" and T_2 is the subset of "Incomplete Educational Attainment". T_1 and T_2 are disjoint. Hence, G: $E_1 \rightarrow T_1$ and G: $E_2 \rightarrow T_2$. In other words, attending school is a pre-requisite for complete educational attainment.

Definition 4: Define H again as a one-one function that maps T onto S, the set of "Employers of Labour". Further $S = N_1 \cup N_2$; N_1 is the set representing "Non – Natural Capital" while N_2 is the set of "Natural Capital" with $N_1 \cap N_2 = \Phi$. Thus, H: $T_1 \rightarrow N_1$ and H: $T_2 \rightarrow N_2$.

With these definitions, we prove four major theorems (see the appendix for proofs), which consist the basis of the empirical model described in the next sub-section. They are as follows:

Theorem 1: Poor children are denied access to a complete level of education, who eventually must be recruited by the natural capital.

Theorem 2: Dropouts or poorly educated children cause greater environmental degradation.

Theorem 3: Positive Externalities of being fully educated reduce the environmental stress.

Theorem 4: Theoretically, tourists bring mixed effects on the environment.

PROOFS

Theorem 1: Poor children are denied access to a complete level of education and who eventually must be recruited by the natural capital.

Proof: Any fisherman *'i'* maximises return in the following manner:

$$\text{Max}\,\pi_i = TR_i - C_i$$
$$= \bar{P}Q_i - C_i$$

The total revenue (TR) function is the product of a constant numeraire price denoted by $\bar{P}$, (as fixed by the cooperative organisation to which the catch is sold) and the quantity produced by that fisherman Q_i. Each fisherman's family, besides its own labour, uses its children as full-time workers and must cater for the latter's subsistence by incurring some costs measured in terms of physical output. C_i represents a vector of all cost associated with fishing. Now Qi = * where n is the family size (including the head of the family and his wife), and j is each member constituting the fisherman's family.

Furthermore, $$Q_j = t_j \times q_j$$

Where t_j is the number of times that individual j fishes in a day or the average number of hours spent at fishing and q_j is the average output generated by individual j per catch or per hour.

Now if we denote s_j as the time to be spent at school, by child j, we can see that for a given family which is obliged to send children to school, its output now will be denoted by:

$$Q_j' = (t_j - s_j) \times q_j$$

Thus the modified profit will be $\pi_i' = \text{Max}\left\{\bar{P}\sum_{j=1}^{n} Q_j' - C_i\right\}$

$$\pi_i' = \text{Max}\left\{\bar{P}\sum_{j=1}^{n} t_j q_j - C_i - \sum_{j=1}^{n-2} s_j q_j\right\}$$

$$< \text{Max}\left\{\bar{P}\sum_{j=1}^{n} t_j q_j - C_i\right\} = \pi_i$$

Hence, the opportunity cost of sending children to school $\sum_{j=1}^{n-2} S_j q_j$ is which is the loss in return to a given fisherman's family.

A further cost of having children as inputs is the private expenditure incurred by the fisherman's family to have access to public education. At this stage, it is essential to observe that despite the prohibitive costs of sending children to school, parents must abide by the law, and have no alternative, as public education is compulsory. To analyse this cost, we may rewrite the profit function as follows:

$$\pi_i = TR_i - C_i$$

$$= \bar{P}Q_i - C$$

$$= \bar{P}\sum_{j=1}^{n} t_j Q_i - \sum_{j=1}^{n-2} q_j s_j - \sum_{j=1}^{n=2}(c_j + u_j) - F$$

The first term represents the total revenue of the family business and the second term as already described is the opportunity cost of sending children to school (or employing child labour). The third term is the total cost of sending children to school (c_j per child) for instance, transport cost, purchase of books and necessary stationery. Altogether, it includes the cost of raising these children (u_j per child). Since children are used as inputs, they do not basically obtain a wage but food instead or at the most a stipend as pocket money. F represents the non-labour inputs used. From our specification therefore the overall cost of sending children to school is $\sum_{j=1}^{n-2} q_j s_j + (c_j + u_j)$.

However, to show that poor children have to be employed by the natural capital, we consider the following:

From the above definitions $\forall x \in A_2$, $F(x) = y \in E_2$; since $E_1 \cap E_2 = \Phi$ then $y \notin E_1$. Also, since G and H are one-one and onto, $\forall y \in E_2$, $\exists z$ such that $G(y) = z \in T$ and $z \notin T_1$. Hence $G(y) = z \in T_2$. Now $\forall z \in T_2$, $\exists m$ such that $H(z) = m \in N_2$. If we define a composite function C such that $C = H[G[F(.)]]$ then C will also be one-one and onto and we can easily show that $C(x) = m \in N_2$. Hence those who have not acquired a complete education have to go back to or remain employed by the natural capital. This completes the proof. We also note that if $G(A_1) = T_1$, $G_1(T_1) \cap G_1(T_1') \neq \Phi$ which implies that students of high income families may also drop out principally due to poor performance at examination but however do not go to the natural capital to seek employment.

Theorem 2: Dropouts or poorly educated children cause greater environmental degradation.

Proof: Consider $\bar{p}_i\,[q^p h^p + kq_k t_k] - k[c_k + u_k] + \pi_F$ which is a modification of the profit function described above and k here represents number of children. If we separate the profit realized by the children from the head of the unit, including his wife, and denote this as π_F, then the left hand side of π_i represents the profit of using child labour. Without loss of generality, assuming that cost, output produced and hours spent at fishing for each child to be constant, differentiation of π_i with respect to k will imply

$$\frac{d\pi_i}{dk} = \bar{p}q_k t_k - (c_k + u_k)$$

$$= MRP - MC$$

The left-hand term is the Marginal Revenue Product (*MRP*) of using child labour by each family and the latter must exceed the Marginal Cost (*MC*) of child labour for $\frac{d\pi_i}{dk}$. This means that each family unit will have an extra child as long as $MRP > MC$ for that child. To maximise the difference between MRP and MC, either t_k or q_k must be raised or $(c_k + u_k)$ be minimized. Clearly this means that the children must increase the marginal output from fishing or their harvest rate by raising the number of hours spent at sea. On the other hand, parents must reduce the subsistence cost of children. These have adverse implications for school attendance, calorie intake and children's performance at school. More importantly, this accentuates environmental stress, as hours of fishing will increase.

Theorem 3: Positive Externalities of being fully educated reduce the environmental stress.

Proof: Let X represent an externality operator. We know that $H: T \to S$ already from Definition 4, but there exists X now and A_3 such that $X: T_2 \to N$, where $T_2 \subset T_1'$ and $F^{-1}[G^{-1}(T_1')] = A_3 \subset A_2$

Hence $\quad X(A_3) \subset N_1$ and $X(A_3) \not\subset N_2$

Thus $$\frac{\partial N_2}{\partial X(A_3)} \leq 0$$

In other words, externalities reduce dependence on the natural capital.

As a famous holiday resort, Mauritius is exposed to high tourist arrivals, and this has a special role to play along the coast, where hotels are mainly located.

Theorem 4: Tourists bring mixed effects on the environment

Proof: In the first place, tourists can generate employment for unskilled workers, thus by the externality argument again there exists A_4 such that

$$X\text{: } A_4 \rightarrow N_1 \text{ where } A_4 \subset A_2 - A_3 \text{ and } \frac{\partial N_2}{\partial X(A_4)} < 0$$

Alternatively tourists also represent a higher population that will tend to increase dependence on fish catch (or on the coastal and marine environment). Also, the construction of hotels greatly interferes with the marine eco-systems and presents yet another threat to fish production.

If we express functionally, $N = f(D)$ where D = dependents, given tourists raise D causing $\frac{\partial N_2}{\partial f(D)} > 0$, environmental degradation accelerates.

Therefore the net effect of tourist (through total differentiation) can be summarized as follows:

$$\text{dN}_2 = \underset{>0}{\frac{\partial N_2}{\partial f(D)} df(D)} + \underset{<0}{\frac{\partial N_2}{\partial X(A_s)} dx(A_4)} \quad \text{(uncertain)}$$

Net effect, which is uncertain, depends on the extent that positive externalities impact on employment created by tourism to offset the consumption of seafood by tourists.

CONCLUSION

This paper emphasises the socio-economic channels through which environmental degradation of the fisheries resources could be explained in Mauritius. We focus particularly on poverty and its by-products, illiteracy and family size, as being among the major causes of the decline in fish catch in recent years. Tourism does not seem to provide a safety net to the coastal community and to reduce significantly the environmental stress due to its mixed effects. The major policy implication emanating from this work is that the government has to establish proper support programmes for poor families and children of the fishing community. Both in kind and in cash methods of intervention would be necessary to reduce fertility and raise literacy rates in an attempt to reduce the environmental stress in coastal Mauritius.

REFERENCES

Cline-Cole, R.A., Main, H. and Nichel, J.E. (1990), On Fuel-Wood Consumption, Population Dynamics and Deforestation in Africa, *World Development*, (18): 513-527.

Dasgupta, P. (1993), *An Inquiry into Well-being and Destitution*, Oxford University Press: Oxford.

Dasgupta, P. and Maler, K.G. (1996), Environmental Economics in Poor Countries: The Current State and a Programme for Improvement, *Environment and Development Economics*, (1): 3-7.

Mellor, J.W. (1988), The Intertwining of Environmental Problems and Poverty, *Environment*, (30): 8-30.

Sobhee, S. K. (2003,) Economic Development, Income Inequality and Environmental Degradation of Fisheries Resources in Mauritius, *Environmental Management*, (34): 150-57.

WCED (World Commission on Environment and Development) (1987), Our Common Future, Oxford University Press: Oxford.

WDR (World Development Report) (1990), The World Bank, Washington D.C.: USA.

11

CUMULATIVE IMPACT ASSESSMENT IN THE SHARAVATHI RIVER BASIN

Ramachandra T.V., Subhash Chandran M.D., Sreekantha, Diwakar Mesta, Rao G.R. and Sameer Ali

INTRODUCTION

Many of the developmental projects in the past have been implemented with little environmental concern in India. The assessment of the projects was based on only technical, economic and political criteria, mainly due to the fact that knowledge of environmental impacts and impact assessment technology were not fully developed at that time. As a result, a number of large-scale development projects led to adverse impacts on the environment. Environmental Impact Assessment (EIA) has gained tremendous importance in nineties and is mandatory for any major developing work in India. An EIA is an effective tool in identifying and evaluating the potential impacts of the project, related to physical, chemical, biological and socio-economic components of the environment and in-turn reducing the likely adverse consequences, by proper mitigation measures, appropriate monitoring as well as auditing programmes. Normally, EIA is vastly implemented prior to project initiation. In absence of such investigations, the scope is expanded to include larger issues, such as biodiversity changes, sustainable development, economic valuation of natural resources and impacts etc., to assess the cumulative impacts. The projects that have not undertaken EIA, during the construction phase can be valued using Cumulative Impact Assessment studies. The Cumulative Impact Assessment or Cumulative Effects Assessment (CEA) considers systematically multiple impact sources, pathways between sources and impact receptors, and direct and indirect, additive and non-additive interactions.

The practice of Environmental Impact Assessment (EIA) was nonexistent at the time of execution of the Linganamakki and Talakalale reservoirs. Naturally these projects had telling consequences on the fragile ecosystems of which no documentation, whatsoever, was ever made. The area affected had substantial tracts of species rich tropical forests with their varied kinds of habitats. These included climax forests, Myristica swamps, gallery forests, grasslands, and ecosystems of varied kinds. Also were lost large tracts of traditional agricultural systems, which harboured priceless heritage of domesticated biodiversity. The submersion of lands and consequent resettlements of the affected people as well as the influx of migrant population brought in for the various developmental activities in the region had tumultuous effects on the entire ecosystems. These included fragmentation of natural habitats, forest encroachments, poaching of wildlife, excessive extraction of forest biomass, disruption of wildlife corridors, diversion of streams for agriculture, over grazing, etc. After the passage of four decades, the need was felt to assess the cumulative impact of the Sharavathi river valley hydroelectric projects on the environment. In this connection, the Karnataka Power Corporation Limited approached the Centre for Ecological Sciences, Indian Institute of Science, Bangalore, which is one of the centres of excellence under the Ministry of Environment and Forests, Government of India, to undertake such a study.

STUDY AREA

The Sharavathi river originating in the central Western Ghats runs through the districts of Shimoga and Uttara Kannada (Karnataka state, India) and is depicted in Figure 1. The river has a catchment area of about 3600 sq. km (Figure 2). The river, and its tributaries and numerous streams run through the rugged terrain of the Ghats. The river, which takes birth at Ambuthirtha in Thirthahalli taluk flows northwesterly for 130 km to join the Arabian Sea at Honnavar, in Uttara Kannada district. At Jog, the river drops precipitously from a height of 253 m into a deep gorge creating, one of the most spectacular scenic places of the Western Ghats. The major tributaries of the river are, Nandiholé, Haridravathi, Sharmanavathi, Hilkunjiholé, Nagodiholé, Hurliholé, Yenneholé, Mavinaholé, Gundabalaholé, Kalkatteholé, and Kandodiholé.

The river basin has been harnessed for hydroelectric projects to its fullest potential. The Mahatma Gandhi Hydroelectric Project with present installed capacity of 120 MW, was commissioned in 1948. This was followed by Sharavathi Generating Station (1035 MW) commissioned in 1964-65, the Linganamakki Dam Power House (55 MW) and the Sharavathi Tail Race Project (240 MW) at Gerusoppa in 2001. These account for about 45 % of the total installed capacity of hydroelectric power in the state. Sharavathi river alone, in its fullest potential, accounts for an estimated electricity generation of about 6,000 million units (kWh) per annum.

The river water is stored in three major reservoirs, at Linganamakki (14° 10′ 24″N, 74° 50′ 54″ E), Talakalale (14° 11′ 10″N, 74° 46′ 55″ E) and Gerusoppa (14° 15′N, 74° 39′E). The areas submerged for these reservoirs are 326.34, 7.77 and 5.96 sq. km respectively. The Linganamakki reservoir

Figure 1
Location Map-Study Area

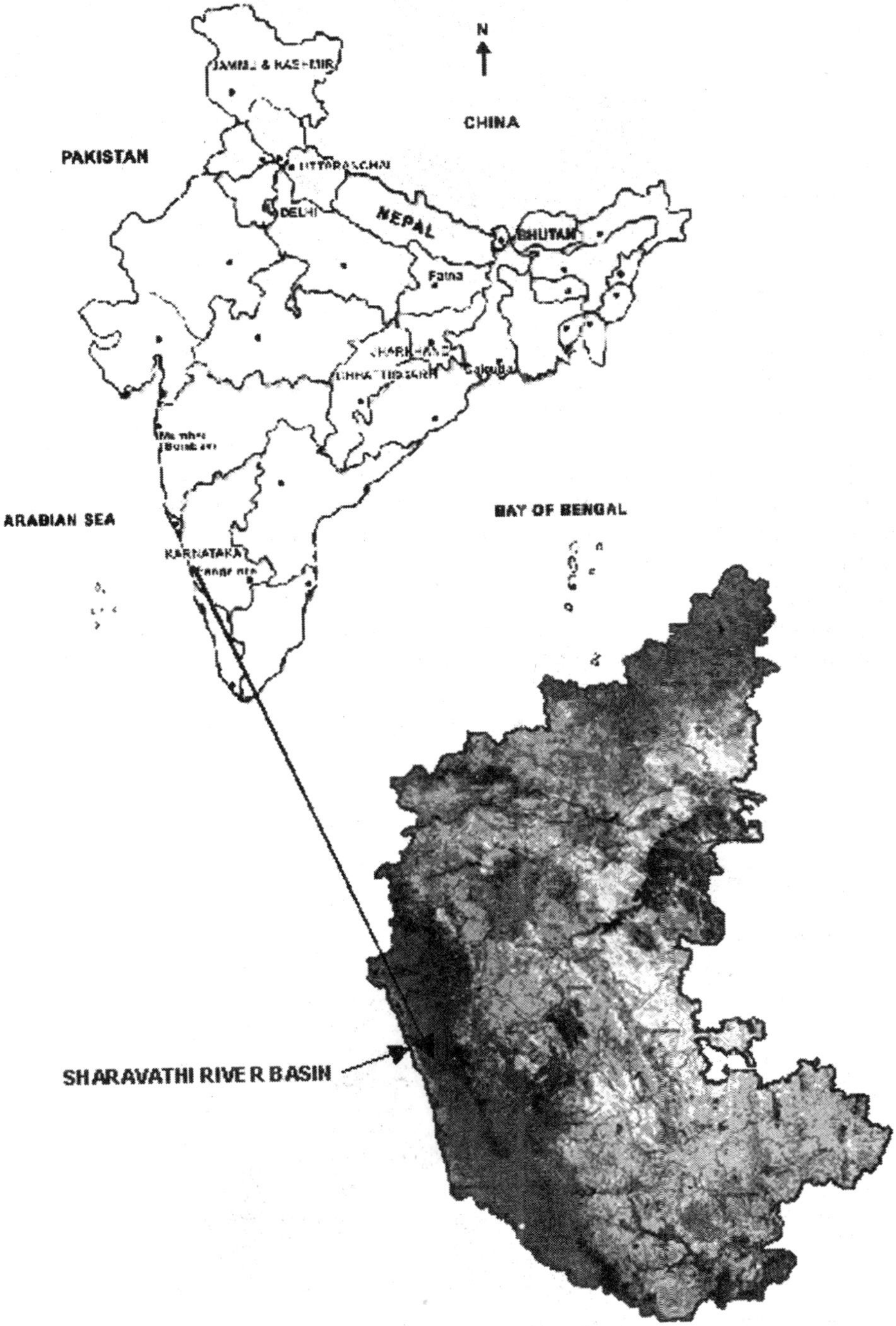

resulted in the full or partial submergence of 99 villages in the Sagar and 76 villages in the Hosanagar taluks of Shimoga district, also causing the displacement of 12000 people. The Talakalale reservoir resulted in the full or partial submergence of 3 villages in the Sagar taluk. Whereas, the Gerusoppa reservoir did not affect any populated village or caused any displacement of humans, but the submergence of 5.96 sq. km of tropical evergreen to semi-evergreen forests. In addition, for the Sharavathi Tail Race project, 4.72 sq. km of forest and 0.08 sq. km of other lands was also acquired for the township, roads, etc.

The current study went a long way in identifying the cumulative environmental impacts of river basin projects. In this regard, the foremost necessity was gathering baseline information on the prevailing ecosystem condition. The outcome of this study, carried out by a multidisciplinary team of scientists and engineers helped in diagnosing the adverse effects of the project on the ecology. This study highlights the need for adopting higher standards in environmental management, which should include conservation, restoration and management strategies for natural ecosystems, like the restitution of corridors of animal migration and improving the quality of aquatic ecosystems and terrestrial ecosystems surrounding the reservoir. The outcome of the study is also a substantial addition to the current benchmark database on existing biodiversity and ecology of the Sharavathi River Valley. The recommendations if implemented properly would certainly help in restoring and conserving the ecosystem in Sharavathi river basin.

Figure 2
Sharavathi River Basin, Western Ghats

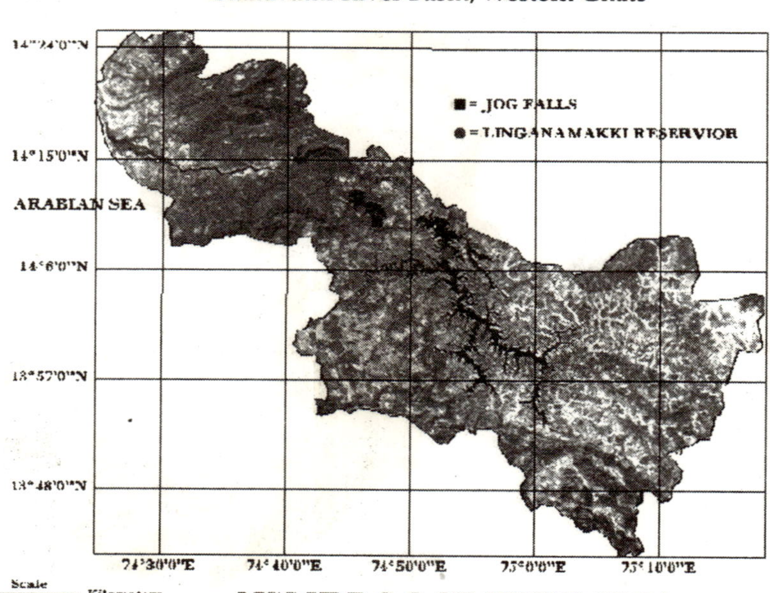

The Linganamakki reservoir has a catchment area of nearly 1991 sq.km. It receives water mainly from rainfall and also from the Chakra and Savahaklu reservoirs, which are linked to Linganamakki through a canal. The water from Linganamakki dam flows to Talakalale Balancing Reservoir through a trapezoidal canal with a discharge capacity of 175.56 cumecs. The length of this channel is about 4318.40 m with a submersion of 7.77 sq. km. It has a catchment area of about 46.60 sq.km. The gross capacity of the reservoir is 129.60 m cu m.

Due to the hilly terrain, the submergence and consequent increase in water spread area has resulted in creation of several islands distributed throughout the reservoir. There are about around 125 islands with sizes ranging from 2-3 hectares to less than 25 sq. km. In most of the submerged areas the highlands and hilltops protrude out of the water-filled surroundings as islands. During monsoon the water level reaches its peak leaving the top of the hills and hillocks on which the vegetation thrives. However, summer season interconnects the Islands as the water level recedes. The remnants of once inhabited villages that got submerged during construction of the dam can be seen in this region as water level goes down. The entire submersion area is rich in relics of varied kinds, which speak of its human habitation once. These relics include temples and idols, asphalted roads , pots and other earthenwares, stone inscriptions etc., lie randomly reminding us of the bygone history and culture. Rows of dead stems of arecanut palms and remains of irrigation channels also surface when the water level goes down, in addition to the diversity of a trunks of dead trees.

LAND COVER AND LAND USE ANALYSES

Land cover and land use analyses was done using remote sensing data as well as collateral data.. Geographic Information System (GIS) has been used for integrating remote sensing data with collateral data. Supervised classification approach using Gaussian maximum likelihood classifier is used for classification of remote sensing data. This is done by collecting a training data (GCP) from the study area and GCP's were uniformly distributed all over the study area. *Normalized Difference Vegetation Index (NDVI)* is the most commonly used Vegetation Index (VI) for land cover analyses. The results show that area not under vegetation ranges from 39 % – 46 % while area under vegetation ranges from 54 % – 61 % in the Sharavathi basin.

Land-use change is significant to a range of themes and issues central to the study of cumulative impact assessment. The alterations it effects in the surface of the earth hold major implications for sustainable development and livelihood systems. Land-use change analyses help to understand (i) the driving forces (exogenous variables) of land use as they operate through the land manager; (ii) the land-cover implications of land use; (iii) the spatial and temporal variability in land-use/cover dynamics; and (iv) impacts on biodiversity and ecosystem functioning. Land use analyses of the upper river basin show that 25% of the area is under moist deciduous, followed by category II (21%, consisting of grass land, scrub, cultivable waste) , evergreen to semi-evergreen (16%), plantations (9.7%), agriculture (8.5%), water body (7.1%), class I (7.14%, habitation, roads, rocky area, etc.) and dry reservoir bed (5.4%).

Human interventions in the ecosystem have resulted in erosion of biodiversity in the recent past in the river basin. The loss in biodiversity has been attributed to habitat loss and fragmentation of the natural landscapes. The analyses revealed extensive fragmentation in the river basin as a consequence of damming the river, conversion of forestlands to agricultural lands, and encroachment etc. Fragmentation has resulted in the remnant areas of native vegetation surrounded by a matrix of agricultural, horticultural and other human impacted lands. These in turn have important influences on the biota within the remnant patches, especially along the edges of the remnant patches as well as in the surrounding matrix. The fragmentation is fuelled by anthropogenic activities, which are degrading the forest habitat of the critical watershed areas. Such human pressures have caused fragmentation of large, unbroken tracts of forests into, smaller isolated patches. This process has lessened the value of the forest as a habitat for many of the plant and animal species native to the Sharavathi river basin, from where, it is feared that many are lost forever. These consequences vary with the distance from the patches and connectivity with the other patches. The physical distances and influences modify the size, shape and the position in the patch and its constituents. The characteristics of the patch along with the spatial and temporal changes in the landscape were done to quantify the extent of damages. The remote sensing data in conjunction with GIS and GPS helped in landscape characterization. Land use changes, Landscape dynamics and Landscape characterization have been analyzed from patch to river basin level to understand the temporal changes due to developmental activities in the river basin. Landscape analysis showed that the indices of shape, richness and diversity provided an additional evaluation of land cover spatial distribution within the complex mountain landscape. The landscape analysis has provided an outline of the degree of propagation of the disturbance from the non-biotic sources and fragmentation. It is revealed that fragmentation has caused loss of connectivity, ecotones, corridors and the meta-population structure.

As a result of the development programme based on ad-hoc decisions, considerable changes in the structure and composition of the land use and land cover in the region have been very obvious during the last four decades. Pressure on land for agriculture, vulnerability of degraded ecosystems to the vagaries of high intensity of rainfall and high occurrence of steep erosion and landslide-prone areas, lack of integrated and coordinated land use planning are some of the reasons for rapid depletion of the natural resource base. These changes have adversely affected the hydrological regime of river basins resulting in diminished river / stream flows. Frequent landslides damage the infrastructure and threaten human lives. In this situation, in order to resolve present problems and to avoid a future crisis, a comprehensive assessment of land use changes, its spatial distribution and its impact on hydrological regime is required and accordingly, appropriate remedial methods should be employed for the sustainable utilization of the land and water resources of the catchment.

BIODIVERSITY

Flora

The vegetation was studied using transect based quadrat sampling methodology in 140 localities of the Sharavathi upper catchment area and in 26 localities in the Sharavathi lower catchment area (*field investigations are in progress in the downstream region of the river basin*). The natural vegetation ranges from the climax tropical evergreen to semi-evergreen forests along the high rainfall areas of the main hill ranges of the Western Ghats to the moist deciduous forests in the undulating plains and low hills along the eastern drier tracts of the river basin. The landscape everywhere is in fact a mosaic of a variety of elements, which are caused by human impacts through historical times to current period.

The number of tree species per hectare in the evergreen to semi-evergreen forests is in the range of 35 – 60. The moist deciduous forests have 15 – 25 tree species per hectare. Basal area per hectare ranges from 35 – 60 sq. mt. in good forests. In the deciduous forests, the basal area ranges from 20 – 35 sq. mt. The evergreen forests are of high conservation value since they are great repositories of endemic species of the Western Ghats (50 – 80 % for trees). On the contrary, the deciduous forests, largely caused by human induced fires have endemism as low as 9 – 15 % for trees. We have observed that there exists a high correlation between vegetation endemism and endemic fauna. Endangered endemic animal species such as the Lion tailed macaque, Slender loris, Jungle striped squirrel, Malabar pied hornbill, Saw scaled viper and the butterflies like Southern birdwing, Budha peacock are associated with the last relics of the climax evergreen patches. *Semecarpus kathalekanensis* an altogether new species of tree has its precarious existence in the endangered Myristica swamps of Siddapur Taluk.

The evergreens to semi-evergreen forests are the major sources of perennial water sources throughout the catchment of Sharavathi. On the other hand in the deciduous tract, the streams mostly dry up in the summer months. Therefore conservation of evergreen forests and restoration of such forests in the eastern parts of the catchment are of paramount importance.

Bulk of the water into the reservoirs comes from natural forests. Unfortunately in substantial part of the catchment area monoculture plantations have been raised causing the drying up of the streams and impoverishment of the ecosystems as a whole. Since the plantations do not yield any fodder or NTFP, the rural population is put to great hardships. Therefore such land uses are not desirable in the catchment area of this important river, which produces substantial part of hydropower for Karnataka state.

The river basin has a variety of habitats (21 listed in the report) that support rich flora of herbs, shrubs and climbers of which we have recorded about 215 species. Temporal seasonal surveys will reveal more number of herbs. Evergreen to semi-evergreen forests and grasslands of the Western Ghats have the largest congregations of endemic herbs. Some of the herbs are exclusive to specialized habitats like tree trunks and wet rocks. The increasing human impact and openings in forest canopy as well as over grazing by cattle are posing threats to many of the herbs.

The numerous streams and the banks of the river and its tributaries in the evergreen to semi-evergreen forest belt are lined with characteristic riparian vegetation of which the notable tree species are *Calophyllum apetalum, Elaeocarpus tuberculatus, Mastixia arborea, Hydnocarpus wightiana, Madhuca neerifolia, etc.* Towards the drier forests of the east the water bodies are lined with tree species such as *Pongamia pinnata, Madhuca neerifolia, Hopea wightiana, Bambusa sp,* etc. The riparian vegetation plays a crucial role in protecting the water bodies from siltation, creating shade conditions to maintain appropriate temperature regime for sustaining populations of endemic fishes, amphibians, phytoplankton, zooplankton and aquatic insects. Of late there has been numerous instances of misuse of the banks of streams and rivers in the catchment area causing severe upsets in the characteristic biota associated with them. Stream waters are often diverted to newly created horticultural farms thereby also affecting the water flow into the reservoir.

Fana

Sharavathi River Basin is rich in animal diversity. The basin also covers part of Sharavathi Wildlife Sanctuary. The region is also adjacent to Mookambika Wildlife Sanctuary of Udupi District and Shettihalli Wildlife Sanctuary of Shimoga District. Many animals from these sanctuaries pass through the catchment area. Due to rising human pressures including creation of numerous monoculture plantations many of the migratory paths of the animals are disrupted. We have therefore proposed the creation of wildlife corridors to facilitate animal movements. These corridors are depicted in the figures 5 – 10.

The study area harbours some of the endemic and endangered species of the Western Ghats. The major mammals found here are Gaur, Leopard, Sambar, Barking Deer, Spotted Deer, Mouse Deer, Common Langur, Wild Boar, etc. The Lion tailed macaque is an endangered endemic species found in dense evergreen forests. Tiger, Leopard, Wild Dog, Civet Cat, Sloth Bear, etc are rare animals in the study area. Some of the minor mammals are Indian Hare, Slender Loris, Flying Squirrel, Giant Squirrel, Common Indian Mongoose, Common Otter, Porcupine and Pangolin. Several species of bats, both frugivorous and insectivorous are also found.

A total of 140 species of birds has been sighted in the study area. Of the birds the order the most abundant are Passeriformes, which include the Flycatchers, songbirds, Warblers, etc. Notable endemic bird species include, Blue Winged Parakeet, Crimson Throated Barbet, Grey Headed Bulbul, Heart Spotted Woodpecker, Malabar Grey Hornbill, Malabar Pied Hornbill and Small Sunbird. The endangered Great Indian Hornbill was sighted near the Malemane village of Siddapur taluk.

The wetland habitats include the reservoirs and their environs, the Sharavathi estuary and several tanks in the study area. The wetlands are dominated by both Passeriformes and the Ciconiiformes, the latter includes the Egrets, Herons, Ducks, Kites, Kingfishers, Coots, etc. While the lands are dominated by resident birds the aquatic habitats have by both resident and migratory birds.

A spate of construction activities in the lower catchment of Sharavathi river and ongoing and threatened fragmentation from the burgeoning human population are perilous to the habitats of several forest birds. There has also been considerable decline in the mangrove vegetation threatening the estuarine birds as well.

The notable reptiles are crocodiles (in the reservoir), turtles, King Cobra, Python, Saw Scaled Viper, Russel's Viper, Malabar Pit Viper, Striped Keel-back, Johnson/s Boa and the Flying Snake.

The notable amphibians and Caecilians observed in the river basin are, *Rana tigrina, Rana limnocharis, Rana curtipes, Philautus sps, Ansonia sps, Ichthyophis sps, etc.*

A total of 134 species of Butterflies were reported from the study area. Some are endangered and some are endemic species. Notable among the butterflies are, Fluffy Tit, Monkey Puzzle, Southern Birdwing (the largest south Indian butterfly), Malabar Raven, , Paris Peacock, Malabar Banded Peacock, etc. The diversity of butterflies is correlated to the diversity of host plants. The varied kind of human impacts are posing threats to several rare plants thereby also affecting their dependent butterflies. Examples are Crimson Rose, Malabar Rose, Southern Birdwing, (dependent on members of *Aristolochia* family), Blue Nawab (host plants unknown), Malabar Banded Swallow Tail (dependent on *Acronychia* and *Euodia*). Therefore while redesigning vegetation in the human impacted landscape of the catchment attention should be paid to the rehabilitation of these rare and endangered butterflies by planting their host plants.

Except for stray references no systematic study has ever been undertaken of the beetles, of which there is an amazing diversity in the Sharavathi catchment area. We could record during a short period about 122 species belonging to 26 families. Many of these beetles are very specific to their microhabitats. The beetles have important role in the ecosystem, as they are pollinators of several plant species including the endangered Myristica of the swamps.

The diversity of common beetles such as the root grubs is high in the monoculture plantations. These root grubs, of late, have become a major threat to the horticultural plantations, which contribute substantially to the economy of Uttara Kannada and Shimoga.

In the absence of any previous studies on the ants of the Sharavathi catchment area, ours is a benchmark study. A total of 84 species of ants representing 31 genera under 5 subfamilies have been recorded from this study. Of the various habitats studied for ants maximum number of 64 species was found associated with deciduous forests. Some of the ants such as members of *Myrmicinae* are associated with all vegetation types.

Ants like *Polyrhachis mayri* are exclusively found inside undisturbed evergreen forest patches. Some species of ants are indicators of disturbances. The islands of Sharavathi frequented by tourists are being inhabited by hot climate specialist ants suggesting disturbances (open canopy, hard soil and absence of litter cover). Such findings highlight the importance of keeping the pristineness of the fragile ecosystems of the Western Ghats from the forays of unplanned tourism.

Lichens are indicators of pristineness as well as disturbances of terrestrial ecosystems. They are pioneers of vegetation and many of them can endure hostile environmental conditions. Some of the lichens have economic importance too. The catchment area is very rich in lichens. Of the 175 species reported from the Karnataka Western Ghats, the study area has 143 species.

The disturbing trend is that the *crustose lichens, with 108 species are found dominating the human impacted areas, especially exposed and fire affected places.* On the contrary, the semi-evergreen to evergreen forests are poor in lichens because of denser shade and humid conditions. Therefore a constant monitoring of such indicator species is essential for the upkeep of the ecology of the catchment.

AQUATIC BIODIVERSITY

Phytoplanktons are the primary producers of aquatic ecosystems. They are also one of the rapid detectors of environmental changes because of their quick response to pollutants and fluctuating light and temperature conditions. Pollution stress reduces the algal species diversity but increases the number of individuals of few tolerant species. We studied seven streams as well as the reservoir waters in the upper catchment and ten streams in the lower catchment for phytoplankton. The study area has a rich diversity of 216 species belonging to 59 genera. Of these the diatoms are more common in streams and desmids in the reservoir. No algal blooms indicative of eutrophication was found during the survey. However, it should be taken with caution that all the waters of streams and the reservoir show light to moderate levels of pollution.

Zooplanktons are the secondary producers of the aquatic ecosystems. They transfer energy from primary producers to higher animals and therefore play an important role in the nutrient cycle of these ecosystems. Zooplanktons were studied in thirteen localities. The freshwater zooplankton of the river is rich and diverse. We could record 39 species. The species richness was low in lotic system compared to lentic water system. No large zooplankton was recorded mainly due to predatory pressures from fishes.

Insect groups like Mayflies, Dragonflies and Caddiesflies are important in the functioning of stream ecosystems. They are important links in the nutrient cycling in forest streams. They process wood and leaf litter and degrade them into forms absorbable by fungi and bacteria. These nutrients are absorbed by plants of the riparian zone. Aquatic insects also form food for fishes. During our study covering 12 sites and 20 sampling localities a total of 37 genera belonging to 27 families were collected. Votehalla in the lower Sharavathi catchment with 18 genera topped the streams in insect communities. In streams subjected to higher temperature due to poor canopy cover the generic diversity was low.

For the first time a monotypic genus i.e., *Phylloneura westermanni,* is recorded in the study area, which was earlier reported from only Nilgiris, Coorg and Wayanad in 1933 by Fraser, from a Myristica swamp of Kathlekan in Siddapur. The presence of Philorus in the Votehalla stream indicates its pristine nature and presence of primeval forests around. Some pollution sensitive genera have also been found in Kathlekan indicating the beginning of human disturbances.

The Western Ghats is renown in the world for the diversity of endemic fishes. A total of 51 species belonging to 32 genera and 16 families has been recorded only from the upper catchment of Sharavathi. From the lower catchment also 51 freshwater fish species have been recorded. The Sharvathi estuary has a total of 44 species of fishes. Altogether, the entire basin has a total of 112 species. The prominent endemic species are *Tor khudree, Labeo fimbriatus, Puntius fasciatus, Puntius filamentosus, etc.* The investigation in the downstream has reported a new species - *Parabatasio sharavatiensis.*

The study of locality-wise species distribution revealed that, western streams and reservoir areas were dominated by native and endemic species and absence of introduced exotic species. Compared to this, on the eastern side most of the native species are gradually becoming rare due to insufficient food availability, disease and over exploitation. Epizootic Ulcerative Syndrome (EUS) affected mainly the, *Channa* spp., catfishes and minor carps. The outbreak of parasitic infection in *Mastacembelus armatus* (havu meenu) substantiates that the native fishes are continually being affected. The practice of targeting the breeding grounds (for fishing) is a threatening factor to native fish species like *Garra, Gonoproktopterus, Puntius, Labeo, Cirrhinus, Mystus, and Pseudeutropius* etc. Silt deposition in the reservoir due to poor catchment conditions has damaged the breeding areas of endangered species such as Mahseers.

AGROBIODIVERSITY

The Sharavathi river basin harbours diverse agriculture systems and the study revealed that the region has treasured 59 paddy varieties of which 29 are traditional. The genetically diverse traditional varieties well suited to the local environment are disappearing due to adoption of high yielding varieties, and greater spread of commercial and horticultural crops. The study also revealed that improper usage of modern techniques in agriculture has resulted in devastating effects on the environment. Extensive use of inorganic fertilizers with the adoption of modern agricultural practices, has resulted in the conversion of self-sustained traditional system to a system depending on external interventions. This is one of the factors contributing to non-point source of pollution in the eastern region.

Soil Erosion and Siltation

Soil erosion is a natural phenomenon even in the pristine ecosystems. However, it gets accelerated due to removal of vegetation cover and disturbance to the upper soil strata. The most severe soil erosion is taking place at the ecotone of upper limit of reservoir level and the adjacent terrestrial ecosystem. Soil-separates (gravel, sand, silt, clay, etc) are transported to varied distances depending on the catchment runoff and ultimately results in silting of reservoir, which reduces its storage capacity. In western region, siltation is arrested by species such as *Holigarna, Madhuca neriifolia, Elaeocarus* and *Pandanus* etc. and by the multi-canopied evergreen forest itself.

There are two paths of escape for surplus water – through infiltration into underground aquifers, and as surface water flows. Natural land cover has various properties that help to regulate water flows both above and below ground. Forest canopy and leaf litter, for example, help to attenuate the impact of raindrops on the earth's surface, thereby reducing soil erosion. Roots hold the soil in place, especially on steeper slopes, and also absorb water. Openings in leaf litter and soil pores permit the infiltration of water, which is carried through the soil into the ground water. Where ground cover is insufficient, sheet, rill and/or gully erosion may result. Such erosion reduces the productivity of the land and may result in sedimentation of watercourses down stream.

Streams eventually carry excess surface water to the ocean, though they may feed intermediate destinations such as lakes and wetlands. In their natural state, the network of streams in a catchment will slow down water flows so that there is a significant time lag between a period of peak precipitation and peak runoff further downstream. Riparian forests can serve as important buffers, reducing sediment loads and keeping runoff from moving too quickly into streams.

Growing vegetation allows retention of moisture and soil are able to absorb more solar energy than the previously sparsely vegetated surface. In addition moisture is available for evaporation, both at the soil surface and in the root zone. These changes affect the temperature and humidity of the lower atmosphere and make rainfall more likely. Conversely, in dry years soils tend to have higher albedos (absorbing less of the sun's energy) and there is less moisture available for evaporation, which can lead to a positive feedback in the opposite direction, yielding lower rainfall.

Water Characterisation

Based on composite sampling, and physico-chemical and biological analyses the river basin is categorized into most disturbed (Sharmanavathi, Haridravathi, Keshwapura, Gazni, Sampakai, Gudankattehole), moderately disturbed (Muppanae, Talakalale Dam, Reservoir, Debbe falls, Hosagadde) and least disturbed (Yennahole, Hurlihole, Nittur, Valagere, Dobbod) zones. The disturbance is due to anthropogenic activities in the catchment, mainly agriculture. Presence of coliform bacteria at Sharmanavathi, Haridravathi, Keshawapura and Nandihole indicates faecal contamination. This study also shows that salinity ingress was found to be absent at Gerusoppa and it decreases with increase in the river discharge.

Soil and Sediment Characterization

Soil samples were collected from 78 locations distributed all over the upper catchment and subjected to physico-chemical analyses. Soils are rich in organic matter and low in phosphate, nitrate and sulphates concentration, while pH ranged between 5.5 – 6.8. The sediments have low sulphate (0.191 – 0.68 mg/gm), nitrate (0.0 – 0.0007 mg/gm) and phosphate (0.00024 - 0.001 mg/gm) indicating close correlation between sediment and catchment soil. The sediment samples are rich in organic carbon and the elements like Na, K, Ca, Mg are found well within the prescribed standards. Bulk density of sediments in streams of the western region indicates porous condition (0.783 – 0.983 gm/cm^3) while in the eastern side they are less porous (1.23 – 1.475 gm/cm^3).

ENERGY AND SOCIOECONOMIC STUDIES

Energy survey carried out in 447 households covering 42 villages in the catchment reveal that the majority of the population is dependent on bioresources to meet the daily energy requirements. The dependence is of the order 85-92% and the per capita consumption corresponds to 1.2 tonnes per year, which is quite high compared to the per captia consumption in Karnataka state.. The major cooking device being used is the traditional Chula. Bioresource status considering the availability and demand indicate that the villages in the eastern parts of the catchment have high levels of consumption of fuel, in the order 40-60%. Most of the households still depend on traditional stoves with efficiency less than 10%, which indicate that there is a scope for energy conservation by just switching over to the fuel efficient devices.

Among the surveyed households, landless labourers in the study area are deprived of many basic needs for their livelihoods. Majority of them are unskilled and uneducated, which make them ineligible for getting any secure jobs. The literacy level among the population ranges from 44% (female) to 64% (male). The annual income of each household with family size of 4.78 is Rs. 9329.67.

THREATS AND MITIGATION MEASURES

- The fragmentation caused by the submersion of vast area and also due to biotic pressures (increasing settlements, agricultural fields, quarrying, fuel wood, fodder and NTFP collection, encroachment, roads in reserve forests, etc.) Land use in the catchment needs to be strictly monitored for changes using satellite imageries and ground surveys. A centralized planning is necessary to maintain landscape in desirable state
- Lopping of branches for fuel wood and collection of leaves for fodder deprives animals of their food. JFM committees to be formed in the catchment area and be associated with future management of forests in respective village territories.
- Overgrazing in the region has resulted in scarcity of resources for wild animals. Village fodder farms to be created to safeguard forests and wildlife and to meet the needs of soil and water conservation as well as for other ecosystem needs.
- Conversion of forests to monoculture in vast areas for commercial purposes has affected the free movement of wild animals and deprived them of food and habitat. Monoculture plantations only serve either small mammals and agricultural pests as hiding places An action plan needs to be prepared urgently for reducing greatly area under monoculture by introducing forest species, an appropriate list of species is given in the report. Moreover conversion of plantations into

natural forests is necessary for meeting other ecological requirements including enhancement of watershed value.

- Fire, within limits, has an ecological role to play in the Western Ghats. But unregulated and frequent forest fires, accidental as well as intentional, have detrimental effects on the flora and fauna, and ecosystems as such. necessitating appropriate management strategies. such as creation of fire lines, restoration of evergreens, which provide greater fire immunity to the forests, adoption of a village centered fire management strategy etc.
- Theft of forest products and poaching of animals have also affected the faunal diversity.Strengthening of JFMs, creation of nature clubs in the village, and spread of awareness can go a long way in controlling hunting menace. The nature clubs may be affiliated to WWF State or NGO sponsored ecological or nature conservation/awareness movements. The village based nature clubs may be associated with animal census activities and other conservation centred activities.

In this regard, there is a need for:

1. A comprehensive conservation strategy should be formulated immediately in consultation with stakeholders such as local villagers, environmentalists, and forest department personnel to conserve the unique flora and fauna of riparian ecosystem.
2. Field observations show that many of the riverine swamps which harbour rare flora and fauna such as *Semicarpus kathalekanensis* and *Phylloneura westermanni are* under severe threat due to encroachment and expansion of agricultural activity. Immediate remedial measures need to be taken by forest department to stop such activities and conserve the endangered ecosystem.
3. A detailed study, focusing on riparian forests is to be initiated immediately to document rare flora and fauna and evolve a conservation strategy for this unique eco-region-the Sharavathi Catchment.

RESTORATION OF HABITATS

The natural vegetation ranges from the climax tropical evergreen to semi-evergreen forests along the high rainfall area in the western side to the moist deciduous forest in the undulating plains and low hills along the eastern drier tracts of the river basin. The evergreen to semi-evergreen forests is the major sources of perennial water sources through the catchment of Sharavathi. On the other hand in the deciduous tract, the streams dry up in the summer months. Therefore conservation of evergreen forests and restoration of such forests in the eastern parts of the catchment are of paramount importance.

Reforestation should not merely be a simple agglomeration of useful fast growing trees, but should stabilize ecosystem with built in dynamics of regeneration of multiplier effect. This would aid in natural balance in climate abetting rich growth of flora and fauna. Trees indigenous to the natural habitats of the region must be given due importance. Local demands of fuel, fodder and medicine and minor forest produce like gum, fruits, spice etc. should also be considered. While planning, emphasis need to be laid on species that would meet fuel, fodder and biomass requirements of artisans like potters, basket makers, mat weavers, wood carvers etc.

List of plants recommended for restoration of forests especially in the drier eastern side of the catchment is given in Table 1.

Table 1
List of Plants Recommended for Reforestation of Dry Areas of Eastern Side

Plant species	*Common name*	*Ecosystem and human value*
Acacia catechu	Khadira	NTFP, MD
Acacia concinna	Seege	NTPP
Bambusa arundinacea	Bidiru	NTFP, Fodder
Buchanania lanzan	Nurklu	EV, FR
Careya arborea	Kavalu	LM, EV, LM
Cassia fistula	Kakke mara	MD
Dillenia pentagyna	Kanagalu	EV, LM
Emblica officianlis	Nelli	NTFP, MD
Ficus benghalensis	Aala	EV
Ficus glomerata	Atti	EV, FR
Madhuca indica	Ippe	EV, NTFP, Leaf fodder
Mangifera indica	Mavu	MD, FR
Phoenix sylvestris	Ichalu	NTFP, FR
Sapindus laurifolia	Antavala	EV, NTFP, MD
Schleichera oleosa	Kindala	LM, Fuel
Semecarpus anacardium	Guddegeru	NTFP, F
Syzygium sp.	Nerlu	EV, F, MD
Zizypus jujuba	Bore	EV, FR

Note: EV:Ecosystem value, FR:Fruit; LM:Leaf Manure; MD:Medicinal; NTFP:Non Timber Forest Produce

SEMIEVERGREEN/*KAN* FORESTS

In the Linganamakki catchment especially in the eastern side there are some isolated small patches of semi-evergreen forests. These forests commonly known as *kans*, were formerly conserved by the local communities considering them as sacred. These forests indicate the local climate to be favorable for sustaining evergreen to semi-evergreen forests. The latter type of forests has greater watershed and ecosystem value than deciduous forests. Therefore all efforts should be made to conserve the existing semi-evergreen forests and also enrich them with more evergreen species. The creation of such forests in all potential areas should be considered seriously. The species recommended for planting in the *kans* and semi-evergreens are listed in the table 2.

Table 2
Species Recommended for the Planting in *kans* and Semi-evergreens

Plant species	*Common name*	*Ecosystem and human value*
Aporosa lindleyana	Salle	Fuel, FR
Artocarpus hirsutus	Hebbalasu	EV, T, FR
Caryota urens	Baini	NTFP, EV
Diospyros assimilis	Karimarlu/ Tumru	EV
Diospyros crumenata	Tumru	EV
Ficus nervosa	Neeruvate	EV
Flacourtia montana	Mullu sampige	EV, FR
Holigarna arnottiana	Sannele holageru	EV
Holigarna grahamii	Doddele holageru	EV
Knema attenuata	Hedaglu	EV
Mimusops elengi	Ranjalu	EV, FR
Olea dioica	Madle	EV
Syzygium cumini	Neralu	EV, FR, Fuel
Syzygium sp.	Neralu	NTFP, MD, FR
Vitex altissima	Nviladi/Barnige	EV, Fuel

Table 3
Plants Recommended for Restoration of Degraded Areas in the Western Side

Plant species	*Common name*	*Ecosystem ana human value*
Olea dioica	Madle	EV
Mimusops elengi	Ranjalu	EV, NTFP
Aporosa lindleyana	Salle	EV, FR
Dillenia pentagyna	Kanagalu	EV
Garcinia indica	Muruga	EV, FR, NTFP
Terminalia paniculata	Hunalu	EV
Flacourtia montana	Mullu sampige	EV, FR
Mangifera indica	Mavu	EV, FR, MD
Syzygium caryophyllatum	Kunnerlu	EV, FR
Syzygium cumini	Neralu	EV, FR
Artocarpus heterophylla	Halasu	EV, FR
Artocarpus gomezianus	Wote	EV, NTFP
Caryota urens	Baini	EV, NTFP
Emblica officinalis	Nelli	NTFP, MD, FR
Strychnos nux-vomica	Kasarga	MD, LM
Pterocarpus marsupium	Honne	Fodder, EV, Fuel

Table 4
List of Plant for Restoration of Semi-evergreen Forests in Western Region

Plant species	*Common name*	*Ecosystem and human value*
Aglaia anamallayana	Kempunola	EV
Artocarpus heterophyllus	Halasu	NTFP, F
Artocarpus hirsutus	Hebbalasu	EV, F
Canarium strictum	Kaidhupa	EV, NTFP
Dimocarpus longan	Kendal	EV
Garcinia morella	Harisina gurgi	EV, F
Holigarna arnottiana	Sannele holageru	EV
Holigarna beddomei	Doddele holageru	EV
Hopea ponga	Haiga	EV
Knema attenuata	Hedaglu	EV
Mimusops elengi	Ranjalu	EV, NTFP
Vepris bilocularis	Mangappe	EV
Polyalthia sp.	EV	
Mangifera indica	Mavu	EV, F
Symplocos racemosa	Chunga	EV
Caryota urens	Baini	EV, NTFP
Garcinia gummi-gutta	Uppage	NTFP
Garcinia indica	kokum	NTFP, MD, EV
Corypha umbraculifera	Talemara	NTFP, EV
Syzygium cumini	Nerlu	Fruit, MD

Table 5
List of Plants Recommended for Restoration of Evergreen Forests in the Western Region

Plant species	*Common name*	*Ecosystem and Human value*
Poeciloneuron indicum	Balgi	EV
Knema attenuata	Hedaglu	EV, F
Myristica malabarica	Rampatre	EV, NTFP
Myristica dactyloides	Patre	EV, NTFP
Persea macarantha	Gulmavu	EV
Calophyllum tomentosum	Surhonne	EV
Dipterocarpus indicus	Dhuma	EV
Palaquium ellipticum	Hadasale	EV
Ficus nervosa	EV	
Mastixia arborea	Niratte	EV
Vateria indica	Saldhupa	EV, NTFP
Elaeocarpus tuberculatus	EV	
Mangifera indica	Mavu	EV, F
Chrysophyllum roxburghii	EV	
Canarium strictum	Kaidhupa	EV
Calamus sp.	Betha	NTFP
Syzygium gardneri	Nerlu	EV, FR

- Compensatory Forest in Non-forest Area: Compensatory forest raised by KPCL to make up for the loss of forest submerged (700 ha) in Sharavathi Tail Race project is undertaken in the non-forest area of 700 ha in Thirthahalli taluk of Shimoga district at an estimated cost of Rs. 77 lakhs. Subsequently these lands have been transferred to the Forest Department.
- Afforestation in Wastelands and Degraded Area: The suitable species for such areas within the different forest types are given in the tables.
- Restitution of corridors of animal migration as depicted with maps.
- Restoration of Disturbed Patches by Planting Native Species – Details on the species to be planted in such patches falling within various vegetational regimes are given in the tables referred to.
- Degraded and vulnerable areas to be considered as priority areas for afforestation and soil conservation program.
- Environmental Safeguards related to Dam Safety and any possible usage of Hazardous or Toxic materials in any activity in the region.

WASTELAND REHABILITATION AND MANAGEMENT

The characteristics and descriptions of land use (agricultural, wastelands etc.) in the eight sub-basins of the Sharavathi upper catchment were given in the *previous sections*.

The main objective of proposed wasteland management is to meet:

1. The domestic requirements of fuelwood, fodder, timber forest produces in rural areas
2. NTFP (non-timber forest produce) for rural population.
3. Soil conservation, to create shelter belts or wind breaks and to reclaim ravine and gullies.
4. Maintenance of ecological balance, biodiversity, ground water recharge and a healthy food web.

These purposes can be met by the selection of appropriate species in the sub-basin based on the land use, soil class or soil capability class, agro-climatic zone, temperature etc.

The factors for species selection includes:

- Silvicultural characteristics.
- Utilization potential.
- Weightage for site-specific native species.

- Avoiding exotic species as far as possible and only considered when the indigenous species are unable to thrive in degraded ecosystem.
- Afforestation should involve a multi-species approach. This would be more resistant to pest and diseases, and equally efficient in utilizing environmental resources. This can also act as a soil cover and serve in regeneration of soil.
- To generate rural employment and also to involve the local population in afforestation, farmer's nursery movement needs to be promoted.
- Wastelands under high-tension line need to be used for productive purposes. Village fodder farms and medicinal plant gardens may be started under the management of village Self Help Groups (SHGs). By bringing such lands under the local community management fire hazards can be prevented. Planting of moisture retaining plants would also assist in reducing fire hazards.
- The plant species suggested for reforestation of open, fallow and degraded lands of low rainfall zones in the eastern part of the catchment are *Acacia catechu, Acacia concinna, Butea monosperma, Pongamia pinnata, Schleichera oleasa, Madhuca indica, Emblica officinalis, Cassia fistula, Strychnos nux-vomica, Odina wodier, Dillenia pentagyna, Xylia xylocarpa, Buchanania lanzan, Careya arborea, Terminalia chebula, Pterocarpus marsupium, Phoenix sylvestris Mangifera indica, Dendrocalamus strictus, Bambusa arundinacea, Azadirachta indica, Cordia myxa, Aegle marmelos, Sapindus laurifolius* and *Spondias mangifera.*

SOIL EROSION AND SILTATION

Solutions to this part of soil erosion problem may come from searching for suitable plants that could grow well at aforesaid ecotone. In estuarine wetlands this problem is even more severe than inland reservoirs because of stronger wave action. However, nature has a solution in the form of mangrove vegetation. Experiments with available and speculated-to-be suitable plant species can be tried.

Species like *Ipomoea carnea, Ipomoea biloba, Ipomoea aquatica, Saccharum sps, Terminalia arjuna, many other aquatic and semi-aquatic plants (creepers-cum-floaters), Cyperus species etc can also be tried. The riverbeds of fluctuating water levels could be looked for suitable plant species. Even mangrove plants like Clerodendron inerme, Ipomoea biloba, Derris sp., etc.* could be tried. The difference between estuarine wetlands and the reservoirs is that mangrove plants have a very long evolutionary history compared to the plants interacting with inland reservoirs. This is because the dams and reservoir (wetlands with highly fluctuating water level) are of recent origin from the point of view of evolutionary time scale. Flood-tolerant species like *Holigarna, Madhuca neriifolia, Elaeocarus* and *Pandanus* etc can be tried. Among the colonizers on the exposed area below the Full Reservoir Level, three species were noticed besides other plants; these were *Cynodon, Eupatorium* and *Cleome*. Even if some species from these plants start growing well in this exposed area, the erosion will reduce. This ecotone zone should be searched whether suitable plants have evolved. Plants suitable for such erosion prone areas may be propagated. *Cynodon* is good for cattle but here it should not be allowed to be grazed. *Eupatorium* is a difficult exotic weed and not good except for its low quality biomass. *Cleome* is a weed as well as a crop. (Please note that no plants can grow in submerged state, except hydrophytes). The others grow seasonally, only when the water level goes down. Grasses are the best option, because of their spreading nature).

ENERGY MANAGEMENT

Lack of employment opportunities throughout the region is one of the main reasons for poverty. Employment programs are important elements for economically sustainable development. These programs should aim at enhancing the employment opportunities by utilizing the technology and thereby increasing the labour productivity. Training camps can be started to train these unemployed and under-employed persons in various non-farm activities. This will diversify the employment opportunities thereby increasing their income. Establishment of self-help groups can be another possibility. Self-employment of rural women can improve their financial situation.

Shifting to improved cook stoves and other energy sources can conserve about 40 % of fuel wood consumption. Low cost and compact designs should be developed in case of biogas plants, by which the technology can reach the poor people, who really deserve it. These alternatives once again decrease the efforts of rural women in fuel wood collection. Reducing the burden of fetching fuel, fodder and water can certainly help the women to engage in productive and income generating activities. Primary education comprising health, nutrition, family planning, water and sanitation is highly essential for women. Providing basic amenities such as basic education, primary health care, low cost water and sanitary facilities is one of the effective methods to improve the quality of life to a large extent. Investing on road networks can be one of the effective methods to improve the poor links between all parts of the catchment. Establishing some of the satellite places to main head quarters, where people can get the basic facilities easily can minimize the transportation problems to certain extent. Providing protection to the degraded forest patches can regenerate them to the original condition and emphasis on appropriate methods of fuelwood collection can drastically reduce the stress on forests. Further, afforestation programmes with preference to native species need to practiced to meet the energy requirements of the region.

Even though the Linganamakki Dam has been constructed with solely power generation purpose, several villages in the catchment (mention some) still lack the power supply. Electricity is the prime amenity required and facilities should be made to reach it to rural areas as well.

ECOSYSTEM MANAGEMENT

Anthropogenic activities due to these developmental projects have caused considerable fragmentation of natural habitats. Encroachments and conversion of forest lands for human habitation and agriculture, etc. have aggravated the fragmentation of otherwise contiguous habitats. Submersion and fragmentation of forests have seriously affected wildlife habitats and wildlife movements considerably. In recent times, it is reported that smuggling, poaching and hunting of rare species of fauna in the catchment. No effort was made to prevent the causal factors of fragmentation. It is noticed during fieldwork encroachments, poaching and smuggling is unabated and sometime it is done at authorities connivance. In addition to this extraction of industrial timber and clear felling of virgin forests to raise teak and eucalyptus plantations are persistent. This has hampered, ecosystem productivity, which is evident from conversion of perennial streams to seasonal streams, reduced biomass productivity and disruption of wildlife movement and migration corridors.

Prior to the Sharavathi River Valley projects, elephants used to move freely to the northern parts of Uttar Kannada. With the commissioning of various projects very little attention was paid to the animal migration corridors, so much so that, one small herd of elephants was trapped in the Gerusoppa and has disappeared now. Similar situation is observed in Kali river catchment small herds still live confined to Dandeli forest patch. In the absence of effective regulatory stipulations on environmental safeguards, and lack of understanding of niche requirement of fauna and flora he Sharavathi River Valley ecosystems have eroded considerably during the late twentieth century.

Some of the families returned to Sharavathi catchment due to lack of proper infrastructure at rehabilitated localities. This along with arrival of migrant workers and traders from neighboring states has led to large-scale encroachments. Other factors that has contributed to devastation of environment are quarrying, unplanned agriculture development and plantation activities. This has caused severe impact on the flora and fauna and their habitats as well as influenced silt movement ending up as depositions in the streams and ultimately in the reservoir.

Establishment of Monitoring team to plan, coordinate, implement and evaluate all the associated aspects as indicated in the recommendations.

Implementation of the Soil Conservation/Afforestation Program:

— Identification of waste, degraded lands and potentially disturbed patches as indicated by the current study.

— Implementation of the afforestation progarmme by the Forest Department and Agriculture Department by planting the species as indicated by the study.

— Bund construction and small check dams to be constructed by the KPCL at places in the reservoir potentially identified for soil erosion and siltation due to landslides.

— Agriculturists and native residents/villagers to be encouraged to venture plantation of leguminous plants and suggested species in patches of wastelands/non-arable lands (to restore these lands).

— Peoples' participation to be extensively advocated in the implementation of afforestation and other conservation programs.

NOTE

1. We thank Karnataka Power Corporation Limited (KPCL), Government of Karnataka and the Ministry of Environment and Forests, Government of India for the financial assistance. We are grateful to Prof. Madhav Gadgil and Prof. Niranjan V. Joshi for useful suggestions during discussions.

12

ENERGY COOPERATION IN THE GANGA BRAHMAPUTRA BASIN: OPPORTUNITIES AND CHALLENGES

Rakesh Tiwary

INTRODUCTION

Situated in the northern half of the Indian sub continent, the Ganga-Brahmaputra basin constitutes the second largest hydrologic region in the world. The total drainage area of the basin region is about 174 million hectare spreading over India, Bangladesh, Nepal and Bhutan in South Asia and Tibet region of China north of Himalayan divide. The Ganga-Brahmaputra basin is a single, sprawling, integrated, transboundary drainage system with a common terminus. The basin consists of two major river systems of the Ganga and the Brahmaputra and one relatively smaller river system of the Barak, which becomes the Meghna in Bangladesh. About 63 percent of the basin is in India, 7 percent in Bangladesh, 8 percent in Nepal, and 2.5 per cent in Bhutan. Almost 10 percent of the world's humanity lives in this region, which contains only 1.2 percent of the landmass.

The Ganga-Brahmaputra basin accounts for about 10 percent of the world population, which contains only 1.2 percent of the worlds land mass. The total population of the basin has reached a level of about 600 million by 1999, of which the share of India is 76 per cent, while the corresponding figure of Bangladesh and Nepal are 21 percent and 3.5 per cent respectively. The total population figures (year 1998) for the different countries are India-987 million, Bangladesh-128 million, Nepal-24 million and Bhutan-1.7 million. The Ganga-Brahmaputra basin is inhabited by about two fifth of the developing world's poor people with a daily calorie intake of less than 2,200-2,400 Kcal. The critically vulnerable groups affected by poverty are the landless and marginal farmers, disabled persons, and destitute with no access to economic activities, rural non-farm workers, urban slum dwellers and rapidly increasing female leaded rural households (Ahmad et. al., 1994, p.14).

In India there has been significantly decline in poverty. In 1983, 37 percent population was living below poverty line, in 1999-2000 the figure declined to 26.10 percent, however, there is considerable regional variation. The population living below poverty line in Bihar, Assam, Madhya Pradesh, and Uttar Pradesh is 42.60 percent, 36.09 percent, 37.43 percent and 31.05 percent respectively, which is much higher than national average. The incidence of poverty is higher in rural areas than in urban areas. Nepal planning commission estimated in 1984 that 42 percent of the population had incomes below the poverty line 95 percent of the poor lived in rural areas of which 82 percent were agricultural laborers of self employed. The per capita GNP for India, Bangladesh and Nepal is (as of 1998) 430 US dollar, 350 US dollar and 210 US dollar respectively; the corresponding global average is 4,890 US dollar.

Table 1
Ganga Brahmaputra Basin: Socio-Economic Indicators

Indicator	*Bangladesh*	*India*	*Nepal*
Population (million) 1998	128	987	24
Annual population growth rate: 1995 – 2000(%)	1.9	1.8	2.5
Infant mortality rate (per 1,000 live births) 1997	75	71	83
Under-5 mortality rate (per 1,000 live births) 1997	104	88	117
Maternal mortality rate (per 100,000 live births) 1990-97	850	437	1,500
Access to safe water (% of population) 1995	84	85	59
Access to sanitation (% of population) 1995	35	29	20
Adult literacy rate (% of people 15 & above) 1997	50(M) 27(F)	67(M) 39(F)	56(M) 21(F)
Female (as % of labour force) 1998	42	32	40
Arable land (hectare per capita) 1994-96	0.07	0.17	0.13
Per capita commercial energy use: annual (Kgoe) 1996	197	476	320
Per capita electricity consumption (KWH\h) 1996	97	347	39
Population below national poverty line (%), early 1990s	48	37	43
Per Capita GNP (US$) 1998	350	430	210

Notes: Kgoe = Kilogram of oil equivalent, KWh = Kilowatt hours
Sources: World Bank (1999, 2000); Human Development Centre (1999); UNDP (1999).

ENERGY SITUATION

Energy is the critical infrastructure for economic development. The energy consumption pattern is used as an index of national or regional level of development. The Ganga – Brahmaputra basin region has one of the lowest per capita energy consumption. The countries of Ganga-Brahmaputra basin have a very low dependence on and utilization of commercial energy. The per capita commercial energy consumption figures in India, Bangladesh and Nepal are 476, 197and 320 Kgoe (Kilogram of oil equivalent) respectively. Per capita energy use in the world is about 1,680 Kgoe on an average while it is around 5,340 Kgoe for high-income countries. The low energy consumption is not due low actual demand. Instead commercial energy supply is very limited and subject to shortages.

Table 2
Per Capita Electronic Power Consumption in South Asian Countries (KWH)

Country	*Per Capita Electronic Power Consumption (KWH)* 1990	1992
India	2340	363
Bangladesh	43	76
Nepal	28	39
Bhutan	-	-
South Asia	228	324
World	1928	2083

Source: World Bank (2000, 2001)

The scenario is despite the fact that Ganga – Brahmaputra basin is vastly endowed with energy resources. India, Bhutan and Nepal have large hydroelectricity power generation potential. India has large reserves of coal. Bangladesh has considerable natural gas reserves. Still, the development of energy resources is abysmally low to raise the people's living standards. The riparian countries of Ganga- Brahmaputra region still depend heavily non- commercial energy, which is derived from biomass (fuel wood, agricultural residues, and animal waste). They not only have low energy efficiency but fuelwood also cause environmental degradation (deforestation, soil erosion etc).

Table 3
Existing Status of Power in GBM

Country	*Total installed capacity MW*	*Thermal*		*Hydro*		*Nuclear*		*Per capita installed capacity Kwh*
		Mw	%	*Mw*	%	*Mw*	%	
India	100,266	71,254	71	24,712	25	2900	3	339
Bangladesh	3770	3584	94	232	6	-	-	57
Nepal	430	57	13	373	87	-	-	39
Bhutan	348	6	2	342	98	-	-	107

Source: Asian Development Bank (1997).

The region faces critical situation in the field of energy as; firstly, there is declining but still high demand for biomass fuels with consequent heavy pressure on forest resources. Secondly, there is less emphasis on diversification of energy sources, development of clean energy resources, conservation and efficient use. Thirdly, there is very low level of energy trade taking place among the riparian countries as compared to the existing potential. Fourthly, energy demand is increasing in the region while coal, oil and gas reserves of the region are exhaustible.

Ganga Brahmaputra basin contains around 15 percent of worlds coal reserves, out of which almost all is located India. India has a dominant position in the basin as far as energy demand as well generation. India accounts for 99 percent of basin coal production, 44 percent of natural gas production and 84 percent of energy consumption. The geopolitical setting also create compulsions where India can bilaterally trade with any nation of Basin but other two nations will have to engage India as third partner. Table number 3 recoverable reserves of primary energy and production potential of the Basin, while table no.4 gives the current rate of extraction of energy resources and installed capacity of power generation.

In such a scenario the hydroelectric energy is the most critical energy resource for the future. Hydroelectric power projects can generate vast amount of energy and they can serve vast areas. There are many benefits from development of hydropower resources. Its operational and maintenance cost are low as compared to thermal and nuclear power. It is clean and renewable source of energy. It is a major component of the multipurpose projects besides irrigation, flood control, navigation etc. It also exhibits a declining unit cost of generation over time, even though initial capital expenditure is very high. However, the large-scale HEP generation has also some adverse impacts like submergence, displacement and threat of seismicity-induced hazards. But benefits substantially out weigh the negative aspects.

The Eastern Himalayan region of the Ganga- Brahmaputra basin has vast potential for hydropower generation. It can not only produce huge amount of HEP but also give additional benefits of irrigation and flood management. The past efforts of HEP development have been within the boundaries of nation particularly meeting the domestic demand of the energy. These efforts should be supplemented with cooperative approach of riparian states of the Ganga-Brahmaputra basin. Such projects can develop complementarities among nations as well as help each other solving problems of flood management, sedimentation and salinity intrusion, migration and displacement. The early endeavors of cooperative development of energy (like Kosi and Gandak projects) are too small as compared to vast potential that exists in the region.

The rivers originating in Himalayan and sub-Himalayan ranges have steep topographical slopes and are almost perennial. Thus they have tremendous hydropower generation potential. Although there are varying estimates, the generally accepted figures are that India has 45,635 MW (economically exploitable), Nepal has 83,000 MW (estimated) and 42,000 MW (economically exploitable) and Bhutan has 21,000 MW potential for the HEP generation. As compared to the potential the developed installed capacities are meager. This may be seen from the figures of installed capacity in India (3,525 MW), Nepal (244 MW), Bhutan (336 MW) and Bangladesh (230 MW). Thus total hydropower installed capacity in the entire basin region is only 4,335 MW; which is about 5 per cent of economically exploitable potential (Shah, 2000, p.23).

The figure suggests that there is tremendous scope for future development of hydropower generations. Nepal and Bhutan have water source as the most precious natural resource for economic growth. These two countries can supply energy to India where large market exists. Thus entire Ganga – Brahmaputra basin can benefit from industrial and agricultural development. This vision can be actualized by regional cooperation.

INDO-NEPAL COOPERATION IN HYDRO POWER GENERATION AND TRANSMISSION

There is relation of geographical interdependences and economic complementarities between the water resources development aspects of India and Nepal. Nepal has large hydroelectricity generation potential. The demand for and consumption of electricity in Nepal is likely to grow at a rate much slower than what its hydropower potential implies. Nepal therefore prefers to view its water resources more as a source of export earnings. On the other hand India has been suffering from power shortage for quite some time with constraint of energy becoming acute every year. The excess hydroelectricity power generated in Nepal may be sold to India. The availability of power in India would enhance the use of ground water for irrigation. The development of water and power resources in Nepal would meet the growing power and irrigation water needs in Nepal and would also bring in revenues from power sales to India. While Nepal can exploit a number of medium and mini schemes to meet its domestic power requirements the more exciting projects that could deliver huge blocks of energy at low cost must find a market in India or through it too Bangladesh. Hydropower represents a renewable source of wealth for Nepal, for it could export this power to India and invest the proceeds in its own further development and prosperity (Verghese, 1992, p.100).

The Kosi and the Gandak are the two projects, which India and Nepal undertook in 1950's. These projects, had agreement over supply of power to the border towns and villages in Nepal leading to development of interconnections between the Nepalese and Indian power systems. After a long gap Mahakali treaty was signed in 1996. The Treaty proposes to build a high dam and two underground powerhouses – one on each bank and share the power generation equally between India and Nepal. The joint detailed project report is yet to be finalized over the Pancheswar project. India and Nepal have also signed a Power Trade Agreement (1996) under which India has agreed to buy power from Nepal and also allow transmission access to the third country. The agreement also facilitates power trading between India and Nepal, by both public and private sectors. Various other project are also under discussion stage like (a) Karnali Chisapani (10800MW) multipurpose project, (b) Sapta Kosi high dam multipurpose project (3000MW), (c) Burhi Gandaki Hydroelectric project (600 MW), and (d) West Seti hydro electric project (750 MW). The power exchange facilities between India and Nepal have been enhanced by the proposed Tanakpur (India) and Mahendra Nagar (Nepal) transmission line.

India and Nepal also have institutional arrangements like India -Nepal Power Exchange Committee (1992) which has functions like (1) monitoring the existing arrangements/facilities for power exchanges, (2) evolving proposals for promoting power exchanges in the future, and (3) recommending principles of tariff for such power exchanges, etc.

Indo-Nepalese cooperation over hydro electricity generation has been very slow. Political mistrust has been the major cause. Nepal complains that it didn't get fair deal in Kosi and Gandak agreements. Further, the issue of sharing the benefits and costs on equitable basis among the regional partners has been the issue of disagreement. Nepal believes that most of benefits (such as flood control, irrigation, navigation) including the availability of surplus clean energy to fuel industrialization from the construction of large dams are bound to accrue to the down stream countries (India and Bangladesh), whereas the associated costs (loss of cultivable land, natural flora and fauna, displacement etc.) of such development projects would be imposed on Nepal (Rijal, 2002, p.101). The hydro electricity projects have been debated hotly in the domestic political as well as academic circles of Nepal. In Nepal, the general opinion is "no deal is better than a bad deal".

Many experts have also questioned the feasibility of huge multipurpose projects. The water resource experts' view that domestic energy needs will be met by utilising small and medium hydropower schemes. The high dam projects come with inherent issues like environmental degradation, silt deposition, seismic activities etc. Further high dam projects are also very expensive and large investments made on such schemes are actually diversion from the other competing sectors like education, health sanitation etc. At the same time Nepal wants to extract its reasonable benefits from the proposed high dam projects.

Nepal has developed West Seti hydroelectric project (750 MW) with an agreement signed with a private developer, the Snowy Mountain Engineering corporation (SMEC) of Australia with the sole intention of exporting the substantial power to the Indian market (Onta, 2000, p.119).

The Karnali project is one of the largest multipurpose projects under consideration since 1960's. The private developer Enron has shown interest in developing the Chisapani dam of the Karnali project. Recently after several discussion and presentation made by Enron, permission was withheld pending more enquiries. The Kosi High dam, a multipurpose project has been under discussion between India and Nepal for a long time.

MAHAKALI TREATY: MAJOR OPPORTUNITY AHEAD

A new chapter in the Indo Nepal relations was opened when the Mahakali Treaty was signed by then Prime Minister of India Mr. P.V. Narsimha Rao, and the then Prime Minister of Nepal, Mr. Sher B. Deuba, in February 1996. The signing of Treaty was preceded not merely by negotiations between the two governments but also by extensive informal consultations covering all parties in Nepal so as to facilitate the process of Parliamentary ratification.

The Mahakali treaty subsumes all other Indo Nepalese agreements relating to downstream projects on the river, thus, it absorbed the regime established by the Sharda Treaty, validated the Tanakpur Agreement, and endorsed the idea of multipurpose Pancheswar Multipurpose Project.

Pancheswar Project

The Mahakali agreement, together with letters exchanged in relation thereto, provide for a joint Indo Nepal project on the basis of a 50:50 cost benefit split. The dam will straddle the border, which lies along the median point of the river. Two power stations are projected, one on either bank, with an overall installed peaking capacity, between 5,500 and 6,480 MW at 20 per cent load factor. A re-regulating dam could be built either at Poornagiri or further upstream at Rupali Gad to hold the waters passing through the Pancheswar turbines and provide regulated back season release to irrigate designated commands in Nepal and India. They have also agreed to have "equal entitlement in the utilization of the waters of the Mahakali River "without prejudice to their respective existing consumptive uses". Besides, they agreed to implement the project in all such sectors as power, irrigation, flood control, etc. as per the Detailed Project Report (DPR) to be prepared jointly by them (Jha, 1996, p.29). The Pancheswar dam will also provide a modest flood cushion. As Pancheswar project is a joint project located on boundary, some general principles applicable to border rivers are laid down, which are further elaborated in a side letter exchanged by two Prime Ministers. The principles commit both sides to design and operate the project as a single, integrated scheme to yield, "the maximum total net benefits accruing to them". The power benefit is to be assessed on the basis of saving in costs as compared with the relevant alternatives available. The treaty envisages basin development under the aegis of a bi-national Mahakali Commission, which shall be guided by the principles of equality, mutual benefit and no harm to either party.

The Pancheswar project is to be implemented by a joint Pancheswar Development Authority to be set up under an agreed time schedule from the date of the treaty's entry into forces to develop, execute and operate the Pancheswar project. The treaty has a life of 75 years. There are provisions for review after 10 years and independent arbitration of disputes, with the chairperson being named, if necessary, by the Secretary General of the permanent Court of Arbitration at The Hague (Verghese, 1999, p.406). The treaty has been welcomed by large number of people on both sides of border. Many Nepalese experts view the Treaty has successfully broken the impasse created by a legacy of entrenched views on national lines on both sides and finds both sides as winners.

India Bhutan Cooperation over Hydro Electricity

The Bhutan forms part of Brahmaputra basin. The major rivers of Bhutan are Torsa, Raidak, Sunkosh, Mao, Manas and Dhansiri. The physiographic characteristic endows Bhutan large potential for hydroelectric power generation. Yet so far only 1.7 per cent of the total estimated theoretical potential of 20,000 MW has been harnessed. Bhutan can generate electricity not only for its own economic development but can also trade it with India and Bangladesh.

Bhutan at present has 11 hydroelectric stations with an installed capacity of 341.75MW. The Chukha station alone generates 336MW. The Chukha hydroelectric project was developed with Indian assistance. The 220 KW lines from Chukha to Birpara provide the link between the Bhutanese system and Easter regional power grid of India. The operation of the Chukha plant is coordinated with the eastern regional grid of India. The 1000MW Tala project on the Raidak River (downstream of Chukha) is the extension of Indo Bhutanese HEP energy cooperation.

The Power Master plan of Bhutan in 1991 listed 89 sites from a desktop study of topographical maps, with a potential of approximately 12,00 MW of hydroelectricity. Further, the Power Master Plan (PSMP) report of 1992 listed 24 sites, which are economically feasible and have a commutative identified potential of 10,988 MW. Most of the schemes are run of the river type with the exception of a few reservoir schemes, mainly along the southern border. The PSMP figure was revised later, as the envisaged capacity of quite a few identified projects have been increased. For example, the Tala project is being constructed with an ultimate capacity of 1,625MW instead of 265MW as planned earlier. Another project under consideration Sunkosh project has also undergone an upward revision. An agreement on the construction of a 1525MW hydel power project on the Sunkosh was signed on January 4 1993, making a notable advance in the Indo-Bhutanese water resource cooperation. The Chukha3 project on the Wangchu River is also under consideration. The experts view that Bhutan can be major supplier of electricity to its energy starved neighbor i.e., India. India has also agreed to purchase all the energy available. The harnessed power from water resources can generate the additional revenue required. In fact, a major portion of Bhutan's revenue comes from export of power from Chukha.

As much as 45 percent of Bhutan's annual revenue now comes from Chukha, with upward tariff revisions since its inception. The availability of electricity has eased the pressure on kerosene imports as well as the deforestation has been controlled. The per capita income of Bhutan is growing as energy intensive industries are being developed at the southern border of Bhutan.

Bhutan has followed cautious approach towards large-scale hydroelectricity generation. The environmental degradation, loss of bio-diversity and displacement of populations are the major concerns of Bhutan. The needs of socio-economic development for growing population in Bhutan are bound to rise in the future. Bhutan can opt for HEP based revenue for it socio economic development. The careful selection of sites and advanced technology for HEP generation can minimize the associated damages.

Indo Bangladesh Power Cooperation

Much possibility exists between India and Bangladesh for power systems interconnection, which have been discussed ever since Bangladesh came into existence. Asian Development Bank conducted a major study under regional technical assistance project called "India Bangladesh Electricity Exchange Project". Following the project report, two 132 KV double circuit lines to interconnect the Indian and Bangladesh have been identified like (1) Durgapur (India)–Bheramara (Bangladesh), Jeerhat (India) – Jessore (Bangladesh) with an aim to export power to Bangladesh from India. (2) Sylhet (Bangladesh) to Kumarghat (India) where Eastern Bangladesh can export power to Northeast India. Both countries have recognized importance of inter connections of their power systems for mutual benefit. The discussion can include Nepal for long term planning for regional energy interconnections.

The Eastern regional grid of India has a power surplus of 3,000 ME and the western zone of Bangladesh suffers from a power deficit of 300 MW. Unfortunately, still there is no power trading between these two countries. Technical level meetings have occurred on very occasions between India and Bangladesh. Agreement has been reached on certain issues like that the interconnections between the two countries would be synchronous and not radial, thus the recipient country will decide on the point of interconnection.

Experts from both the countries recognize that interconnection of their power systems as being postulated would be beneficial to both .It will be beneficial at this stage for the two countries to develop alternative scenarios of power development for a 10-15 year time frame and undertake studies of interconnection for different levels of power exchange.

Bangladesh Gas Resources and possibility of exchanges

Bangladesh has major reserves of natural gas. There are 15 gas fields of which 14 are onshore and one offshore, together account for a total reserve of 37.08 trillion cubic feet. Of the total reserves, the recoverable [proven and possible] reserves are estimated at 25.66 trillion cubic feet and proven reserves account for 16 trillion cubic feet. Bangladesh plans to intensify the assessment of reserves in known gas fields and undertake the exploitation. Current stance of Bangladesh regarding export possibilities is very cautious one. Based on supply demand forecast experts argue that the existing gas reserves will be sufficient to meet the domestic requirements for another thirty years or so provided adequate investments are made in the development of additional gas fields and creation of adequate gas distribution facilities. In other words, Bangladesh current gas reserves do not provide for possibility of export to neighboring countries. Bangladesh view that even if the entire recoverable gas reserve, estimated at 26 trillion cubic feet is accounted for, the only known non-renewable energy resources of Bangladesh may not be available for exports.

India's Demand: Key Driver

Size of Indian demand is much larger than other members of GBM. India presently requires 480 billion units annually for its economic activities but only 450 billion units are available resulting in an energy shortfall of 30 billion units i.e 6.2 percent. India's present peak demand is 73,000 MW but it can avail only about 64,000 MW, which result in a 9,000 MW shortage i.e 12.4 percent. In order to overcome this power shortage, India needs to invest about 10 billion US dollar. Of the five Regional Grids of India, The Eastern Regional Grid is the only one that has an excess capacity of about 3000 MW. This has been attributed to lower load growth than that projected by the forecast. Hence the power shortage of India presents great opportunity embark upon regional power trading in GBM region.

CHALLENGES IN COOPERATION OVER INTERCONNECTIONS OF ENERGY SYSTEM IN GANGA-BRAHMAPUTRA REGION

Various barriers exist at policy level, technical level, institutional level and financial level in the development of interconnection of energy sector in the region.

The Politico–Strategic Issues

The hitherto prevailing ideology of economic self-sufficiency hinders the energy cooperation and trade among riparian countries. The major economic power of the region, India focused in meeting the demand through inters- regional power transfer within the country. The basin state governments have inward looking energy policy based on Electricity Acts, Regulations and regulatory mechanisms. While cross border power trading has been merely a marginal interest. Bhutan is the only country where cross border power trade is the national priority. Nepal-India cooperation has been victim of mistrust and historical baggage. There is a need to gradually build up confidence among the riparian countries along with the gradual harmonization of macro policies in the energy sector. The psyche of hydroelectricity being perceived, as "strategic" commodity should be shed away if the region aims at collaborative development and management of water resources. Further, the historical debates of bilateralism/multilateralism, augmentation, upper riparian and lower riparian rights should not cast unnecessary shadow on the vision of regional power trade (Pun, 2000, p.34).

Technical Issues

The countries have adopted their own norms and standards in planning and design of their power systems. Differences exist in their standards which may act as impediment in their interconnections. The sizes of the systems, voltages, adopted, inadequacies in generation and transmission facilities, varying structure of the electricity supply and operating system poses the technical challenges in interconnections. Different countries have different mix of public and private sector participation and they are going through decentralization, unbundling and deregulation in the power sector. Thus integration on various technical issues will be required.

Institutional and Governance Issues

Different institutional regulation and control has been developed over the period to oversee the implementation of treaties/ negotiations over water resource development for example – Mahakali Commission, Power Trading Corporation, etc. Their integration and smooth operation will be required. Further, the individual countries have their own load dispatch and system operations, monitoring and control facilities and operations. An institutional mechanism to coordinate the operation and control of inter connected system will be required to ensure smooth operations confirming to principles and ethics of integrated operations.

To initiate any power interconnection proposal, the structure of project management, ownership and the relationship among investors, operators, dispatchers and regulatory entities is vitally important. Lack of appropriate rules on these matters, which take care of the mutual interests of the parties involved, or lack of transparency of such rules will affect the confidence of all potential developers of interconnection projects. Regional trading has as its foundation the regulatory framework of the region and that regulatory framework has as its foundation the authorizing legislation and the underlying commercial codes and laws of the various countries or regions. Parties to regional trading will decide which laws govern those arrangements. Laws dealing with market dominance, self-dealing, anti-competitive mergers and acquisitions, industry ownership restrictions and cross ownership provisions should be examined to ensure their existence is consistent and support the open and competitive trade.

Financial Issues

The financial barriers in terms of tariff (both generation and transmission) limit the ability to identify and evaluate the real delivered cost of power. The riparian countries of the Ganga- Brahmaputra region are developing countries, which are facing resource crunch. Besides internal resource generation they depend on foreign direct and indirect investments which many times complicates the issue. The power trading is often treated on political lines rather than on commercial considerations. The deal should be left to commercial decision making specifically to ensure its long-term sustainability and for ensuring transparency and accountability.

REGIONAL ENERGY GRID IN THE GANGA-BRAHMAPUTRA REGION

Though cooperation on interconnection of power system development in the Ganga-Brahmaputra region is at present at preliminary level, a long-term vision of establishing a regional energy grid in the region can be realized. The riparian countries Bhutan, Bangladesh, Nepal and India can take advantage of resource complementarities, demand diversity and geographic proximity. Three of the four countries have border around a narrow 20 to 40 km wide stretch of India territory. This unique geography provides a viable basis for establishing inter-connection among all four countries. This unique geography provides a viable basis for establishing interconnections among all four countries. Many opportunities and challenges exist for realizing the regional grid.

The following factors will justify the establishment of inter-connections, which could be located within the corridor:

1. A 220 KV transmission line originating from Bhutan (Chukha Power Scheme).
2. The nearest 132 KV terminal grid points in Nepal (Anarmani) and Bangladesh (Thakurgaon) are within a distance of 35 KM:
3. Transmission line originating from Bhutan, Chukha, Anarmani (Nepal), Thakurgaon (Bangladesh) can provide the linkage. The

implementation of proposed HEP projects for example, Arun, Kosi high dam project, Sunkosh project, say by year 2015 can create a hydroelectric power pool (banking) of up to 6,000MW which can be supplied to the markets of India and Bangladesh.

4. All these generating stations and inter connecting ties will become an integral part of India's Eastern Regional Power Grid – the prime beneficiary. The resulting trans-country power grid will substantially improve the peaking capacity of the Eastern Grid of India and also help in meeting the demands of the Western Bangladesh.
5. The regional grids of India has already been inter connected. The Power Grid Corporation of India can play a major role in transportation and wheeling of the large quantum of electricity and establishing interconnections.
6. The major challenge lies in the implementation of the projects although institutions such as Power Grid Corporation of India, Bangladesh Power Development Board and Nepal Electricity Authority are enthusiastic. The major initiative needs to be taken by the respective governments.

The regional collaboration on energy based on economical and reliable interconnections promises multiple benefits like savings accrued from load diversity, hydrologic complimentarity, transfer of periodic surplus of power, economies of scale in building large HEP projects with sharing of investments, savings on power generation investments and operation costs etc. The efforts in joint development activities can facilitate mobilization of financial resources.

CONCLUSION

The basin states are going to face huge energy crunch in coming decades. There are multiple energy sources available, which show high interdependence and complimentarity among the nations for their development and utilization. Though there are numerous constraints: the lack of conducive geopolitical environment, mistrust, absence of regulatory mechanisms, financial and technological gaps etc, but there are huge prospects and opportunities of power trading among riparian states. The interconnections of energy systems of the riparian countries of Ganga-Brahmaputra basin can become crucial agent of economic and social development of the entire region .There is need to take political initiatives at highest levels then create institutional structures for sustained mutual interactions. There is a need to build up interconnections step by step. India, being largest nation in the region has larger role to play without hurting the sensitivities of smaller nations. Gradual opening up of the region and sectoral reforms can help in filling up the institutional, technical and investment gaps. We need to have the vision to seize the opportunity.

REFERENCES

Ahmad, Q.K., Ahmad Nirlufer and Rasheed K.B.S. (1994), *Resources Environment and Development in Bangladesh*, Dhaka: Academic Publishers.

Asian Development Bank (1997), *Electric Utilities Data Book 1997*, Manila.

Human Development Centre (1999), *Human Development in South Asia 1999*, Islamabad.

Jha, Hari Bans (1996), "Mahakali Treaty and Its Implications" in Hari Bans Jha ed., *Mahakali Treaty: Implications for Nepal Development*, Lalitpur: FESC Publications.

Onta, Iswer R.(2000), 'Harnessing the Himalayan Waters of Nepal; A case for Partnership for the Ganges Basin', in Asit K Biswas and Juha I. Uitto,eds., *Sustainable Development of the Ganges Brahmaputra Meghna Basins*, Tokyo: United Nations University Press.

Pun,S. B. (2000), ' Electric Power Trading in the SAARC Region', in Sridhar. K. Khatri (ed.) *Energy Policy: National and Regional Implications*, Kathmandu: Modern Printing Press.

Rijal, Kamal (2002), "Role of Nepalese Water Resources for Meeting Energy Needs in South Asia", in Sridhar Khatri, ed., *Energy policy: National and Regional Implications*, Kathmandu: Modern Printing Press.

Shah, R.B. (2000) "Ganges Brahmaputra: The Outlook For the Twenty First Century", in Asit K. Biswas and Juha I. Uitto, eds. *Sustainable Development of The Ganga Brahmaputra Meghna Basin*, Tokyo: United Nations University Press.

UNDP (1999), *Human Development Report 1999*, New York.

Verghese, B.G. (1992), 'Learning to Say Open Sesame' in David J. Eaton, ed., *The Ganges Brahmaputra Basin: Water Resource Cooperation between Nepal, India and Bangladesh*, Austin: Lyndon B. Johnson School of Public Affairs Publication.

Verghese, B.G. (1999) *Water of Hope: From Vision to Reality in the Himalaya-Ganga Development cooperation*, New Delhi: Oxford & IBH Publishing CO. Pvt. Limited.

World Bank (1999), *World Development Report 1999*, Washington D.C.

World Bank (2000), *World Development Report 2000*, Washington D.C.

World Bank (2001), *World Development Report 2001*, Washington D.C.

13

ENVIRONMENTAL COST OF DRINKING WATER IN ANGUL-TALCHER IN ORISSA: A CASE STUDY

M. Mishra and N. C. Sahu

INTRODUCTION

Water is one of the most precious gifts of Nature without which no life can exist. In early days with very limited number man could afford to contaminate and abandon one after the other sources considering water bountiful. Population growth, rapid industrialisation and faster urbanisation accompanied by inadequate infrastructure investment and management capacity contributed to widespread problems of water scarcity and water pollution throughout the world. With the exponential growth of humans and increasing farming and industrialisation drinking water has become scarce and costly, and the supply has been reduced to minimum per capita daily requirement of 23 liters. Though the provision of clean water and hygienic means of sanitation are basic to human life, 20% of world population lack access to safe drinking water and 50% do not have adequate provision of sanitation. According to the WHO study more than 5 million people die each year for the lack of water and sanitation facilities (CSO, 1997).

Water is polluted by four kinds of substances: traditional organic waste, waste generated from industrial processes, chemical agents from agriculture and silt from dams. Of these though industries contribute smaller volume but share half of the pollution load. Industrial units located in densely populated areas discharge their untreated or inefficiently treated effluents to the surface water channels. Industrial pollution is a worldwide problem that reflects the conflict between economic progress and the environment. Since independence, economic growth through industrialization in India is widely contested as less than satisfactory. In specific localities industrial agglomerations with high intensity of pollution have seriously injured the natural environment. Robust industrial activity is the economic basis of all other activities. However, economic prosperity should not be at the cost of the quality of natural environment. The objective of economic progress should be to provide desirable services to improve the quality of human existence. But the environmental problems like water and air pollution caused by industrial activities leave far reaching consequences on human health and well being through changes in the epidemiological environment that needs quantification and valuation. The cost of supplying alternative sources and the loss in hauling safe drinking water needs to be compared for environmental policy making.

PROBLEM OF STUDY

Water is a prime natural resource, a basic human need and a precious asset, in the absence of which no socio-economic developmental activities can sustain. The most significant environmental problem and threat to public health in both rural and urban India is inadequate access to clean drinking water and sanitation facilities. Almost all the surface water sources are contaminated and unfit for human consumption. Many of the rivers and lakes are getting contaminated from industrial effluents and agricultural run-off, with toxic chemicals and heavy metals, which are hard to remove from drinking water with standard purification facilities. Water pollution is a major environmental concern in India. The main sources of water pollution are discharge of domestic sewage and industrial effluents, which contain organic pollutants, chemicals and heavy metals, and run-off from land-based activities such as agriculture and mining. The increasing discharge of domestic and industrial wastes and the increasing use of fertilisers and pesticides in agriculture has also led to the contamination of groundwater at many places making it unfit for human consumption. An analysis of water quality during twelve years, 1986 to 1997, undertaken by the Central Pollution Control Board (CPCB) has brought out that water quality has deteriorated over the years. Discharge of industrial effluents is another major source of water pollution. The degradation of natural resources, or loss in the quality of the environment imposes a burden on present and future generations. Pollution leads resource scarcities, decline in tourism etc. A need clearly exists for a realistic assessment of the costs of environmental pollution and ecosystem degradation. The costs include not only a human health cost but also cost involved in the loss of efficiency or production. The rapid paces of urbanisation, industrialisation and commercialisation of agriculture have led to significant changes in the stock and the quality of the environment and natural resources. Mitigating widespread environmental problems requires a broad array of actions.

One of the most important threats that human kind faces today is the scarcity of drinking water and its impact on human health. The problem of drinking water remains an urgent need because only 70.5% of the urban and 8.7% of rural households have safe drinking water through piped supply and rest depends on surface and ground water, which is contaminated and untreated. The approach paper to ninth plan (1997–2002) claims that 85% of urban populations have access to safe water and the provision is still inadequate in slums, rural and backward areas (CSO, 1997). Water quality in most stretches of major rivers is below the CPCB and WHO prescribed designated best use standard. The progress during 1991–1997 is quite unsatisfactory. Even in the urban areas recommendations of the National Commission on Urbanisation is not adequately met. The distribution of water in India is extremely inequitable, unjust and biased against the poor. Rich people use the resource relentlessly while poor people spend hours in fetching and filling of water, which is even not sufficient to meet their minimum needs. Hundreds of millions of people suffer from ill health because of contaminated drinking water. The use of polluted water for drinking and bathing is one of the principal routes of infection that kill millions and sicken more than a billion people each year. The water borne diseases particularly of children and poor are huge. Unsafe and contaminated water is implicated in many cases of diarrhea diseases, which as a group kill more than three million people mostly children and cause about 900 million episodes of illness each year (World Bank, 1990). The common effects of organic pollutants present in water are a threat to public health. Outbreak of dysentery and typhoid in the world in the past was associated with organically polluted water supply. Ways the environmental factors affect human health and well-being have attracted growing attention in recent years with much illness and disease now linked with environmental factors. The direct impact of pollution on health is reflected in illness and premature mortality. The patterns of disease and death exhibited in Indian health data reflect the importance of environmental factors particularly quality and composition of drinking water in human health. Two of the important pollutants such as particulates in air and water borne bacteria in surface drinking water are particularly associated with large health losses in the developing countries. World Bank (1993) estimates 30% of total DALY losses in India are due to pollution related diseases. Such health problems are very much costly in terms of premature mortality, medical care and associated pain and suffering, which need quantification and evaluation from the point of view of policy making.

Industries as important point sources spew a huge quantity of toxic and hazardous pollutants that leave irreversible damage to the air, water and other environmental resources. In specific localities, on account of industrial agglomerations, there occurs manifestation of excessive pressure on the natural environment. Such areas are known as problem areas or pollution hotspots and attract growing attention of policy makers. This study has taken up one such problem area Angul-Talcher in Orissa, India and examines the negative link between development and environment, more particularly the impact of growing industrialisation on water quality and stress on drinking water supply in a rural area. Further the work intends to quantify the environmental damage in terms of water quality through extra labour time spent on fetching drinking water.

SCOPE AND OBJECTIVES

The territorial scope of the study covers the Angul-Talcher industrial agglomeration in Orissa, one of the 24 pollution hot spots of India. The area is famous for rich coal deposits and presently endowed with seven large scale public sector undertakings mainly coal based and are highly polluting in nature. The industries release a huge quantity of pollutants toxic and hazardous in character. Extensive mining, power plants, increasing industrialisation and growing urbanisation have seriously injured the natural environment of the study area. Presence of several pollutants, heavy metals, fluoride etc. compelled local people to walk distant places to fetch a pot of drinking water. Water once the free goods have become precious and costly especially for the women folk. Such an environmental damage urgently calls for valuation and quantification. The study aims at assessing the environmental damage cost of water pollution through valuing drinking water that involves extra labour time involved in fetching water from a distance place.

METHODOLOGY

The provision of safe drinking water in a country leaves significant impact on the quality of life of its citizens. People in the developing countries, women particularly spend significant portion of their labour time towards fetching and filling of drinking water from sources to their home. Enhanced access to environment improving services in the provision of potable drinking water yield substantial benefit in the form of productive time saved. Attaching a value to the time people spend in collecting water thus indicates the economic benefits of improved access to water services which than can be compared with the estimated costs of such services, to assist decision making on appropriate service provision (Zyl *et al., 2000).* The Inter American Development Bank study considered drinking water pricing to be made by time saving at a rate of 50% of the existing wage rate for the unskilled labour. But estimates of the value of time on the basis of travel mode choice in developing countries typically value travel time as the price of alternative arrangements though at a less than market price (Bruzelius, 1979; Yucel, 1975).

Industrial pollution severely contaminated the surface and ground water and made it useless and disease-borne. Therefore, either it requires alternative arrangements or involves heavy labour cost to fetch clean potable drinking water from a less polluted distant place. The damage in this context is calculated through labour time lost in fetching water, which could have been used for other productive purpose. In the study area, the rural women walk miles distance devoting more of time to fetch and fill drinking water. So the problem is approached through a modified travel cost method. Hence the economic value of drinking water hardship is measured by the number of working women (main female workers) times the number of hours of labour lost multiplied by the wage per hour, which can be expressed as

$$W_D = N.\ H.\ Wh$$

Where

W_D = Damage to drinking water facilities

N = number of working women (main female workers) engaged in fetching drinking water

H = hours of labour lost in fetching water

W_h = wage per hour

SOCIO-ECONOMIC AND POLLUTION PROFILE OF STUDY AREA

Angul-Talcher is one of the two largest industrial agglomerations of Orissa in India. Talcher, which is famous for its coal deposits, has attracted rapid industrialization being supported by other factors like cheap labour, plentiful water supply, good transport network, abundant space for construction and other locational advantages. Industrialization has taken a big leap during 1960s to 1980s(OEP, 1994) and the area has emerged as a big source of coal and thermal power in the country (SAC, 1990).

The area spreads over the Angul district in the central part of Orissa, stretching over $20^0 48'$ to $20^0 58'$ N latitude and 85° 53′ to 85° 28′ E longitudes (CGWB, 1993). The area is 160 Kms. away from the Bay of Bengal. It occurs at an altitude of about 100 mtrs above the mean sea level. The pollution impact zone covers an area of 2829 Sq. Kms. encompassed by 30 Kms. radius circle on both sides of river Brahmani inhabited by a population of 8.4 lakhs. The area has a rural agro-based economy with 83% of rural population and 65% them depending on agriculture (GOO, 1993). However, there is a recent trend towards urbanisation due to phenomenal growth of industries. Since traditional and low productivity agriculture is the main occupation, majority of people are poor and backward (GOI, 1991).

There exist seven large-scale public sector undertakings (PSUs) in both central and state sectors. The industries produce urea and ammonia, aluminium, thermal power, chemicals and heavy water (Table-1). The area is also well known for coal mining, which dates back to 1921 (SAC, 1990). At present there operates seven coalmines in the region under the

Table 1
Industry Profile of the Region

Name/Year	*Output Produced*	*Input used*	*Investment (Rs.in crores)*	*Employment*	*Annual average output (MT/MW)*	*Annual average value of product (in Rs. Crores)*
FCI, Talcher (1980)	Urea, Ammonia	Coal	225.52	1380	1,12,585	79,20
NALCO ASP Angul (1986)	Aluminium,	Alunima	2408	5450	1,81,365	1120,00
NALCO (CPP) Angul (1986)	Thermal Power	Coal	3,519			
TTPS, Talcher (1967)	Thermal Power	Coal	750	1200	12,83	91.30
ORICHEM, Talcher, (1982)	Sodium Dichromate & Basic Chrome Sulphate	Chromite Ore/Soda Ash/ H_2SO_4	08	200	2,600	7.60
H.W.Plant, Talcher, (1982)	Heavy Water	Chromite Ore	NA	NA	NA	NA
TSTPP, Kaniha,	Thermal Power	Coal	NA	NA	NA	NA
NBFA(1997)	High Carbon Ferrochrome	Chromite Ore	NA	NA	NA	NA

Note: Annual Average refers to an average of 6 years, 1990-91 to 1995-96.
Source: OSPCB, (1990); Annual Reports of concerned Industries, Government of Orissa, 1992; and Through Questionnaire cum Schedule.

Table 2
Details of Coal Mining in the Area

Name of the mine and type of mining	*Year of Operation*	*Area (in Sq.Km.)*	*Coal reserve in million tones*	*Average daily Production metric tones*	*Employment*
Deulbera (UG)	1926	8.11	17	450	1870
Handidhua (UG)	1929	8.51	3.80	100	320
Talcher (UG)	1927	6.02	69.91	500	1410
Nandira (UG)	1962	6.00	33.26	325	700
South Balanda (OC)	1960	8.27	36.60	1900	1100
Jagannath (OC)	1985	4.89	177	700	900
Bharatpur (OC)	1985	12.05	263	—	1070
Total		53.85	600.57	3975	7370

Note: UG = Underground; OC = Open Cast
Source: OSPCB (1990)

Table 3
Pollution Profile of the Region
(Pollutants in Kg/day; Wastewater m³/day and Solid waste tones/day)

Name of the Industry	*SPM*	*SO_2*	*NO_X*	*Waste Water*	*Solid Waste*
FCI	53,740	5,664	3,792	4,757	459
NALCO(ASP)	200	2,208	897	12,485	18
NALCO(CPP)	8,980	33,591	19,297	71,455	3,879
TTPS	65,700	29,400	21,250	15,076	1,836
ORICHEM	154	03	NIL	NIL	13
TSTPP	13,219	19,056	23,640	11,616	NA
TOTAL	1,41,993	89,922	68,876	1,15,389	6,205

Note: Apart from the above FCI emits 530 kgs. Of ammonia (NH_3) AND 1,843 of urea dust daily and NALCO smelter release 76 Kgs of Fluoride daily.
Source: OEP, (1994)

Mahanadi Coal Fields Limited (MCL), the basic information of which is presented in Table-2. Extensive mining, power plants, industries and growing urbanisation has seriously damaged the environment in the area (ORSAC, 1994). Main air pollutants in the area include SPM, SO_2, NO_x, CO, CO_2, Urea dust, ammonia, fluoride, fly-ash, dust, soot, smoke, fume, hydrocarbons, etc. These pollutants remain suspended for a long period of time in the ambient air and cause serious health hazards and other damages over a large stretch of area. These air pollutants from the industries cause the biotic and a biotic component of the ecosystem degraded in the region (Das, 1990). Large quantity of water drawn from the rivers and water bodies for various processes return back with contamination and degradation. Water bodies receive 1,15,389 m³ of industrial effluent daily, containing heavy metals like lead, chromium, cadmium, cobalt, zinc, thorium, aluminium, etc., with high adverse effects on living creatures. An amount of 6,205 MTS of solid waste per day is also a matter of serious concern. The polluted water /industrial effluents include various heavy metals, fluoride, hexavalent chromium, ammonical nitrogen, ash, suspended solids, oil and grease etc. Water in the area is so polluted that it is not fit for direct use as drinking or bathing. Also a huge quantity of solid waste in the form of ash, ash slurry, scrap metals etc. is generated which adds to the pollution problem of the locality. Table-3 contains information regarding the nature and extent of pollution of ambient water of the Angul-Talcher area. Since majority of industries are thermal power plants and coal based, all the industries found in the area come under the list of 17 highly polluting industries classified by the Central Pollution Control Board (CPCB) and the Government of India. Therefore, the extent of pollution is so high that Angul-Talcher area of Orissa is reeling under severe environmental pollution caused by the industries. The pollution problem has become so high that the area has attracted attention from within and outside the state and has been branded as a problem area and one of the twenty-four important hot spots of industrial pollution in the country.

STATE OF WATER QUALITY

It is observed that the parameters like pH, BOD, COD, SS, and oil and grease are at higher concentration than the specified range as prescribed by EPA, 1986. The presence of such parameters in excess renders water unfit for direct use. It is observed that all the industrial effluents in the study area ultimately find their way into rivers and other water bodies rendering them replete with both organic and inorganic pollutants. Discharge of industrial effluents/waste water into the water bodies have converted them into private industrial drains. They are heavily silted due to large deposits of suspended solids. The impact is clear by analysing water quality of Brahmani and Nandira rivers. Brahmani, the second largest river of the state passes through the study region. All the major industries depend on river Brahmani for water resource and discharge their waste-water and effluents into it. Besides the industrial effluents, the sewage water of the urban bodies and rural settlements enter into the river directly. This Brahmani is subject to heavy stress of pollution (OSPCB, 1990). For all the parameters of water quality, downstream figures have remained high. The water pollution situation is worst at Kamalanga where all industrial effluents of Angul - Talcher area joins the river. Kamalanga represents the turning point, below which the water quality is very bad. The OSPCB is regularly monitoring the water quality of Brahmani at several stations, including Kamalanga, which also comes under the Monitoring of Indian National Aquatic Resource (MINAR) programme. The CPCB has classified the water of Brahmani from Kamalanga onwards as class 'C' water on the basis of designated best use, where class 'C' water indicates that the water should be suitable for public water supply after conventional treatment followed by dis infection (OSPCB, 1990). The level of SS remains objectionably high throughout the year. The level of TDS is very high during monsoon months because of soluble matters and silt coming through runoffs (OSPCB, 1990). The other parameters like BOD, COD, DO also do not comply with the prescribed limit for designated best use. A significant decrease in BOD value leads to marked increase in DO level which are not good from the point of view of aquatic life. Comparison of data of Brahmani with the water quality criteria (ISI0500, 1991) confirms that Brahmani water is not suitable for direct drinking purpose. However, the stretch of the river, which is classified as class 'C', also becomes below that level with regard to pH and BOD. Nandira rivulet, a tributary of river Brahmani is the most polluted river of Orissa. It is used like a private industrial drain (OSPCB, 1990). The average concentration of SS, BOD, COD and oil and grease are well beyond the acceptable limits prescribed by EPA, 1986 along the entire stretch of Nandira during all the seasons for which FCI, TTPS and NALCO (CPP) are largely responsible. Nandira water has a very high pH value. Since water of Brahmani is classified as class 'C' water from the point of view of designated best use by CPCB, Nandira water also belongs to class 'C' water. In case of TDS the tolerance limit for class 'C' water is 1500 mg /l. But at downstream of FCI' s industrial discharge TDS is as high as 4780. So also for DO, as against the limit of 4 mg / l, at times it becomes 7 mg / l. While for BOD, the tolerance standard is 3 mg / l, Nandira water at the extreme attains a figure of 24 mg /l. Similarly for all the required parameters the observed value is far higher than the prescribed tolerance limit. Therefore Nandira is biologically a dead rivulet for the people and for aquatic life (Das, 1990).

In and around Angul - Talcher industrial area ground water is also not safe and not suitable as potable drinking water due to industrial pollution. The CGWB has carried out an extensive study of ground water quality in the study area. The result of the experiment indicates that pH level ranges from 5.92 to 8.25. Very high values are also found in case of chloride, fluoride, phosphate, magnesium, alkalinity, hardness etc. Hence, it is not potable at all (OEP, 1994). The presence of the Aluminium Smelter plant of NALCO further worsens the situation as at least 220 tonnes of fluoride is annually discharged from the smelter plant through emissions and effluents. Most of the fluoride naturally settles in the immediate vicinity of the factory and finally finds it way into the nearby water bodies and ground water through leaching and percolation. At places, as is evident from the monitoring of ponds and wells (both dug wells and tube wells) by the CGWB the fluoride concentration has exceeded the standard prescribed for drinking water (OSPCB, 1997). The presence of heavy metals in higher quantity in the mine seepage and industrial discharge water contaminates the ground water quality.

Water Quality of River Brahmani

Brahmani, the second largest river of the state passes through the study region, having a total catchment area of 39,033 square kilometers and an annual average discharge of 18,310 million cubic meters of water. All the major industries depend on river Brahmani for water resource and discharge their wastewater and effluents into it. Besides the industrial effluents, the sewage water of the increasing population of growing urban centers and rural settlements enter into the river directly. This Brahmani is subject to heavy stress of pollution (OSPCB, 1990). For all the parameters of water quality, downstream figures have remained high. For different parameters such as BOD, COD, suspended solids, oil and grease fluoride, MPN and turbidity the values are very high in the downstream area as compared to upstream points. On the other hand, for the other parameters like pH and DO the figures lag behind in the downstream indicating an adverse environmental state.

Kamalanga at which all industrial effluents join the river below which the water quality is very bad (Table-4). The OSPCB is regularly monitoring the water quality of Brahmani at several stations, including Kamalanga under the monitoring of Indian National Aquatic Resource (MINAR) programme. The CPCB has classified the water of Brahmani from Kamalanga downwards as class 'C' water on the basis of designated best use, where class 'C' water indicates that the water should be suitable for public water supply after conventional treatment followed by disinfections (OSPCB, 1990). The level of SS remains objectionably high throughout the year. The level of TDS is very high during monsoon months because of soluble matters and silt coming through runoffs (OSPCB, 1990). The other parameters like BOD, COD, and DO also do not comply with the prescribed limit for designated best use. A significant decrease in BOD value leads to marked increase in DO level which are not good from the point of view of aquatic life. Comparison of data of Brahmani with the water quality criteria (ISI0500, 1991) confirms that Brahmani water is not suitable for direct drinking purpose (Tables 4&6). However, the stretch of the river, which is classified, as class 'C' also becomes below that level with regard to pH and BOD.

Table 4
Water quality in Brahmani River

Parameters / *Year*	*pH*	*TDS*	*DO*	*Total Alkalinity*	*Total Hardness*	*BOD*	*COD*
(1)	(2)	(3)	(4)	(5)	(6)	(7)	(8)
1985	7.5	52	3.8	105	72	7.6	25.9
	8.8	256	7.6	107	72	7.7	27.8
1986	7.5	52	3.8	70	72	3.9	16.6
	8.2	256	7.6	104	82	6.4	20.0
1987	7.9	52	3.8	92	62	3.8	18.3
	8.4	256	7.6	106	72	5.7	28.0
1988	7.5	52	3.8	92	89	3.5	14.6
	7.7	256	7.6	108	84	11.2	41.3
1989	7.9	52	3.8	85	69	3.2	16.7
	7.9	256	7.6	87	72	6.5	28.9

Note: (1) All figures are in mg / l except pH

(2) Upper figures represent water quality at Samal barrage (upstream) and lower figures represent water quality at Kamalanga, (downstream) the confluence point of Nandira and Brahmani.

Source: OEP (1994).

Table 5
Pollutant Concentration of Nandira along its Entire Stretch

Parameter	*Average values 1992*	*Average values 1993*
pH	7.7	8.2
SS	4246.6	3157.33
TDS	593.3	349.3
COD	302.6	258
BOD	126	—
DO	6.4	—
Oil and grease	32	—
Fluorides	0.5	1.4
Phosphates	5.3	—
Nitrates	8.3	—

Note: All values in mg / l except pH.
Source: OSPCB (1993)

Table 6
Tolerance limits for Inland Surface Water subject to Pollution
(IS – 2296-1982)

Class of Water *Parameters*	*A*	*B*	*C*	*D*	*E*
(2)	(3)	(4)	(5)	(6)	(7)
pH	6.5 - 8.5	6.5 – 8.5	6.5 - 8.5	6.5 - 8.5	6.5 - 8.5
DO	06	05	04	04	—
BOD	2	3	3	—	—
COD	—	—	—	—	—
TDS	500	—	1500	—	2100
Total Alkalinity $CaCo_3$	300	—	—	—	—
Oil and grease	—	—	0.1	0.1	—
Fluoride(F)	1.5	1.5	1.5	—	—

Note: All values in mg /L except pH
Source: OSPCB (1990)

Water Quality of Nandira

Nandira is the most polluted river of Orissa. It is used like a private industrial drain (OSPCB, 1990). The average concentration of SS, BOD, COD and oil and grease are well beyond the acceptable limits prescribed by EPA, 1986 along the entire stretch of Nandira during all the seasons for which FCI, TTPS and NALCO (CPP) are largely responsible (Table). Since Nandira is the most polluted water body and it is a rivulet to Brahmani, the pollution load from Nandira to Brahmani is well understood (Table-5). Water quality and water pollution problem can be judged by comparing water quality of Nandira with the Tolerance limits of Indian Standards

IS: 2296 – 1982 presented in Table: 6. A careful analysis of Tables 4, and 5 in comparison to Table 6 (IS- 2296 – 1982) shows that Nandira water has a very high pH value. Since water of Brahmani is classified as class 'C' water from the point of view of designated best use by CPCB, Nandira water also belongs to class 'C' water. For all the required parameters the observed value is far higher than the prescribed tolerance limit. Therefore Nandira is biologically a dead rivulet for the people and for aquatic life (Das, 1990).

Impact on Ground Water Quality

In and around Angul – Talcher industrial area ground water is also not safe and not suitable as potable drinking water due to industrial pollution. The CGWB has carried out an extensive study of ground water quality in the study area. The result of the experiment indicates that pH level ranges from 5.92 to 8.25. Very high values are also found in case of chloride, fluoride, phosphate, magnesium, alkalinity, hardness etc. Hence, it is not potable at all (OEP, 1994). The presence of the Aluminium Smelter plant of NALCO further worsens the situation as at least 220 tonnes of fluoride is annually discharged from the smelter plant through emissions and effluents notwithstanding the fluoride emission control measures adopted by the industry. Most of the fluoride naturally settles in the immediate vicinity of the factory and finally finds it way into the nearby water bodies and ground water through leaching and percolation. At places, as is evident from the monitoring of ponds and wells (both dug wells and tube wells) by the CGWB the fluoride concentration has exceeded the standard prescribed for drinking water (OSPCB, 1990). The presence of heavy metals in higher quantity in the mine seepage and industrial discharge water contaminates the ground water quality.

ECONOMIC VALUE OF DRINKING WATER

The industrial agglomeration at Angul-Talcher acts as a growth pole for the nation. The products produced have strong propulsive effects inducing further industrialisation and higher economic activity. But the people of the region suffer heavy social cost in the form of environmental damages. It is clearly evident from the previous section that all the industrial units discharge effluents directly to the rivers and other water bodies. Both ground water as well as surface water has become completely unfit for direct use. Since the water is totally contaminated woman folk do not feel it secure for their family to consume water from the local points and walk miles distance for a pot of safe drinking water.

The economic value of the hardship in fetching drinking water is assessed through labour time spent or wage lost in the activity. In this work the scarcity problem is focused only to the rural people because the industrial colonies and urban pockets are provided with piped supply of drinking water. It is observed that within a radius of 10 kms 50,000 rural households in the neighboring villages suffer the scarcity of water. Since both ground water as well as surface water is contaminated and tube wells have become defunct, as the level of ground water has gone deep, women either have to walk more distance or have to wait long in a queue. Traveling a long distance or waiting in the queue on an average consume two hours daily which they could have used in a direct productive work or in household chores. The cost of fetching drinking water involves time, which can be valued at the existing wage rate. Using the method outlined in Section.4, damage cost of water comes to Rs.5 lakhs per day (2 hours of 50,000 females @ Rs. 5 per hour). Since industries injure the aquatic ecosystem they need to pay for alternative arrangement (through piped supply or hand pumps) of drinking water to the victims as recommended under polluter pays principle.

CONCLUSION

Water is a basic natural resource. It not only sustains human life but also promotes socio-economic development and quality of life. Apart from the problem of distribution and continuous availability of fresh water resources, pollution is a major threat to the quality of water that is available in a local area. In this paper we attempted the impact of industrial and mining activities on the quality of water in one of the pollution hot-spots of India, namely Angul-Talcher industrial agglomeration in Orissa. The study goes into the economic damage suffered by the local community due to industrial expansion. A travel cost method is adopted to estimate the cost borne by the local community by way of arranging sate drinking water from a distance.

The water pollution outcome of seven large scale industries and mining activity in the region is that Brahmani and Nandira rivers are polluted, ground water is contaminated and local ponds and other sources of drinking water are unsafe. It has been observed that use of polluted water has resulted in health problems (morbidity and mortality) for the villagers. Thus, the people (particularly the women folk) travel on an average two kilometers to fetch water that is considered safe. On the basis of the time spent for this activity and the prevailing wage rate it has been estimated that environmental damage cost of polluted drinking water is about Rs. 5, 00,000 per day . This implies a loss of income to the tune of Rs.10/- per day per household, which is not small given the state of poverty in the region.

In environmental economics there is a clear insight that this is an external cost of industrial and mining activity. The best way to deal with this problem is to make the industries responsible to internalize the by providing for deep bore wells and/or through supply of piped water to every affected village.

REFERENCES

Bruzelius, N. (1979), *The Value of Travel Time*, Croom Helm Ltd, London.

Central Ground Water Board (1993), *Study on Fluoride Pollution of Ground Water around Angul*, Ministry of Water Resources, Eastern Zone, Bhubaneswar.

Commission on Sustainable Development (CSD) (1997), Freshwater Resources, Ch 18 *in Progress Review Report on Agenda 21*, New York, UNDPCSD.

Das, R. C. (1990), *Water Resources: State of Environment in Orissa, Vol. I*, Bhubaneswar, Orissa State Pollution Control Board.

GOI (1991), *Census of India*, Series 19, Orissa, Provisional Population Totals, Paper 1 of 1991 Supplement, Directorate of Census Operations, Orissa, Bhubaneswar.

GOO (1993), *District Statistical Hand Book, Angul* (1993), Bhubaneswar, Directorate of Economics and Statistics, Orissa.

OEP (1994), Environmental Management Plan for Angul-Talcher in Orissa (Prepared by Kirloskar Consultants, Pune), Bhubaneswar, Orissa Environment Programme.

ORSAC (1994), *Monitoring and assessment of Eco-degradation and Environmental Pollution of Angul-Talcher Region* by Remote Sensing, Bhubaneswar, Orissa Remote Sensing Application Centre.

OSPCB (1990), *Status of Polution in Angul-Talcher Area, Orissa and Recommended Action Plan*, (Report submitted to the CPCB), Bhubaneswar, Orissa State Pollution Control Board, Orissa, Bhubaneswar.

OSPCB (1997), *Parivesh Samachar*, The Official News Letter of Orissa State Pollution Control Board, Orissa, Bhubaneswar.

SAC (1990), *Impact of Mining Activities and Super Thermal Power Stations on Environment: A Study of Talcher*, Ahmedabad, India.

World Bank (1990), *World Development Report*, Washington D.C., Oxford University Press. World Bank (1993), *World Development Report: Investing in Health*, Washington D.C., World Bank, Oxford University Press.

Yucel, N. C. (1975), *A Survey of the Theories and Empirical Investigations of the Value of Travel Time Saving*, World Bank Staff Working Paper No. 199, The World Bank, Washington, D.C.

Zyl Hugo van, T. Store and A. Leiman (2000), Valuing Time Spent Collecting Water in A Kenyan Town, in Mc Cracken, J. R. and Hussein Abaza (eds) *Environmental Valuation: A Worldwide Compendium of Case Studies*, Earth Scan, London, pp. 23-33.

14

SOCIO-CULTURAL IMPACT OF ARSENICOSIS ON RURAL COMMUNITIES OF BANGLADESH

Maheen Sabreena Karim

INTRODUCTION

Arsenic poisoning[1] has been detected as an acute health problem in Bangladesh, where as many as 70 million people are living in a possible risk of contamination. Arsenic pollution is causing a serious health hazard, and also at the same time, it has been creating acute socio-cultural and environmental problems in the country (DCH, 1998). It causes widespread social disorganization where the rural people in our country, mostly due to their ignorance, superstitions and lack of proper knowledge often confuse the skin lesions caused by arsenicosis with leprosy (World Bank, 1999). As a consequence of it, the arsenic victims are neglected from the medical point of view and the villagers treat it as a contagious disease, which according to them, is a curse of nature. For that reason, they fully isolate them from the community and do not allow them to come in public.

Arsenic exists in the underground water, and it enters the human body through the drinking water from the tube-well. As we know that arsenic is a water-borne disease, and since 95 per cent of the people in our country now have some sort of access to the tube-well water, it is quite obvious that almost a similar number of people have now the possibility of being contaminated by it (Karim *et al.*, 2003)[2]. In Bangladesh, ground water becomes contaminated because of its over-use, and also because of the people having too much dependence on it[3]. As compared to many other countries of the world, the amount of ground water use in Bangladesh is quite extensive, and absolutely for that reason, the possibility of arsenic contamination becomes higher in this country[4].

Arsenic contamination of ground water was first discovered in the Bengal belt of West Bengal and Bangladesh during 1990s, because of a similar geographical location and physiographic condition (Karim, 2003). From global context, the situation in Bangladesh is perhaps the worst in the world. Whatever may be situation however, arsenic issue has now been dealt with very seriously, and a multi-disciplinary approach has been employed to conduct various types of research in this context. Dhaka Community Hospital, ICDDR-B and NIPSOM have started conducting important research from their respective viewpoint. But to my knowledge, very few research have been undertaken to study the social impact of it, which seem to be very much essential at this moment. This research thus, is an outcome of the field-based data collected from a peri-urban village in the northern part of the country. The main focus of this paper is to explain analytically as to how arsenicosis has been causing serious socio-cultural problems in rural Bangladesh. At the same time, the paper also finally suggests a few recommendations, which the policy planners and researchers may consider to tackle this social crisis.

PHYSIOGRAPHIC LOCATION OF STUDY VILLAGE

Arsenic contamination of ground water was widely reported in Bangladesh in the beginning of 1990s. It was discovered mostly in the Bengal deltaic belt of West Bengal and Bangladesh, perhaps because of the similar geomorphological affinity and socio-economic condition. But as compared to West Bengal of India, the situation is much worst in Bangladesh. The availability of 'injurious arsenic' in the ground water was found in the western bordering districts of Ganges deltaic plane. Recently, it has been discovered in many districts of Bangladesh (Quamruzaman *et. al.*, 2000)[5]. Out of a total 64 districts in the country, 44 of them have been found affected by arsenic contamination. And among them, at least 17 districts have been identified as severely affected areas.

The most vulnerable physiographic regions of Bangladesh which have been identified as severely contaminated are: (i) the Ganges-flood Plane; (ii) The Atrai flood plane, (iii) the tidal regions; the coastal plane; and (v) the Meghna flood plane. By adding the high probability zone, moderate probability zone, and low probability zone, it is suspected that total of 60 per cent areas have now been identified as arsenic-prone areas of Bangladesh (see DCH 1997). A map on the level of arsenic contamination in different parts of the country was prepared to find out the extremely contaminated districts of Rajshahi Division (see Karim, 2003 for details). In Rajshahi District, the incidence of arsenicosis is found in some neighbouring peri-urban villages of Rajshahi city areas. Among other areas, Charghat Upazila in Rajshahi District is the most vulnerable zone of this region. Among a number of such villages, Manirampur, Mianpur are most prominent. In this research, Mianpur has been selected as study area, because of its widely known presence of a number of arsenic victims (for details see Karim, 2004). In Mianpur, 24 arsenic patients were identified, and among them, two died of arsenicosis in different times in the past years. The tragic death of one of the victims named Pinjira in village Mianpur became a kind of legend in the field of arsenic research in

Bangladesh is recent times. It has been incorporated in the following section as a kind of case study for explaining the socio-cultural context.

METHODOLOGY

Based on author's own participation at the field-level, the research has employed entirely an anthropological technique for collecting data. In total, 22 arsenic patients were identified from the village Mianpur. Apart from an intensive interview of the sample survey, detailed case studies were also conducted. The method of the study also included the conduct of two focal group discussions (FGDs), through a community participatory approach. A total of 468 households were surveyed most extensively to provide an over-all socio-economic condition of the villagers.

SOCIO-DEMOGRAPHIC PROFILE OF ARSENIC VICTIMS

The discussion in this section deals with the socio-demographic data of the arsenic victims in Mianpur village of Rajshahi District. As part of this anthropological investigation, a survey was conducted in 468 households at the initial stage to find out 24 arsenic victims in the village. The village contained a total of 1872 persons, of which, 1010 (53.95%) are males, and the remaining 862 (46.05%) are the females. In Mianpur village, 24 patients reside in two neighborhoods, of which two women died of arsenicosis very recently.

Among the twenty-four patients, 50% each are females and males. Out of 22 arsenic patients, 29.17% are found to be in mild condition of their suffering. This is a positive aspect of it, which suggests that a proper medical attention may allow them to get recovery from their disease. The suffering of 33.33% patients is of moderate in nature, which might go worst, if they are not brought to proper medical care at the moment. And the conditions of at least 37.50% patients seem to be quite serious. They require immediate medical care at an intensive level.

Arsenic victims are classified mainly into three types. These are: (a) melanosis (b) leucomelanosis and (c) keratosis. A person having contaminated by arsenic is first struck with menanosis where the patients have some black dotted spot. When the white spots are seen, it is called leucomebosis. Long term ingestion of inorganic arsenic in drinking water leads to keratosis (for details see Karim, 2003)[6]. The patients, whose monthly income is less than Taka 3000 are classified as poor. A middle-income person is he who earns between Taka 3001 to Taka 10,000 a month. And, rich are those, who earn more than Taka 10,000 a month.

CASE STUDY OF A VICTIM

Pinjira Begam, a 26 years old woman was the worst arsenic victim from village Mianpur, who died of it, in the year 2001. Her cruel husband Masood tortured her both physically and psychologically before her death.

I could not however, see Pinjira, but I had the opportunity to see her children, who initially were reluctant to talk to me. When I asked Shapla (12 years old daughter) as to how her mother died, she pointed directly showing a red-marked tube-well at the premise of her house. While talking to the neighbours, I learned that Pinjira once predicted that she would die because of the drinking water from the tube-well when she found enormous black spots all over her body, hands and feet. It may be noted here that Pnjira had another four years old small daughter named Juthi, who died after her death. It was not even investigated as to what were the causes of Juthi's death, as I suspect that she might also had been infected by arsenic.

However, Pinjira was married to Masood, who was a daily wage labourer in the village having an earning of around taka 2000, a month. As Masood could not maintain the family economically with such a low amount of money, Pinjira had to substitute it, by working as maid to other rich families in the village. Pinjira had to perform all type of heavy works for those rich families from boiling paddy to drying them on the sun, and finally grinding them to rice. Yet, it was difficult for her to maintain her family. Most of the time, Pinjira remained half-fed or un-fed, as she usually distribute all foods to her family members. From the very day Pinjira was married to Masood, she had a struggling life, economically and socially.

But her ill-fate was further worsened during 1996 when all on a sudden, Pinjira noticed some spots on parts of her body. As she has shown it to her family members and neighbours, they initially took it as a kind of skin disease, which would be banished soon. But by the beginning of the rainy season of the same year, her spots started increasing, which also turned blackish. The cracks started to appear on her toes and fingers. By this time, she felt unbearable pain at the time of doing her household work. With this, the villagers became frightened and suspected that Pinjira had contracted tuberculosis, which was totally incurable. Her mother-in-low shouted at her exclaiming, 'you have brought this dangerous disease from your parents house, and as you will die, soon you will also destroy my son and other family members'.

Pinjira's mother-in-law and her husband used to beat her frequently because, they thought that Pinjira had brought with her an incurable disease which might transmit among other family members. Her husband Masood started avoiding her and fully stopped taking food and water from her. While her condition grew worst, Pinjira lost her work as maid at a rich man's house. She was denied to enter these houses and had no longer any access to her neighbours' houses as well. Villagers also forbade her to fetch water from their tube-wells, and also disallowed her from bathing in the village pond. In two incidences of marriage ceremonies, Pinjira was simply driven out from the festivals without any food. Once, a few religious leaders from the village denied giving her sweets in a religious gathering. The religious leaders rebuked and warned her not to attend such gatherings in future.

Once a village volunteer named Nargees came to her rescue who took Pinjira to a local hospital. After the careful investigation, the doctors diagnosed her both tuberculosis, and poisonous scabies on the skin. The doctor prescribed her vitamins and also advised her to eat nutritious food everyday. But it was absolutely impossible for Pinjira to afford it economically.

In June, 1998, a group of researchers from NIPSOM came to Mianpur and tested the tube-well that Pinjira was using. They were shocked to find that the water they have been using from this tube well for the last 16 years contained arsenic at an alarming level. The team however, suggested Pinjira to get admitted at a hospital. The village volunteer Nargees along with a few ladies took Pinjira to a Chest Disease Hospital. A dermatological specialist was also consulted. After a thorough check-up, it was found that her liver had been seriously impaired by arsenic. This time, many people sympathized her and came to rescue her, though at that time, it was too late. But interestingly, her husband Masood remained altogether indifferent enjoying conjugal life with her newly married wife through all these ordeals, as Pinjira was a victim of arsenic.

SOCIO-CULTURAL ASPECTS OF ARSENICOSIS

Based on data and case studies, it has been found that the arsenic victims in Mianpur face extreme miseries and problems. When I moved around the village, I found that the villagers were all doing their regular activities and everything seemed to be so normal until I realized that the village had been covered with a curse of arsenic. The curse of arsenic caused unpleasant events in the lives of villagers which never have had occurred, if it was facing acute arsenicosis. Through an informal talk with the villagers, I learned that there occurs a number social problem like unemployment, family breakdown, divorce, marital dissolution, and many other social incidents. Young children are being debarred from their educational institutions. The arsenic victims are sometimes forbidden to attend marriage ceremonies, annual festivals, public function and even religious gathering. As I talked to the victims, I came to know that people often refuse to take food and water from the arsenic affected families. They further told me that people also keep a social distance with the arsenic victims.

Social Isolation

Social isolation is the first social problem among the arsenic victims and there is a visible gap between the affected and unaffected people in Mianpur. While taking to some of the victims of the village they reported to me that they prefer staying alone rather than becoming an issue of discussion among the villagers. For example, a victim named Murshida told me that as she has arsenicosis and for that reason, she was restricted to go to her friend's house during a festival, as she was afraid of teasing and insulation. In her own words, "it is better to be alone than being a pile of hatred".

Many of the victims reported to me that they were told indirectly to remain absent from schools with an apprehension that arsenic might be transmitted to their fellow classmates and teachers. Consequently, some victims had to leave their friends and schools and lead a lovely life and home. From my observation, I saw a few good students are not attending classes, being the victims of circumstances. Among the 22 arsenic victims, Shapla and Arif (daughter and son of Pinjira) have already faced warnings from their school teachers. Students have complained in several occasions to keep a reserve bench for those children who come from the arsenic victims.

Marriage Dissolution

The effect of arsenic not only resulted in serious health and environmental problems, but also it often breaks up marriage tie having a fear of arsenic transmission. For example, Pinjira who died of arsenicosis from Mianpur had serious contention with her husband. As she became crippled day by day, her husband Masood tortured her to death. One informant from Mianpur also told me that he went to a neighbouring village to see a bride, but refused to accept the proposal as the girl had been affected with arsenic. The girl was beautiful but her fingers were so badly affected that they had no other alternative but to reject the proposal.

Social Disorganization and the Impact on Inter-personal Relationship

Another impact of arsenic in Mianpur can be understood through analyzing the total network of relationship between the affected and unaffected people. Those families who are the victims of arsenicosis are avoided. No matter how wealthy and highly educated they are, if they are the victims of arsenic, the villagers do not hesitate to dishonour them. Chronic exposure to arsenic has an adverse impact on the magnitude and strength in interpersonal relationships. I found that the relationships between individuals were becoming disrupted due to having affected by arsenic. Inter-village conflicts and disputes have arisen in the study area, because of arsenic contamination. Misunderstanding has been created in the inter-personal relationship among the villages and different households. A kind of conflict between some families started on the collection of arsenic free water in Mianpur, especially among the women. Though initially the conflict started between two women, it eventually involved a number of families.

Prejudices and a Wrong Conception in Regard to Arsenic

The victims of arsenicosis bear some wrong conceptions. Many villagers regard it as leprosy, contagious and hereditary. Therefore, problems arise in free mixing, marriage, service and business because of their wrong conception of its infectiousness. The older persons and religious leaders often attribute the disease as act of their bad works, having wrath of Almighty Allah or due to ill desire of the ghosts.

CONCLUSION AND RECOMMENDATIONS

The study clearly shows that most of the arsenic patients are illiterate and poor. Many of them are not at all aware of its consequences. Although gradually the villagers have started to realize the consequences of arsenic yet, they often think it as a contagious disease. The arsenic patients are socio-culturally isolated where; people often do not want to mix with them. Many villagers hesitate to accept food and water from them. If a child is attacked by arsenic, he or she always remains at home. The married women become worried because, if she is attacked by arsenic her husband may marry again. This is what happened in the case of Pinjira who died two years ago.

The arsenic victims also feel psychologically upset. They feel shy to come in front of the people. They suspect that if people know that he or she is attacked with arsenic, they will maintain a distance from him/her. Based on my discussion, I may suggest the following recommendations. Due to ignorance, superstitions and lack of proper motivational programmes, the arsenic problem in Bangladesh has been taking a serious shape in the rural areas. The victims even do not know as to what extent, they are affected, for what fault they are isolated from the community and why they are ignored by the people in the villages.

i. The Government and the local NGOs should make a regular follow up to check the arsenic situation in the village.

ii. The villages should be surveyed properly with the help of the social scientists to identify the arsenic patients and to know about their socio-economic conditions.

iii. The tube-wells which are found to be contaminated by arsenic should immediately be sealed. The authorities should inform the villagers about the bad effect of drinking arsenic contaminated water from these tube-wells.

iv. A kind of social awareness should be created among the villagers through social mobilization programs. The educated people, the school teachers and the local leaders should convince the villagers saying that arsenic is not a contagious disease. One should not have any social distance for the arsenic victims, instead, one should be careful in using arsenic contaminated water.

v. The anthropologists should come forward to conduct in-depth socio-cultural studies on arsenic victims and suggest remedial measures. I also feel that the technical experts and medical scientists should include anthropologists in their research team to cover the socio-cultural aspects of their research.

vi. Mass media like newspapers, televisions, radio, posters, leaflets, books, cassettes, video movies, drama, and cinema should act as agents to increase awareness to make the people to realize the effect of using unsafe tube-well water. The victims at the very early stage of their infection should consult the doctor, if they find any pigmentation or irritation of skin. This awareness can be created through the mass media.

vii. Since the professionals are now suspecting arsenic contamination in agricultural produce, I suggest that we should immediately go for rain water harvesting as practiced by 'Jaldisha Project' in India (see Khanum et. al. 2003). We can also go for preserving rain water for future consumption.

viii. The NGOs and the Government should check the tube-wells very frequently through a follow-up monitoring system.

NOTES

1. Arsenic is a naturally occurring tasteless and odorless element, which exists in the earth's crust with an average of two to five thousand microgram per liter of water (Sarkar and Mohiuddin, 2002). The fatal dose of arsenic is 125 mg, and it is four times poisonous than mercury. Arsenic has been identified as a cause of cancer. Considering the potential carcinogenicity, the International Agency for Research on Cancer (IARC) arsenic has been placed in the first category of poison (DCH, 1998).

2. DPHE, UNICEF employed several strategies to provide tube-wells for pure drinking water. And in our research, it has been found that now 95 per cent of the people have somehow access to tube-wells.

3. In Bangladesh, the extensive use of ground water for drinking was started at least three decades ago, and prior to that people mostly depended on the surface water. At that time, there were lot of ponds and wells in the country and the number of tube wells was fewer in number. Most of the people during that time were drinking water from rivers, ponds and wells. As part of a filtration process, water was kept in the pitcher for the whole night to allow the sediments to fall at the bottom of the pitcher.

4. The extensive use of ground water in Bangladesh was started during 1960s, when the Comilla Model of Rural Development was instituted for introducing farming based on High Yielding Variety Seeds (HYV). This type of HYV cultivation necessitated the extensive exploration of ground water through irrigation (Khan, 1973). Many villagers at that time, did

not like the idea of such extensive use of ground water for such mechanized cultivation. After four decades however, we have now started realizing the effect of lifting too much ground water.

5. From several researches, it has been learned that out of 64 districts in Bangladesh, at least 44 have been affected with arsenic contamination and many more villages in Bangladesh are in the vulnerable stage (Khan and Akhtar, 1997; Quamrruzzaman *et al.*, 2000).

6. The clinical manifestation of chronic arsenic toxicity develops slowly after six months to two years or more, depending on the amount of arsenic intake. Taking nutritious diet and vitamins on daily basis, may often delay a person being contaminated with arsenic. Based on the clinical manifestations, arsenic mainly is of three types: melanosis, leucomelanosis and keratosis. A person drinking arsenic water is first struck with melanosis, where black dotted spots are seen on the palms, chest, tongue, feet and even on the gum of the teeth. When melanosis is evenly distributed all over the body, it is called diffused melanosis. When white spots are seen on the black dotted ones, it is called leuco melanosis. It also shows 'rain drop pigmentation'. And for long term ingestion of inorganic arsenic in drinking water, leads to keratosis. In the infection of keratosis, the skin of the hands and feet becomes hard and rough. When keratosis becomes, serious, it is called hyper-keratosis (details see Karim, 2003; Saha, 2002, Tondel, 1999; Ahmed *et al.*, 2001). Based on the level of sufferings, I have categorized the patients in my village broadly into three categories. As I am quite ignorant about the medical investigation, I recommend the medical practioners to work on it in details.

REFERENCES

Ahmed, Sk. Akhtar *et. al.* (2001), Arsenic in Drinking Water and Urinary Level of Arsenic, *Environmental Sciences*, 8(5): 417-423.

Dhaka Community Hospital (DCH) (1998), The Arsenic Problem in Bangladesh: An Introduction to the Country and its Arsenic Situation, In DCH & SOES, *International Conference on Arsenic Pollution of Groundwater in Bangladesh: Causes, Effects and Remedies*, Dhaka: DCH.

Karim, A. H. M. Zehadul, Abu Bakar Siddique Bhuiyan and Muhammad Mahmudur Rahman (2003), A Situational Analysis of Water Use, Sanitation Practice and Personal Hygiene Awareness the Rural Bangladesh: A Baseline Study, *South Asian Anthropologists*, 3(1).

Karim, Maheen Sabreena (2003), The Socio-Cultural Aspects of the Arsenic Victims in Rural Bangladesh: An Anthropological Study of Mianpur Village in Rajshahi District an unpublished BSS (Hons.) Research Report. Department of Anthropology, Rajshahi University, Rajshahi, Bangladesh.

Khan, Akhter Hamid (1973), *Tour of Twenty Thanas: Impressions of Drainage Road, Irrigation and Cooperative Programmes*, Comilla: BARD.

Khan, A. Wadud and Sk. Akhter Ahmed (eds.) (1997), *Arsenic in Drinking Water: Health Effects and Management*. Dhaka: NIPSOM.

Khanum, Sultana Mustafa, Kazi Tobarak Hossain, Muhammad Hasa Imam, M. Zulfiquar Ali Islam and Wardatul Akmam (2003), Access to Primary Health Care Services of Rural Women in a Bangladesh Village, an unpublished Result Report Funded by DFID and Managed by British Council, Rajshahi University, Department of Sociology.

Quamruzzaman, Quazi; Mahmudur Rahman and Allisor Quazi (eds.) (2000), *Arsenic in Bangladesh: Report on the 5000-Village Rapid Assessment Project*, Dhaka: Dhaka Community Hospital.

Saha, K. C. (2002), *Arsenicosis in West Bengal: Environmental Problems and Solutions*, Kolkata: Sadananda Prakashani.

Sarkar, Prafulla. C., and M. Mohiuddin (2002), The Socio-Economic and Cultural Life of the People of Bangladesh, *South Asian Anthropologist*, 2(2): 97-102.

Tondel, Martin *et al.* (1999), The Relationship of Arsenic Levels in Drinking Water and the Percentage Rate of Skin Lesias in Bangladesh, *Environmental Health Perspectives*, 10(3). World Bank (1999), *The Bangladesh Arsenic Mitigation Water Supply Project: Addressing a Massive Public Health Crisis*, Washington: South Asia Brief.

15

THE SOUTHEASTERN ANATOLIA PROJECT AND THE QUESTION OF LARGE DAMS: THE DILEMMA OF DEVELOPMENT IN SOUTHEASTERN TURKEY

Eddie J. Girdner

INTRODUCTION

The Southeastern Anatolia Project Gunay Anadolu Projesi (GAP) is the largest state development project in the history of the Turkish Republic. The plan includes 22 dams and 19 hydroelectrical plants and is projected to provide irrigation for 1.7 million hectares in the Tigris and Euphrates river valleys when completed. The major dam-building phase is likely to be completed in 2010, according to the Master Plan. There are major dilemmas both in terms of fulfilling the technical promises of the project and in terms of wider socioeconomic development in Southeastern Anatolia. GAP is projected to promote industrial development, creating some 3 million jobs in the region and spurring broader socio-economic change. Clearly, a major thrust behind GAP has been the attempt to deal with the problem of Kurdish separatism. The further success of the project, beyond water management, irrigation for agriculture, and the production of electrical power, may depend upon greater political decentralization and regional autonomy. Land reform is needed as a basis for reforming the feudal social relations in the Southeast region.

On the other hand, the state has continued to rely upon local tribal chiefs and landlords who are generally the most socially and politically conservative among the population. Without land reform, the project may tend to strengthen feudalism and increase existing inequality. The latter considerations may ultimately prove to be the Achilles heel of the project, in spite of the massive state capital investment of about $32 billion, over some 30 years. It has been estimated that the dam will not pay for itself in terms of the benefits of electrical power and agriculture before 2045. Benefits from the production of electricity and from agriculture are estimated at $1.8 billion a year between 2000 and 2010 and $3.3 billion per year after all electrical generators and irrigation facilities have been completed (Yackley, 2001). The Turkish Government argues, however, that the project will provide a broad foundation for more comprehensive development, including the development of human factors and a sustainable pattern of development. It is claimed that the project has realized a 51 per cent return on investment ratio.

This article will first look at some broader issues of large dam projects in the context of changing perspectives on development. After a brief historical overview of water development in the Tigris-Euphrates Valley and some legal and political considerations, the article will provide an overview of the Southeast Anatolia Project. Finally, the article will evaluate the project on a number of critical dimensions, in light of recent recommendations of the World Commission on Dams. This article argues that the historical attempt to contain Kurdish separatism since the early decades of the Turkish Republic has contributed to the backwardness of the Southeastern region. GAP was a late response to the need for development in the area and the feudal social relations, which have threatened the success of the GAP project. It appears that the large state investment in infrastructure will continue to benefit the west of the country disproportionately. While the project receives much effusive praise as a grand achievement in Turkey, and has provided a psychological lift to many Turkish citizens, whether the massive investment of state funds represents the best use of development funds remains highly debatable among a broad range of analysts (Mortan, 1998).

WATER, LARGE DAMS AND DEVELOPMENT

A number of established notions about dams have been challenged in the report of the World Commission on Dams (McCully, 2001). The World Commission on Dams (WCD) set out basic guidelines, such as the opening to public debate of the performance, benefits, and negative impacts of dams; and the goal of "equitable and sustainable development" (Parasuraman and Sengupta, 2001, p. 1881).

The WCD report notes that while dams have been used for thousands of years, the large dams built in the twentieth century, almost one per day on average, are qualitatively different as they are associated with development and modern economic progress. These dams, in fact, destroy traditional communities. The thrust for pursuing large dam projects came from the US Army Corps of Engineers, first in the US and Europe. This approach was then extended to development projects around the world.

There are more than 45,000 large dams in the world and nearly one-half of the world's rivers have at least one large dam, defined as a dam at least 15 meters high or 5-15 meters high with a reservoir volume of more than three million cubic meters. Three-quarters of all large dams are in only five countries. These are China, the US, India, Japan and Spain. Two-thirds of large dams are in developing countries with 45 per cent in China alone. Large dams generate one-fifth of all electrical power in the world (Parasuraman and Sengupta, 2001, p. 1881) with one-third of all countries getting half of their electricity supply from such dams. Some one-half of large dams are built exclusively or mainly for irrigation, irrigated agriculture accounting for 12 to 16 per cent of world food production. These figures suggest that the world would be significantly poorer without large dams. However, there might have been cheaper and better ways to produce this energy, food and fiber (Allan, 2002). In many cases large dams have not, in general, proven to be cost-effective.

Dams promote urbanization, population growth, modern agriculture, and industrialization. They are a part of the public infrastructure, which facilitates the accumulation of capital, but they represent the pattern of growth envisioned by planners and engineers rather than a decentralized pattern of development (Parasuraman and Sengupta, 2001).

Dam development peaked in the 1970s with some two or three large dams being commissioned per day. Since then there has been a decline and movement toward the decommissioning of old dams in North America. In the US, dams on Butte Creek in California and dams in the Florida Everglades have been decommissioned (Reisner, 1997). Some 467 dams have been removed to date in the US, 28 of them large dams. This decline is also seen in the developing world. Four major factors have contributed to the trend to seek alternatives to large dams: shifts in national economic priorities; the drying up of financing; rising conflicts around social and economic issues; and the decline in large-scale state-driven public infrastructural projects (Parasuraman and Sengupta, 2001, p. 1882).

In earlier periods, the benefit of dams was seen as "self-evident," with no need for post-facto review and public accounting. Environmental considerations were largely ignored but eventually displaced and affected groups began to organize and the professed objectives of dams, in relation to actual performance, became an issue. Consequently, opposition to big dams has emerged in many countries.

On the global level, the debate over dams has become polarized. The downside of dams include the debt burden; cost over runs; displacement of residents and their impoverishment; destruction of ecosystems, sometimes including fishery resources; and the inequitable sharing of costs and benefits. Until the last few years, opponents of large dams were largely crying in the wilderness, with the political and economic momentum on the side of proponents. Advantages of dams include added power and water resources, favorable public opinion, enhancement of state power, profits and prosperity for the dam industry, and more business for multilateral banks. Opponents have generally been accused of stalling progress and trying to protect minority groups over national interests. On the surface, it was easy for states and firms to tout the benefits of dams as the alleged disadvantages could be argued away. Those negatively affected were generally the marginalized with little political representation. This is, in fact, a more general pattern seen in most environmental struggles against large development projects that have major ecological consequences and where local communities become sacrifice zones (Girdner and Smith, 2002).

With the global rise of environmentalism, green movements emerged along with human rights campaigns, networking, and research, and these gained a good deal of public support. The global debate today revolves around development strategies and choices, and the question of what best promotes the public good. Scholars have moved to consider the broader issues and impacts of disrupted river flow; uprooted existing settlements; destruction of culture; destruction of sources of livelihood; depletion and degradation of environmental resources; whether dams are the best use of public funds and resources; the track record of dams; and what direction to take in future.

The larger debate hinges fundamentally on the gap between promised benefits and actual outcomes. To some extent, these can be quantified and measured. Secondly, the challenge of water and energy development, from the perspective of "nation-building and resource allocation," is important (Parasuraman and Sengupta, 2001, p. 1882). Proponents argue that dams have performed well in over 140 countries. Opponents say that there are better, less costly, and more benign options for meeting water and energy needs, but these have generally been ignored. These options include small-scale decentralized water supply; small-scale electricity options; large-scale end-use efficiency; and demand-side management options. Demand side management includes the use of more efficient appliances and energy conservation. Critics argue that these options may be cheaper, more sustainable, and more equitable. While common ground has been found, there are "still deep fault lines between critics and proponents" (Parasuraman and Sengupta, 2001, p. 1882).

The fault lines emerge around financial, economic, social and environmental issues similar to other big development projects, with proponents arguing that negative social and economic impacts can be dealt with through compensation and restoration. The opponents say that this has not succeeded because the impacts are simply too large (Parasuraman and Sengupta, 2001, p. 1883). It is widely recognized today that community participation is necessary for successful watershed management (Kolavalli and Kerr, 2002).

HISTORY OF WATER DEVELOPMENT IN THE TIGRIS-EUPHRATES VALLEY

Irrigation has been practiced on the rivers in Iraq for thousands of years. In the early part of the twentieth century, the Hindiya Barrage on the Euphrates River in Iraq was completed in 1911 for flooding, and water canals. The water schemes in Turkey can be traced back to Mustafa Kemal Ataturk in the 1930s. The Electrical Works Studies Agency was established in 1936 (Akmensoy, 1996) and the State Hydraulic Works Administration (DSI) was set up beginning in 1953. From 1950 to 1960, there were drilling works on the Euphrates and Tigris rivers. In the 1950s, Iraq built another dam, Ar-Ramadi, for flood control and irrigation. The Samara Barrage on the Tigris River was built in Iraq in 1955. The Tabqa High Dam in Syria was begun in 1966 and completed in 1973, later named al-Thawrah (The Revolution). The Keban Dam, the most northern dam on the Euphrates River, was started in 1966 and completed in 1974. On the Euphrates River, the Karakaya Dam was built in 1986; the Ataturk Dam was started in 1983 and completed in 1992 (Kaya, 1998). The fourth major dam on the Euphrates River in Turkey is the Birecik Dam.

THE HYDRO-SECURITY COMPLEX

Only the most critical international problems in the Tigris-Euphrates Basin can be briefly mentioned here. While the watershed should be seen as linked together in terms of national security, in reality, the dams have created international tensions. A "regional security complex" is defined as "a group of states whose primary security concerns link together sufficiently closely that their national securities cannot realistically be considered apart from one another" (Schultz, 1995, pp. 91, 97). The Tigris-Euphrates river Basin should be seen as this sort of regional security complex.

Historically, Iraq had developed 58,000 hectares for irrigation by 1917. This was increased to 1.44 million hectares by 1950 and 2.6 million hectares by 1992. By 1992, Syria had developed some 279,000 hectares for irrigation. Turkey was relatively late, developing only 65,000 hectares by 1992. Future plans in Iraq, under the former Baathist Regime, called for the development of up to four million hectares for irrigation. Plans for irrigation in Syria included up to 358,000 hectares; The Turkish Government has plans to irrigate another 1.7 million hectares. Until the end of 2004, GAP had produced 222,617 hectares of arable land with irrigation facilities for another 123,500 hectares under construction.

The major potential for conflict is over the Euphrates River. The likely future pre-emption of river water by Turkey has been put at 16.9 billion cubic meters (BCM). This is 55 per cent of the flow as it leaves Turkey. Syria's plans call for using four BCM. Planning in Iraq, under the previous regime of Saddam Hussein called for the use of 17 BCM. As the average annual flow at Shatt al-Arab is 32.7 BCM, this would leave a shortage of some five BCM. Further, the flow is extremely uneven over the years. Past conflicts have involved protests by Iraq when Syria was filling the Tabqa Dam (Lake Assad) in 1974 and between Syria and Turkey, when Turkey began filling the Ataturk Dam in 1992.

The tri-national Joint Technical Committee (JTC), established in 1980 between Turkey, Syria, and Iraq, produced no tangible results and because of the lack of a negotiated settlement, the World Bank and other international lending agencies have been unwilling to extend credit for GAP. As there are no binding international laws on the resolution of international water disputes, the resolution must be by treaty. While the principle of "equitable and reasonable utilization" was provisionally accepted in 1987, this is subject to varying interpretations such as geophysical and socio-economic characteristics of the riparian states, optimality, prior appropriation, and mutual impact of existing and potential usages.

There has also been a wide gap between the positions of the riparian states. Turkey wants both the Tigris and Euphrates rivers to be treated as "transboundary" rather than "international waters." Syria and Iraq have argued that the rivers should be treated as "international waters" to give legitimacy to their claim to "share" the waters. Turkey prefers "allocation" on the basis of negotiation. For Turkey, "the allocation should be based on the assessment of water needs for irrigation, that is, on economically developable irrigation of the riparian states and on optimal irrigation methods. For Turkey, the Euphrates and the Tigris form a single basin, rather than separate basins, and so the "allocation" of waters among the riparian states should be on the joint total flow of the two rivers, rather than on the flow of each river considered separately (Mutlu, 1996). In the absence of an international agreement, Turkey has forged ahead on its own.

THE SOUTHEASTERN ANATOLIA PROJECT

In Turkey, the developmentalist state mentality and centralization remains strong. While projects such as GAP are becoming sunset industries in developed countries such as the United States, as mentioned above, they are still seen as part of the national agenda in many developing countries such as Turkey, which still looks to the state to ensure development. The historical developmentalist agenda is especially salient in relation to GAP.

In Southeastern Anatolia, the issue of Kurdish separatism has historically been the underlying issue, rather than the issue of development *per se*. "...GAP is first and foremost a product of concerns over internal problems... Turkey's objective with GAP is to fundamentally transform the socioeconomic environment of the region, and to create a climate less hospitable to future secessionist movements" (Mutlu, 1996, p. 59). It became clear that poverty in the region was making it easier for the Kurdish Workers Party [Party Karkaren Kurdistan (PKK)] to recruit new members. Consequently, the Kurdish secessionist insurgency in the 1980s put pressure on the Turkish Government to accelerate development in the area. Secondly, there was strong bias at the state level against the concept of regional planning because it was considered to have implications for the Kurdish question (Mutlu, 1996). Third, the first five-year development plan "set forth the principle that resources should be invested in the most productive areas, and that projects in the backward regions should be preferred over other areas only if they were at least as beneficial as the others with respect to economic investment criteria" (Mutlu, 2001, p. 110). This approach contributed to the differentiation of the economic landscape into a core and a periphery (Mutlu, 2001). Disparity between the Southeast and the rest of Turkey has since grown worse, as explored below. Fourth, while the construction of the GAP project, the biggest project ever attempted in Republican Turkey, has poured billions of dollars into the Southeast, hydro-electrical power from the project has been consumed primarily in the west of the country. Fifth, the second phase of GAP for the primary purpose of expanding irrigation facilities in the Southeast of Turkey, has been badly handicapped by the language problem. This may change with new laws for cultural rights for Kurds, passed in 2004. Instruction was not allowed in the Kurdish language up until 2004. Some have argued that there is no empirical basis for the claim that Kurds have been economically exploited, "...[t]he pattern was produced entirely by actors behaving in conformity with economic logic. The process was impersonal and, at the aggregate level, unintentional" (Mutlu, 2001, pp. 113-114). Nevertheless, it could be plausibly argued that the decision to invest development resources in the West was as much a political decision as an economic one and contributed to the out-migration of population from the Southeast to population centers in the west.

The utility of large dams, such as those on the Tigris and Euphrates Rivers in Turkey, is widely recognized in terms of water management. Regulation of the water is desired by states when there is much seasonal variation of flow. This volume ranges from 150-200 cubic meters per second in the summer up to 5000 cubic meters per second in the spring floods. Turkey points out that the dams provide more stable flows. Some of the dams are termed "after-bay dams" or regulation dams. These dams regulate the water released from large-scale dams, which is highly variable. The variation from the Ataturk Dam is from 2000 cubic meters per second to 200 cubic meters per second, depending upon how many of the eight electrical generating units is being operated at any one time. The Birecik Dam is an after-bay dam for the Ataturk Dam and the Al-Baath Dam is an after-bay dam for the Tabqa Dam in Syria. The Euphrates River is some 3000 kilometers long while the Tigris River is 1800 kilometers long.

SCOPE OF THE GAP PROJECT

As mentioned above, GAP is the largest and most ambitious project ever carried out by the state in Turkey. Planning for the project actually got underway in the 1960s, with the objective of addressing the Kurdish Problem, producing electrical power, and providing irrigation. Subsequently, the "integrated development project" was introduced to encompass the comprehensive development of the entire region, including industry, transportation, mining, telecommunications, health, education, tourism, urban and rural infrastructure, and the "human factors of development" (Akmensoy, 2001; Unver, 1994). The 1989 GAP Master Plan was prepared by the Japanese Nippon Company. The total cost of S32 billion, as claimed on the official website, is somewhat misleading, as the estimated real cost, over the life of the project, has been estimated to be around $128 billion dollars. To complete the project by the projected date of 2010, Turkey needs to put at least $2 billion a year into the project. This would almost double the rate of spending so far, which has been around $3 million a day (Yackley, 2001). In 2002, GAP received 5.4 per cent of the total state budget of Turkey, a decrease due to the 2001 economic crisis. Over the period 1990-2001, the project has absorbed, on average, about 7 per cent of the budget. The GAP enterprise covers nine provinces, encompassing about 10 per cent of the total land and about 10 per cent of the population of Turkey (6.1 million people: 64 per cent urban, 36 per cent rural) in the provinces of Adiyaman, Batman, Diyarbakir, Gaziantep, Kilis, Mardin, Siirt, Sanliurfa, and Sirnak. The population of the GAP area is about 66 per cent Kurdish, 25 per cent ethnic Turks, and 10 per cent Arab. There are numbers of Assyrians, as well. About two-thirds of the population is engaged in farming (Hillel, 1994; Mutlu, 1996). The development area covers some 73,000 square kilometers.

The GAP area includes the least developed part of Turkey. As noted, the project envisions bringing about 1.7 million hectares of land under irrigation to add to Turkey's current 8.5 million irrigated hectares. The southeastern area has an urban population of 56 per cent and rural population of 44 per cent, compared to an urban population of 59 per cent and rural population of 41 per cent for the entire country. The population growth rate in the region, of 3.4 per cent, is considerably higher than the national average of 2.4 per cent. Officially by 1994, 95 per cent of villages in the area had electricity and 98 per cent were linked to roads. Some 70 per cent of rural settlements had water supply. In 1990, the per capita GDP of the region was 47 per cent of the national per capita GDP. While the Southeast makes up about ten per cent of the area of Turkey, it contributed only 4 per cent of total GDP and 2 per cent of national value added in the industrial sector, in 1994. The Tigris and Euphrates rivers provide 28 per cent of the entire national water supply (Unver, 1994).

GAP also set a goal of increasing the population of the southeastern area to nine million by 2005. This was to be achieved by attracting people from outside the region. As noted, the pattern of development in Turkey has resulted in the economic marginalization of the Southeast since 1935 and per capita regional income, as a percentage of per capita national income, has fallen. However, there was a marginal increase from 1935 to 1960. The figures are as follows – 1935: 72.7 per cent of per capita national income; 1960: 85.8 per cent; 1965: 64.4 per cent; 1975: 64.2 per cent; 1980: 52.4 per cent; 1985: 49.0 per cent; 1990: 53.2 per cent; 1994: 53.8 per cent. Total regional income as a percentage of national income has also fallen. The figures are as follows: 1935, 4.18 per cent of national income; 1960, 5.02 per cent; 1965, 3.81 per cent; 1975, 3.97 per cent; 1980, 3.23 per cent; 1985, 3.27 per cent; 1990, 3.79 per cent; 1994, 3.60 per cent (Mutlu, 2001, p. 102). Among other countries, only Brazil has a higher regional disparity. The population-weighted coefficient of regional per capita incomes was 0.5812 in 1965 and 0.5892 in 1986. The average per capita income ratio from the richest to the poorest region was: 5.54 (1975); 9.76 (1980); and 14.68 (1986). This illustrates the rapidly widening gap between the Southeast and the rest of Turkey in spite of the GAP development efforts (Mutlu, 1996, p. 60).

THE MAJOR GAP PROJECTS AND STATUS

The GAP Regional Development Administration (RDA), a ministerial body with offices in Ankara and Sanliurfa, was set up in 1989 to oversee the project (Yackley, 2001). However, all projects must be finalized by the GAP Higher Council, consisting of the Prime Minister or a state minister in charge of GAP, the state minister in charge of the State Planning Organization, the Minister of Public Works, and the Minister of Agriculture and Rural Affairs (www.gap.gov.tr). This arrangement means that the administration remains highly centralized.

As noted above, 22 major dams and 19 hydroelectric projects are included in the project. Including the smaller projects, some 80 dams, 66 electrical plants and 68 irrigation systems are projected to be completed (Schultz, 1995, p. 99). The largest of the dams, the Ataturk Dam completed in 1992, is the sixth largest dam in the world in terms of volume. The dams include:

ON THE EUPHRATES

Keban Dam: Completed 1974, but officially not a part of GAP. The lake is 640 sq. km. with a storage capacity of 30 BCM and it is primarily for electrical production. Its capacity is 1240 megawatts feeding Ankara and Istanbul.

Karakaya Dam: Completed 1988. The lake is 300 sq. km. has an installed capacity of 5300 megawatts of power. By 2001 it had produced 107 BKWH of power (cumulative).

Ataturk Dam: Completed 1992. The lake is 877 sq. km. Cost was more than $3 billion. It irrigates some 200,000 hectares around Harran. Storage capacity is 48.4 BCM. Provides 10 BCM a year for irrigation. Installed capacity is 1800 megawatts. By 2001 it had produced 75 BKWH cumulative power. Attached to this is the Sanliurfa Hydroelectric station with a capacity of 50 MW power.

Birecek Dam: On the Euphrates at Birecik. Installed capacity is 672 megawatts of power. By 2001 it had produced 1.6 BKWH of power, cumulative.

Karkamis Dam: Has an installed capacity of 190 megawatts.

ON THE TIGRIS

One-third of the water of the Tigris (5-7 BCM per year) is pre-empted by Turkey. The GAP plan projects some 2200 megawatts of power generation at completion. These dams are projected to irrigate some 550,000 hectares.

Kralkizi Dam: Installed capacity is 94 megawatts of power. By 2001 it had produced 0.3 BKWH power.

Dicle Dam: Installed capacity is 110 MW. By 2001 it had produced 0.1 BKWH of power.

Batman Dam: Installed capacity is 198 megawatts. In operation but not yet completed.

Silvan Dam: Projected capacity of 150 megawatts. Status: Not yet constructed. Proposal being studied.

Kayser Dam: Projected capacity of 90 megawatts. Status: Not yet constructed. Proposal being studied.

Garzan Dam: Projected capacity of 90 megawatts. Status: Not yet constructed. Proposal being studied.

Ilusu Dam: Projected capacity of 1200 megawatts. Status: Not yet constructed. Planned.

Cizre Dan: Projected capacity of 240 megawatts. Status: Not yet Constructed. Planned.

POWER PRODUCTION

By the end of 2004, cumulative energy production from GAP reached 234 million kWh equivalent to 14 billion dollars worth of energy. With around 12 billion kWh being produced by GAP in a year, from a total installed capacity of 7476 megawatts, this represents about half of all hydraulic electrical production in Turkey and about a quarter of total electrical production. The projected annual power production from GAP, when completed, is 27 BkWh per year. It is projected that Turkey's electrical usage would rise to 290 billion kWh in 2010 and 547 billion kWh in 2020. Future production from GAP might allow power to be exported to neighboring countries (www.gap.gov.tr).

IRRIGATION

At the end of 2004, 222,617 hectares of land was under irrigation from GAP dams. This represented only 13 per cent of the planned irrigation capacity. Another 7 per cent of capacity is under construction, 25 per cent at the contract stage, and the other 55 per cent is planned. The value of irrigation realized until the end of 2001 was about $3 billion.

The large Sanliurfa irrigation works conducts water to the Harran Valley through a 24 km tunnel. In the Harran Plain, south of Sanliurfa, cultivable land has tripled and agriculture output per hectare has doubled (Yackley, 2001). The Turkish Government plans to increase irrigated areas by 50 per cent, increase the gross regional product by fourfold, and double per capita income in the region (Unver, 1994). About 20 per cent of all irrigable land in Turkey is in the GAP region.

INDUSTRIAL ZONES AND TRANSPORT INFRASTRUCTURE

Industrial development in the region is a major goal of GAP. Five organized industrial zones (OIZ's) have been established, three in Gaziantep, one in Sanliurfa, and one in Kilis. About 550 enterprises have been established in these zones, employing 43,500 persons. There are 18 small industrial sites, which have been established, including six in Diyarbakir, six in Gaziantep, three in Sanliurfa, and three in the Batman-Siirt-Mardin region. These include about 5500 enterprises with 33,000 employees (www.gap.gov.tr).

Another dimension of GAP is the development of the transportation infrastructure. While a wide range of main arteries and village link roads are included, a major project has centered around the 196 km. Sanliurfa-Gaziantep Motorway. A $53 million loan was obtained from the US to finance this project. The seven airports in the region are being upgraded and the GAP International Airport is under construction north of Sanliurfa.

EVALUATION OF GAP

Similar to other large dam projects, GAP is conceptualized as necessary for water management in the Euphrates-Tigris basin and for development in Southeastern Anatolia. The major dams on the Euphrates in Turkey, however, have not been without controversy. Especially the Birecik Dam came under international protest because it inundated major historical and archaeological sites, the ancient city of Zeugma. A similar debate arose over the Ilusu Dam on the Tigris River, which will inundate the ancient site of Hasankeyf. From the perspective of the state, on the other hand, these dams are a necessary and integral part of water management in the basin. They produce a significant portion of the electrical power for Turkey and are modernizing agriculture in the area, particularly cotton and fruit production.

It is still too early to evaluate the long-term results of GAP. In Sanliurfa, in the wide, flat plain near the Syrian border, the production of cotton and maize has greatly increased with much improved yields. While this development is producing a great deal more wealth, the feudal system of land tenure is still largely in place. Although there is some effort to give plots of land to the landless, genuine land reform is lacking. Further, there are problems in gaining the cooperation of local landowners in the proper use of water. Farmers tend to believe that the more water used, the better.

Consequently, much of the water is not being used efficiently. In some areas, overuse of water has damaged crops. The most serious problem with over-use is salination, which could render the land infertile in future. While much of the local labor is currently absorbed by the hand tilling and picking of cotton, mechanization and greater use of chemicals in future, will pose further serious problems. The toxic runoff will affect downstream waters. Moreover, the area is extremely socially and politically conservative. Production of cotton in such a land tenure system is not likely to be conducive to dynamic economic development and progressive social change. The need to elicit the cooperation of landowners in sound farming practices is important, but to date this has not been very successful in terms of irrigation methods. Many drainage canals are becoming clogged with the growth of weeds and resistance to change is seen in the practice of burning straw off wheat fields, which farmers are continuously advised against.

It is interesting to compare Sanliurfa to Gaziantep some 135 km to the west. Gaziantep is much richer in both trade and agriculture. Pistachio nuts and olives are produced on relatively dry and hilly land and the crops are well cared for. The people of Sanliurfa sometimes say that Gaziantep does not need GAP, as the people are already wealthy, but it is clear that part of the basis of that wealth is a more socially progressive attitude and greater economic dynamics. It can be argued that the lack of land reform in the GAP region has been a major failing in terms of laying the foundation for economic development and, along with terrorism, has tended to be another factor in discouraging investment in the area. Some argue that greater local autonomy and more privatization is essential (Yackley, 2001).

A serious contradiction in the GAP project is that Turkish Governments have relied upon the loyalty of the landlords (*agas*), since the early days of the Republic, who are extremely socially and politically conservative, to help stem the threat of Kurdish separatism. GAP continues this historical policy of relying upon the local feudal notables (*agas*) who support the state but resist progressive social change. The efforts launched for "women's education" in GAP seems to be largely a token effort. Indeed genuine progressive social change is not welcomed by certain sections of the local conservative population, particularly men. The GAP project itself represents top down, centralized, control, whereas genuine development could possibly bring pressure for democratization. It appears that the conservative land tenure system is a serious constraint on achieving the major goals of the GAP project. So far, GAP has not created the conditions for more equitable circulation of wealth and democratization (Yackley, 2001).

Consequently, regional socioeconomic development, a major rationale of GAP, is problematical. While the project is officially scheduled to finish around 2010, it appears that 2025 is a more realistic date. At the rate of construction observed in the mid 1990s, however, it has been estimated that it would take another 137 years to complete the irrigation works alone. Under strict economic criteria, a number of the planned irrigation facilities would never be built. For example, 15 out of 17 irrigation projects cannot be met on economic criteria, based on an internal rate of return of 12 per cent return. There are additional other reasons why these projects may not be justifiable. *First*, the Master Plan assumes a land use intensity of 134 per cent, which is clearly unrealistic. *Second*, the plan assumes that on-farm works will have been completed by the time irrigation starts, which is also unrealistic. The capacity of the related Government agency, dealing with the land preparation, had a capacity of only 40,000 hectares per year and so cannot possibly proceed parallel with the irrigation works. This situation is very serious in terms of the likely resulting damage to the land. The land must be properly leveled for economic water usage, to prevent accumulation of salts in the soil. With gravity flow, as is presently used in Harran, salinity will build up rapidly and productivity will fall, making current assumptions irrelevant.

Third, the plan assumes that farmers will have been trained in modern farming methods and proper irrigation techniques and that a properly functioning extension system will be in place. This is implausible because of difficult personnel problems, the extremely dispersed rural settlement pattern, and the language problem. The training has so far been given in Turkish, while the mother tongue of 75 per cent of the population is a language other than Turkish. *Fourth* the problem of erosion is a major concern as the rate of silting of the dams is very high. It was found that in Kartalkaya Reservoir, the rate of accumulation of silt was 15 times higher than originally estimated. What is needed is massive works for afforestation of the area, but erosion control works is very low. In a typical year, there were only 1700 hectares of erosion control completed and 550 hectares of afforestation. Related to this problem is that of grazing on public lands. This must be strictly controlled, but this is nearly impossible due to the "sacrosanct" nature of traditional grazing rights. *Fifth*, is the constraint of extremely skewed land distribution in the absence of land reform? As irrigation expands, a "bimodal" form of land exploitation will likely develop. This promises to intensify social conflicts, rather than lessen them. Location, climate, topography, social structure, and birth rate are all major factors, which contribute to the backwardness of the region (Mutlu, 1996 and 2001).

Additionally, prospects for industrial development in the area are also problematical. Previous "incentive programs" to attract capital investment, have been "singular failures." The major reasons, in addition to terrorism and economic crises in Turkey, are that firms can reap protective tariffs, or rents, where they are currently located. With the improvement of roads, the area is likely to be increasingly supplied from the core. Secondly, firms can be more competitive in the core. Third, there is a lack of network of producer services, except in Gaziantep. Fourth, there is little capital accumulation in the area. Finally, there is a lack of skilled workers and need for trained managers. Capable managers, however, hesitate to relocate to the area due to lack of education facilities and the conservative local culture (Mutlu, 1996). Overall, the prospect for the scope of development envisioned by the GAP project is not great. Furthermore, the allocation of necessary funds is not likely due to economic problems under deepened IMF austerity following the February 2001 economic crisis in Turkey.

As a result of the realization of these problems and many studies, which have been conducted in the area, a number of projects have been organized around the goal of "sustainable human development." These social projects were launched with the GAP Social Action Plan in 1994. Among these projects are the CATOM (Multi-Purpose Community Centers), intended to increase local skills and raise incomes, provide medical services, and provide social and cultural activities for local women. Some 15,000 women have participated to date. Another project, to resettle and assist people displaced by the Birecik Dam was carried out in the late 1990s. Another project, Social Progress for Youth, began in 2001 funded by the Swiss Government and the United Nations Development Program (UNDP). This project has established youth centers to promote social, cultural and sports activities. Another project called "Rehabilitation of Children Working in Streets" was opened in Diyarbakir and Adiyaman (www.gap.gov.tr).

Another project, "Enhancement of Employment and Capacity in Urban Informal Sector," targeting the rural poor, was begun in 2000 in Diyarbakir in cooperation with the UNDP. The GAP Region Public Health Project" is designed to assess the public health effects of likely ecological and environmental problems resulting from the change to irrigated farming in the region. The UNDP has also participated in the "Program for Sustainable Development in the GAP Region" since 1997. The 28 sub-projects include social services such as education, health, housing; gender equality; urban management; environmental problems; and strengthening civil society organizations, with a small budget of 5.2 million US dollars (www.gap.gov.tr).

Other projects in the region include the "Sanliurfa-Harran Plains On-Farm and Village Development Project" and the "GAP Urban Planning and Sanitation Project." These projects are funded with a $650,000 grant from the World Bank. Additionally, the "GAP-FAO Rural Development Program" involves participation between the GAP administration and the United Nations Food and Agriculture Organization. The "GAP Regional Development Program" is funded with a grant from the European Union of around 50 million Euros. Other projects and "training programs" are being carried out in the GAP area with other countries, such as Bangladesh,

Bosnia, Canada, China, Egypt, France, Israel, and Macedonia. A number of US universities are also involved in projects and training programs in the area (www.gap.gov.tr).

DOWNSTREAM AND ENVIRONMENTAL CONCERNS

The international dimensions of GAP are also relevant in terms of the greater Tigris-Euphrates water basin. Kolars (1994) has looked at the implications of GAP for the ecology of the entire Tigris-Euphrates Basin over the next 50 years of development. Addressing such issues depends upon nations being able to take a supra-national point of view. It should be understood that hydrological development is likely to go forward, in spite of the protests over dams. The concept of "sustainable development" is relevant. However, it is difficult to take into account the possible impacts of fertilizers, insecticides, and herbicides from large-scale agriculture. In terms of the shared impacts of each of the three riparian states in relation to the ecological degradation of the Persian Gulf, Iraq's impact has been greatest, but that from Turkey will increase significantly toward 2040 as more land is irrigated. It is clear that "...downstream ecological impacts of dams, reservoirs, and irrigation...can be severe." International cooperation is needed in ecological studies and planning and "upstream reservoirs are absolutely necessary" to regulate the seasonal water flow. The dams in Turkey must be part of any regional management scheme (Kolars, 1994, pp. 129-136). Roughly, in terms of ecological impact, in 2020, the share of each country will be: Iraq 77 per cent; Turkey 16 per cent; and Syria 7 per cent.

The dams in Turkey are the most efficient in the basin in terms of evaporation. This largely depends upon the depth of water in the lakes. But Turkey must practice correct use of fertilizers and insecticides to preserve the water quality downstream. As far as the impoundment of silt that happens in upstream dams, this "does not, of itself constitute a threat to the ecology of the Gulf. A threat is the drainage from the Main Outfall Drain (MOD), which carries roughly 9 BCM. The major danger is the increased salinity, but also the impact of pesticides and herbicides is a problem. These could well destroy the habitat of the fish in the Persian Gulf. Not enough is known, however, about the likely ecological consequences (Kolars, 1994).

PROTEST OVER DAMS

Protest over dams in Southeastern Turkey has involved a number of dimensions, such as environmental impacts, women's rights, cultural destruction, displacement, rehabilitation and political repression. Some 250,000 persons have been displaced in the area of the dams (Hillel, 1994). Exploration of these dimensions is beyond the scope of this article so will only be briefly mentioned.

One dam project that has been the target of much protest is the Ilusu Dam, which is 65 km from Syria on the Tigris River. This is the second largest dam in the whole the GAP project, second to Ataturk Dam and the largest dam on the Tigris. (Dorsey) The project involves the submerging of 183 villages and ancient city of Hasankeyf. Some 78,000 people, mainly ethnic Kurds have been affected.

Turkey has been accused of lacking a consistent policy on the Dams. The British Government, Department for Trade and Industry, demanded four conditions for funding: independently monitored resettlement plans for the displaced, saving as much as possible of the archaeological ruins of Hasankeyf, water flow controls, and water quality standards (Dorsey, no date).

The Ilusu Dam, costing an estimated $2 billion US dollars, is projected to produce 1200 megawatts of electrical power. Finance was arranged mostly from Europe; The Union Bank of Switzerland; and export credit agencies of Austria, Germany, Italy, Japan, Portugal, Sweden, Switzerland, the UK and the USA. The Swedish construction company Skanska, however, pulled out of the program presumably because of the protests. The environmental report in 1997 from Sulzer Hydro was not released to the public. The contract was obtained by Switzerland Sulzer Hydro and ABB Power Generation. Construction was carried out by Britain's Balfour Beatty, PLC, Italy's Impreglio Spa, and three Turkish firms.

A report by Swiss-based Berne Declaration in 1998 stressed "major political, social, environmental and archaeological problems with the proposal." It is clear that the project did not meet international standards for several reasons. It is also claimed that the project violated core provisions of a UN convention, which aims at preventing water wars between states. Also abrogated were World Bank guidelines and OECD's Development Assistance Committee's guidelines. A Kurdish human rights organization called the dam a potential "major human rights disaster."

In the case of Birecik Dam, near Gaziantep, a part of the archaeological finds from the ancient city of Zeugma have been preserved in local museums and a portion of the ancient site is now under some 372 meters of water (www.gap.gov.tr).

Given the argument that the development of the Southeast is likely to fall far short of that envisioned in GAP and that it is not a rich area except for land and water, it has been argued that it is a tragedy that the great wealth of cultural heritage, primarily of the Kurds, is being destroyed primarily for the purpose of producing electrical power for the west of the country. Mutlu has shown that this power could have been produced in the west for only a very marginally greater cost (Mutlu, 2001). While all 19 power plants will contribute an estimated $1 billion per year in terms of oil for energy, the marginal gain for Keban, Karakara, and Ataturk dams is only some $445 million per year in 1993 prices (Mutlu, 2001).

MINORITIES AND WATER CONTROL IN THE MIDDLE EAST

The dynamics of Dam construction problems seen elsewhere have been compounded in the case of GAP, with the area under military rule during much of the 1990s, and given the threat of Kurdish separatism. Therefore, guidelines that have been set out for the desirable approach to water resources development have often not been followed. During the construction of GAP, terrorism and the war with the PKK has resulted in the death of more than 30,000 people. While GAP is intended to bring benefits to the Kurdish population, in significant ways, the Dams have sometimes increased Kurdish resistance among some who see the program as a violation of human rights and destruction of Kurdish heritage (Schultz, 1995).

To a considerable extent, countries in the Tigris-Euphrates Basin demonstrate how water development projects and policies can be directed in negative ways in ethnic conflicts. Turkey has used water development to stem the Kurdish resistance. Iraq used water development against the Marsh Arabs under the regime of Saddam Hussein. Israeli water policy has been used against the Palestinians. In Syria, "local peasants and farmers have not benefited from the Euphrates dams as was expected. Irrigation schemes are not working as intended, and local populations have been forced to accept governments' development strategies without any concern for their demands and wishes." Rich lands overuse water and poor peasants lose crops (Schultz, 1995, p. 116). Saddam Hussein's draining of 57 per cent of the 15,000 square km. of marshland was for the purpose of subduing the south of the country. This has been called eco-genocide or a "hydro-jihad" (Schultz, 1995, pp. 109-110).

In general, a number of these approaches demonstrate some of the worst downsides against dam projects that have been addressed by the World Commission on Dams. Yet the issues are perhaps harder to address in the Middle East due to the nature of politics and the lack of viable and legitimate civil societies. The same applies to Turkey, as the southeastern area experienced an extended period of military rule. Therefore, neither human rights nor environmental dimensions can easily be addressed in the region. The Southeast has been the site of recent wars and ongoing military operations and conflicts. Clearly, then, the obsession to press on with the projects involves more than just developmentalism. For Turkey, as noted earlier, GAP involves national security and overcoming the problem of Kurdish separatism.

THE WORLD COMMSSION ON DAMS APPROACH

The World Commission on Dams has worked to come up with an alternative framework for decision-making on large dam projects. This would establish, "a set of internationally applicable criteria and guidelines for future decision-making on assessment, design, planning, implementation, operation, monitoring, and decommissioning of dams..." Non-dam options for development also need to be explored (Parasuraman and Sengupta, 2001, p. 1883). The WCD approach was meant to be as objective as possible, embracing anti-dam groups, as well as large dam-building industry groups. The study involved at least 1400 individuals from 80 countries. Considered were social and distributional issues, environmental issues, economic and financial issues, options assessment, and institutional processes. One of the ten case studies was the Aslantas Dam in Southern Turkey.

Questions considered included how the actual costs and benefits of dams compared with the original projections; who gained and who lost as a result of the overall project; whether the project complied with the existing criteria and guidelines; how key decisions were made; and whether such a project would be undertaken in the current context. Difficulties were encountered due to the lack of systematic post facto information on many dams, which indicates that there is often a failure to evaluate the performance of dams satisfactorily.

In terms of technical, financial, and economic performance, the results showed that many dams fell short of net benefits, physical requirements, and economic targets. This was particularly true of irrigation dams, compared to hydropower dams, which were generally more economically viable. Also dams for municipal and industrial water supply have generally fallen short of their goals (Parasuraman and Sengupta, 2001). Sometimes dams for flood control have actually led to greater vulnerability to flood hazards. Multi-purpose dams have generally pursued over-optimistic targets and so have greater shortfalls. Also schedule delays and cost over-runs are typical of large dam projects. Some 32 large dam projects had cost over-runs of more than 500 per cent.

Another problem is that as a dam ages, safety problems arise, so that dam safety requires additional investment. Other problems are the change in hydrology with climate change; the high rate of water logging; and salinity of agricultural land (some one-fifth of irrigated land globally). This has severe, long-term, often permanent, impact on land, agriculture and livelihoods. Sedimentation leads to loss of storage capacity (Parasuraman and Sengupta, 2001).

Impacts on ecosystems include loss of forests and wildlife habitat, loss of species population, and degradation of upstream catchment areas. There can also be a loss of aquatic biodiversity and upstream and downstream fisheries. The dam is a barrier to species migration, but sometimes fishing has also been improved. Also the loss of wetlands, riverine, estuarine, and adjacent marine ecosystems has been seen. This is worse when there are several dams on a single river, such as in the Euphrates complex, and can go undetected for long periods of time. There can also be loss of species and diverse ecosystems. Another problem is the emission of greenhouse gases from reservoirs, which affects global climate change. The rotting of vegetation and carbon inflows is sometimes equivalent to that of thermal power plants. Also "it is not possible to mitigate many of the impacts of reservoir creation on terrestrial ecosystems" (Parasuraman and Sengupta, 2001, pp. 1885-1886). In general, the WCD report notes a continued disregard for the environment in large dam projects.

In terms of social impact, there is a considerable range of effects that generally are neither adequately assessed nor accounted for. There are severe impacts on communities that depend upon the river environment and these social impacts have generally not been considered in decisions that displace some 40 to 80 million people worldwide. Many people downstream depend upon fishing and flooding for a livelihood. Many displaced persons have not been resettled or compensated. It has been shown that the larger the community, the less likely that the livelihoods can be restored. Another problem is the lack of a legal framework to enforce compliance (Parasuraman and Sengupta, 2001). The most vulnerable groups generally bear most of the risks, such as indigenous, tribal, pastoral communities, fishers, and flood plain agriculturalists. Often these groups may suffer impoverishment and social decline. There is generally a lack of commitment and lack of capacity to deal with displacement.

Another aspect is the adverse affect on cultural heritage; submergence of archaeological sites burial sites and so on, as in Kurdish areas mentioned above. It is argued that future generations, therefore, bear more costs of dams but do not get an equal share of the economic benefits. Indigenous, tribal, vulnerable ethnic minorities, and women are likely to suffer more through displacement and suffer loss of livelihood, loss of culture, and adverse health, while reaping fewer benefits.

The true economic benefit of dams is not known because of the failure to account for environmental losses, social costs, and possible other beneficial human factors. Sacrifice zones are often created which are not compatible with equity and social justice. Some populations, it is claimed, must be sacrificed for the good of majorities such as the urban populations and elites that matter most in the political equation. In some cases, there is a failure and even refusal of states to enter into a dialogue on these matters. This is clearly seen in Turkey.

The WCD has set out options for water and energy resources development. When dams already exist water efficiency can be improved. The Aslantas Dam study, in Turkey, for example, found that irrigation efficiency was only 40 per cent while field level efficiency varied from 25 to 55 per cent. It is possible to increase surface irrigation to 60 per cent with proper techniques and micro irrigation methods such as drip and sprinkler irrigation, to 70-90 per cent. While this is initially capital intensive, the costs can be recovered over several years.

The major uses of dams are irrigation, power, flood management and water supply and it is not clear that all of these could have been met with traditional techniques, particularly the need for flood control. However, it is also not clear whether four major dams were actually needed on the Euphrates in Turkey. A combination of fewer dams and use of traditional methods might have served the country better. A good deal of the land, for example, where wheat is grown, produces without irrigation. So it remains unclear whether GAP is actually justified in terms of a number of serious considerations.

SOCIAL AND POLITICAL INDICATORS

The GAP project has clearly fallen short to date in the area of adequately advancing social and political change. The statist tradition, the continued weakness of civil society, and the perceived threat of Kurdish separatism in the Southeast are major impediments. But, on the whole, this pattern of statist top-down development is largely accepted in Turkey by the mainstream elites. The World Commission on Dams, on the other hand, has made several recommendations, which encourage more decentralization and grass-roots development.

From a more grass-roots point of view, stakeholder groups, including women and minorities, should have a right to participate in decision-making on dams. The authorities should be required to obtain the free, prior, and informed consent for projects from those who will be displaced or otherwise affected. The local population must have access to information, legal support and financial assistance (Parasuraman and Sengupta, 2001). In addition, there should be comprehensive options assessments for water, food and energy and choices selected from a "range of possible options." Social and environmental aspects should be accorded the same high priority as economic and social factors. The efficient use of existing water sources should also be addressed, along with problems, with the option of even decommissioning dams if the criteria cannot be met. It is important that the health and integrity of the river system be sustained, along with the livelihoods of the people. Benefits must be shared equitably with people both upstream and downstream. There should be binding arrangements that ensure compliance of authorities with regulations, criteria, and guidelines. Finally, rivers should be shared for peace, development, and security. Mechanisms for dispute resolution must be put in place.

For future projects, the WCD has made the following recommendations. There are five critical decision points when water and energy options may be considered. These are needs assessment; selecting alternatives, project preparation, confirming compliance before commissioning, and the project operation. It is a serious and problematical question as to what extent states will be willing to abide by these criteria, which are designed to protect the rights of the local population.

CONCLUSION

In Turkey, there has been a strong impulse for modernization as a result of the emphasis on science and technology by the founder of modern Turkey, Mustafa Kemal Ataturk. This has become a strong pillar of the national secular religion of Kemalism. Civil society, on the other hand, has remained relatively weak while statism remains strong. Further, the notion that the Kurdish population constituted a minority has officially been viewed as anathema and "separatist." This remains the case today. Within this context, the GAP project was launched, in part, to ensure the territorial integrity and national unity of the state in Turkey.

GAP is representative of the state development pattern and intended to transform the Southeast, economically, socially and politically. Nevertheless, today there are major problems and contradictions in the GAP program. The Turkish Government has sometimes failed to adhere to international guidelines in building the dams and has not addressed a number of the fundamental social and ecological issues stressed in the WCD report. It appears that rather than promote democracy and needed grass-roots social change in the region, the project is likely to shore up the power of the traditional landowning elites as well as further erode the fragile ecology of the region.

REFERENCES

Akmansoy, S. (1996), Southeastern Anatolian Project, *http://www.ce.utexas.edu/prof/maidment/grad/akmensoy/gap/paper.html*. (Accessed March 19, 2001).

Allan, J. A. (2002), *The Middle East Water Question: Hydropolitics and the Global Economy*. London: I. B. Tauris.

Dorsey, J. M. (no date), Controversial Dam Plan May Spur G-8 to Action, *Wall Street Journal* (no date), *http://www.eca-watch.org/artedorseywsj.html*. (Accessed March 23, 2001).

Girdner, E. J. and J. Smith (2002), *Killing Me Softly: Toxic Waste, Corporate Profit and the Struggle for Environmental Justice*, New York: Monthly Review Press.

Kolavalli, S. L. and J. Kerr (2002), Mainstreaming Participatory Watershed Development, *Economic and Political Weekly*, 37(3): 225-242.

Hillel, D., (1994), *Rivers of Eden: The Struggle for Water and the Quest for Peace in the Middle East*, New York: Oxford University Press.

Kaya, I. (1998), The Euphrates-Tigris basin: An Overview and Opportunities for Cooperation under International Law, *Aridlands*, (44), Fall/Winter, (online) *http://ag.arizona.edu/OALS/ALN/aln44/kaya.html*.

Kolars, J. (1994), "Managing the Impact of Development: The Euphrates and Tigris Rivers and the Ecology of the Arabian Gulf – A Link in Forging Tri-Riparian Cooperation," in A. I. Bagis, ed., *Water As An Element of Cooperation and Development in the Middle East*, Ankara: Hacettepe University and Friedrich Naumann Foundation.

McCully, P. (2001), *Silenced Rivers: The Ecology and Politics of Large Dams*. London: Zed Books.

Mortan, K. (1998), GAP, *Private View*, 2(6).

Mutlu, S. (1996), The Southeastern Anatolia Project (GAP) of Turkey: Its Context, Objectives and Prospects, *Orient*, (37): 59-86.

Mutlu, S. (2001), Economic Bases of Ethnic Separatism in Turkey: An Evaluation of Claims and Counterclaims, *Middle Eastern Studies*, 37(4): 101-135.

Parasuraman, S. and S. Sengupta (2001), World Commission on Dams: Democratic Means for Sustainable Ends, *Economic and Political Weekly*, 36(21): 1881-1891.

Reisner, M. (1997), Deconstruction the Age of Dams, *High Country News*, www.hcn.org/servlets/hcn.urlRemapper/1997/dir

Schulz, M. (1995), Turkey, Syria and Iraq: A Hydropolitical Security Complex, in L. Ohlsson ed, *Hydropolitics: Conflicts Over Water as a Development Constraint*. London: Zed Press.

Unver, O. I. (1994), Innovations in Water Resources Development in the Southeastern Anatolia Project (GAP) of Turkey, in A. I. Bagis ed., *Water As An Element of Cooperation and Development in the Middle East*, Ankara: Hacettepe University and Friedrich Naumann Foundation.

Yackley, J. N. (2001), The Southeastern Anatolia Project (GAP) and the Imperative of Regional Development, *Journal for Studies on Turkey*, (14): 255-266.

www.gap.gov.tr, GAP, T. C. Basbakanlik Guneydogu Anadolu Projesi Bolge Kalkinma Idaresi Baskanligi, 2000 Yili Faalihet Raporu (2001), Ankara.

16

ASIA'S LARGEST FRESHWATER LAKE VANISHING: PROBLEMS AND REMEDIES

F. A. Shaheen

INTRODUCTION

Lakes and wetlands have inestimable value and are more important and often easily available water resources. Lakes supply drinking water, food, irrigation and recreation. These are cradles of vital biological diversity and provide the water and primary productivity upon which numerous species of plants and animals depend for survival. Preservation of lakes and wetlands is a life necessity for people in many regions of the world, given their cultural, ecological and socio-economic value.

Jammu and Kashmir (J&K) state is famous world over for its lakes and wetland systems, ranging from largest natural freshwater Wullar lake to largest brackish or salt water Pangong lake in Asia. There are about 1248 water bodies / lakes in all the three provinces of state. These lakes and wetlands are known all over world for their unique flora and fauna. Thousands of migratory birds that throng to these water bodies are a special attraction to the tourists. The lakes and wetland system of state is, however, under serious threat because of human greed, onslaught of mindless urbanization and encroachments.

Among all these lakes, Wullar lake has a distinct position and status. It is the largest and probably most picturesque fresh water lake of Asia. The lake serves as an important life support system regulating hydrological regime of the region and as a repository for a variety of avifauna, fish and aquatic life. Wullar lake, in view of its importance as a vibrant heritage aquatic ecosystem and one of the internationally recognized wetlands has been recently brought on the Ramsar sites map. The lake is situated in Baramulla district of Jammu and Kashmir state to the north of the capital city Srinagar. It acts as a flood absorption basin for river Jhelum, the life line of Kashmir.

The lake has a historical, ecological, aesthetic and economic significance. The adverse change in the ecological scenario of the state for last more than five decades has a deleterious impact on the normal functioning of the dynamic ecosystem and consequently lake has witnessed tremendous deterioration in its size, shape, flora, fauna, water quality and the entire lake dynamics. It presents a very dismal picture as also the fact that the water surface area of the lake has been reduced by two third over last few decades from around 273 sq. km to just 100 sq. km as at present (peak period).

The lake sustains the livelihoods of a large section of population of valley. It provides fish, waternuts (*Trapa natans*), lotus (*Nelumbo nucifera*) rhizomes, fodder, timber, water etc. Besides this, it acts as an important flood absorption basin and natural reservoir for power plants located downstream. It also performs vital ecological functions. The lake was declared as a Ramsar site in March, 1990 due to its importance as a permanent winter habitat for waterfowl.

Once spread over 273 sq. km, the emerald green lake has effectively shrunk to 24 sq. km today. If the reclamation process continues with same pace, are we able to see the magnificent lake after four to five decades? Is the Asia's largest freshwater body going to disappear for ever? What are the major problems with the lake and is there any remedy to these? The paper will try to discuss all these problems along with their remedial measures.

METHODOLOGY

The paper is based on the review of past studies/literature and also on some old chronicles as well as the primary observations/visits and author's personal experience/knowledge with regard to lake. The paper will try to put forth and discuss all the problems/major environmental issues with which lake is encountered and will be followed by the probable remedial measures to overcome these problems.

PROBLEMS AND MAJOR ENVIRONMENTAL ISSUES

Wullar Lake is manifested with a large number of problems ranging from anthropogenic, siltation, environmental, catchment to trans-boundary issues. These problems are discussed as follows:

Siltation

The lake has experienced considerable deterioration and decline for the last century, especially, the last few decades and has virtually been reduced to less than half of its size. The principal factor responsible for its shrinking area is siltation. The only perennial source of water inflow into Wullar is through Jhelum river. The river originates from Verinag in south Kashmir and traverses through middle of the valley, drains into Wullar Lake near Banyar after traveling more than 100 kms.

So the major deposit of silt into Wular comes from Jhelum river. The Jhelum basin lies in oval trough between 33° – 25' to 34° – 32'N and 74° – 0' to 75° – 30'E, between Himalayas and Pir Panchal mountain ranges. The river passes through the middle of the Kashmir valley, through the alluvium of its own depositions and the depositions of its catchments. Out of the entire catchment area of Jhelum consisting of 24 major codified catchments, 21 catchments drain directly or through numerous streams and nalas into the river that ultimately drains into the Wullar Lake. The other 3 catchments drain into the Jhelum after its outflow from Wullar Lake. The 21catchments comprising of 123 watersheds, and 204 micro watersheds are predominantly in (their most part) extremely degraded forests and sloping agricultural and horticultural lands where mostly unscientific farming practices are practiced.

The river passes through most of the densely populated towns, including Srinagar and receives the entire thrust of sewage and sewerage disposal from these urban areas. Since, Wullar Lake functions as the settling basin of river Jhelum; it is the recipient of the entire silt load, both the loads in traction and suspension that the river Jhelum carries from its degraded catchment. Unless efforts at stabilization, through afforestation and soil conservation of Jhelum catchment are made in real earnest, efforts at conservation of the Wullar and other lake bodies will be only half hearted an approach.

A catchment area of 9, 61,362 ha of the river Jhelum drains directly or indirectly into Wullar lake. The six directly draining catchments of Gundar, Ningli, Wullar-l, Madhumati, Erin and Wullar-ll comprising of a total catchment area of 1,14,501 ha discharge lakhs of tones of silt in form of

Map 1
Wullar Lake and its Surroundings

load in traction and suspension into the lake. These factors are changing the very dynamics of this live ecosystem.

Shrinking Area

The area of the lake according to recent survey is just 79 sq. kms as compared to 189 sq. miles (272 sq. kms in flood time). This is highly alarming and if things are allowed to proceed as such and no timely remedial measures are taken, the lake will get converted into a piece of agricultural land within around 50 to 60 years time as per the visible rate of shrinkage of lake in past. The average depth of water has considerably diminished.

Table 1
Chronology of Lake Area

Source	*Flood time*	*Normal time*
Kalhana's Raj Tarangni (one of the oldest Chronicles of Kashmir, see Stein, M.A. 1898)	273 sq. kms.	(72 sq. miles or 189 sq. kms)
Other historic chronicles	Not given	12 miles × 8 = 96 sq. miles or 252 sq. kms.
Valley of Kashmir Lawrence (1996)	Not given	72 sq. miles (189 sq. kms)
Directorate of Environment J&K State with the help of Satellite Imageries (see Government of Jammu and Kashmir, 2001)	100-130 sq. kms.	79 sq. kms (Water surfacearea only)

Source: Directorate of Ecology and Environment (2001), J&K.

The satellite imagery of 1996 has shown 11.10 sq. kms wetland area, and 10 sq. kms of water chestnut area in addition to the clear water surface of the lake of 65 kms^2 with maximum depth of over 5.8 mtrs. Evidently, the area of the lake has shrunk by more than half of its size.

Encroachments

The lake body has been facing a lot of encroachment on many fronts. As the periphery of the lake gets silted up, the villagers reclaim the same either by plantation or by paddy cultivation. In this process big villages and human settlements have mushroomed up around the lake or the erstwhile lake body, over a period of time in the past. The lake body within has been brought for almost more than half of its area under Ningali plantations by government departments which if carried unabated, can pose a severe threat to the lake body. It looks as if entire area west of Jhelum river including Hajan has been a part of lake, where big towns/ villages have sprung up. Paddy cultivation is being made, in this area and in the area on the eastern fringes of the Lake.

The Ningli plantations in the middle of the lake on its eastern, southeastern and western part have compartmentalized the majestic lake into small fragments especially in its eastern areas. The main lake body is now limited to western and northern part, free from such big human disturbances.

Innumerable pretty long chains of embankments and roads have been build in the interior of this magnificent wetland disturbing its very configuration, on account of racing human settlements. The population of these villages/towns in the direct catchment of the lake has risen to several lakhs, whose entire pressure is being borne by this lake directly or indirectly.

It needs political will and coordinated legal, administrative and scientific measures coupled with earnest management and conservation strategies on a comprehensive scale, to restore the lake to its pristine glory.

Willow Plantations

Willow plantations have been raised around and within the lake which deteriorate water circulation and enhance siltation besides adding a lot of biomass from its foliage. These plantations *interalia* act as barriers for heavily silt waters, check the velocity of the current and force it to discharge the sediment load completely into the lake, thus causing siltation at an alarming rate. Besides the bio-matter resulting from the leaf and other parts of the plants after death and decay add to the organic content of the

Photo Showing Willow Plantations in the Inside Lake Body

lake, deteriorates water quality and invites weeds. It is high time not only to stop further plantations but to remove them gradually in a scientific manner.

Sedimentation

Hills once compact forests have been reduced to highly degraded areas by wanton deforestation, overgrazing, encroachments, faulty agricultural practices so much so that a single shower of rain is enough to bring down thousands of tonnes of silt and soil into the lake body directly, thereby, posing a severe threat to this vast lake ecosystem in near future in the absence of comprehensive restoration and remedial program.

Map 2
Erosion Intensity Map of Wullar Lake and Surrounding Areas

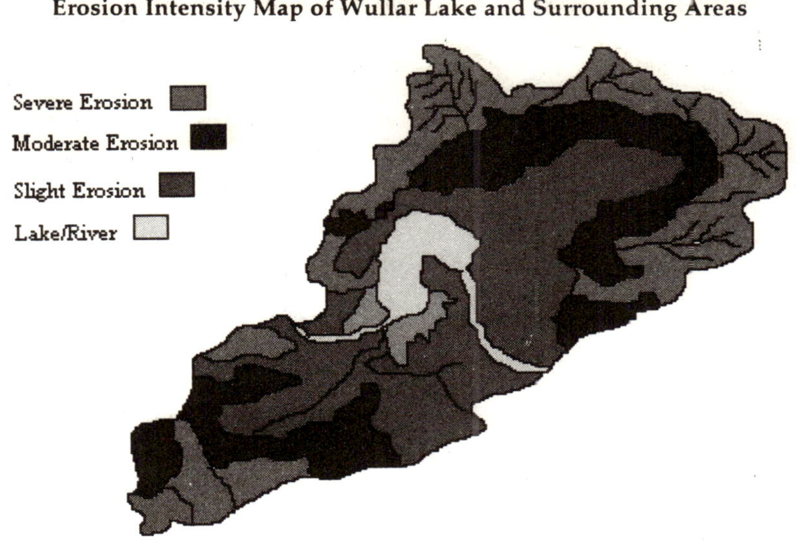

Sewage/Sewerage Disposal

All the human settlements around the lake body have conventional raw sewage and sewerage disposal systems. Most of the effluents from houses are directly drained into the Wullar or her feeding channels/nallahs. Even the municipality authorities of adjoining towns are dumping the tones of refuse, particularly the used polythene bags on the banks of Wullar Lake which poses a great threat to its health. This increases the chloride content of lake water and is an indicative of organic pollution of animal origin. Unless scientific alternatives for the sewage and sewerage disposal are provided to these settlements, the eutrophication of the lake on account of nutrient inflow especially of Nitrogen and Phosphorus will continue at an alarming level.

Photo Showing Dumping of Refuse on the Lake Periphery by Town Municipality

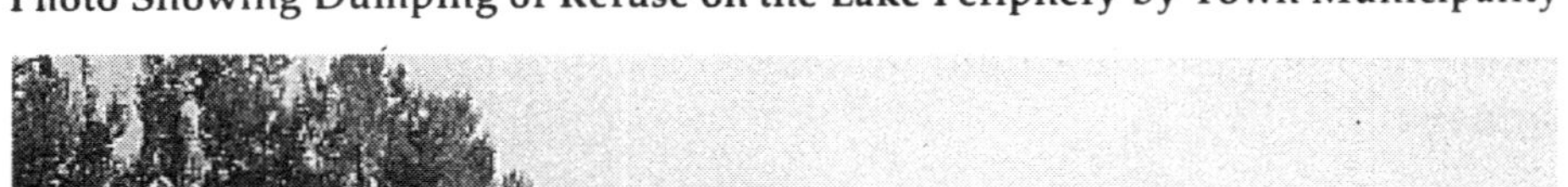

Catchment Problem

Out of the total catchment area of 12,75,696 ha of river Jhelum, 9,61,362 ha drain into the Wullar. More than 55% of the catchment area is demarcated highland forests, predominantly, coniferous. The cold months extend to around six months, with a minimum temperature as low as –12°C. In the event of insufficient alternatives of energy sources, viz., Kerosene, LPG, etc., the forests have been subject to heavy pressure for fuel and charcoal manufacture from the human settlements all around. The root cause of the degradation in the catchment is tremendous increase in population, both human and livestock. Impact of population growth has reflected in the form of excessive pressure on forests, encroachment, over grazing, faulty land use, unscientific agricultural and horticultural farming practices. Unmindful of the ecological and environmental consequences, these forests forming the principal catchment of the lake have been over exploited on account of;

(i) Increased demand for timber to satisfy the greed & need of the people, as the forests were considered only as a revenue generating source.

(ii) Encroachment of forest land.

(iii) Over grazing as the direct pressure on the forests & grass lands increased proportionate to the increased uneconomic cattle population.

Besides, the tilling of slopes across the contours and other faulty agricultural practices, changing the land-use status, excessive use of fertilizers/pesticides in the agricultural/horticultural fields, most of which ultimately washes down into the lake, thereby, changing the chemical composition of lake water. The ruthless deforestation coupled with all afore said adverse factors put together exposed the slopes and enhanced soil erosion to alarming levels. The entire surface runoff with all the slit and sediment accompanying it, finds its direct entry into the lake. Resultantly the lake water quality has deteriorated considerably thereby speeding up eutrophication of the lake. The lake has become shallower by 4 – 5 metres in a matter of just 50 years of time.

Map 3
Land Use Land Cover Map of Wullar Lake

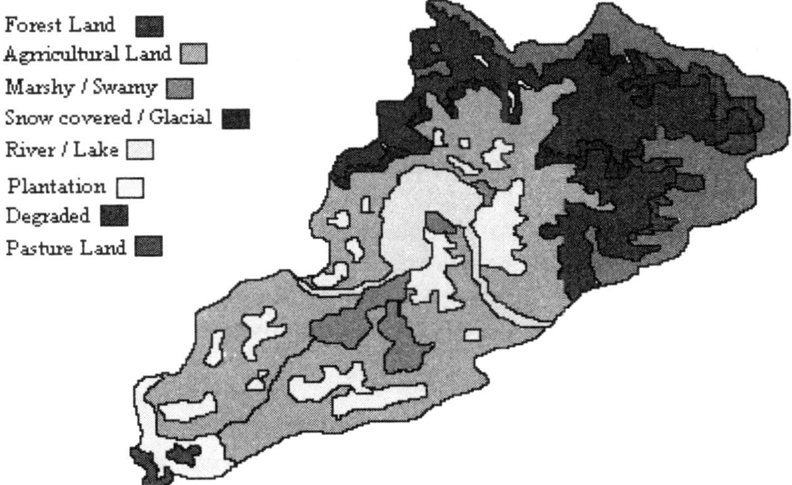

Deteriorating Water Quality

In the past, depth and area of clear water surface of lake was more with better quality. Due to human interferences in the catchment as well as lake body, more particularly deforestation, grazing, use of fertilizers etc, the depth and extent under freshwater has got reduced considerably. As a result, the physico-chemical composition of the water changed considerably. The chloride content of the lake is much higher in comparison to other water bodies which are an indicative of organic pollution of animal origin. Specific conductivity values, ammonical nitrogen, nitrate nitrogen and total phosphorus contents are quite significant and suggests the enriched status of the lake.

Weed Infestation

Siltation/sedimentation and an extraordinary increase in the NPK level have significantly changed the composition of the lake due to invasion of many weeds. The marshy and shallow zones of the lake are infested with weed growth of which more important are *Nymphoides peltat*, *Salvina natans* and *Potomogeton species*; the comparatively deeper invaded by *Ceratophyllum demersum*, *Myriophyllum spicatum* and *Trapa bispinosa*.

Economic Loss

Wullar Lake serves as a flood absorption basin of the river Jhelum, a natural reservoir for irrigation, drinking water and hydro-electricity projects downstream. Any disturbance in the water holding capacity can cause unimaginable loss to the state. Besides, Wullar Lake is a major source of fish, water and chestnuts, nadru etc., therefore, any change in its size and structure will not only cut off these special gift supplies but also jeopardize the livelihood of more than half lakh population who directly depend on the Lake.

Loss of Biodiversity

Wullar Lake is an inseparable component of the ecological and environmental heritage of Kashmir. It is the biggest fresh aquatic ecosystem of India. It is home to innumerable phytoplanktons, zooplanktons, hydrophytes, fishes and other aquatic animals. Besides it acts as a feeding and breeding ground for numberless avifauna both domestic and migratory. It is a heaven for visitors, ornithologists, naturalists, researchers, academicians and scientists. Any ecological degradation of the lake is bound to endanger the biodiversity of this freshwater ecosystem.

REMEDIES

The Wullar Lake which is the flood basin of river Jhelum, the life line of Kashmir valley, is very critical component for the well being of Kashmir in general and the people living in the surroundings of the lake in particular.

The lake which was over 200 sq. km in area only 100 years back has shrunk to merely 50 sq. km today. The vast sweet water lake of Asia is turning into a cess pool where all sediments, fertilizer and pesticide residues, raw sewage and sewerage and mining wastes besides other pollutants are being dumped. The well being and upkeep of the lake is not the baby of any specific department. All the departments are doing activities of their own without any coordinated effort. The management of lake calls for a combined effort of the departments like forest, wildlife, soil conservation and social forestry, agriculture, horticulture, pasture land development, Irrigation and flood control, animal husbandry, power development, rural development, geology and mining and the tribal welfare department.

It is therefore imperative that the eco-development and conservation of lake and other linked sectors are directly or indirectly benefited by managing the lake on scientific lines. The development activities of all the departments in these sectors will help in the eco-restoration and development of lake. It is not only the lake body which needs eco-restoration but also the core sector of 1, 14, 000 hectares of the direct inflowing catchments in various watersheds like Erin, Madhumati, Gundar, Ningli, Wullar-I and Wullar-II etc. in particular and also the rest of indirectly flowing catchments forming the tributaries of Jhelum which passes through the lake.

The sustenance of the lake depends on the survivability, sustainability and better coordination of these inter-linked sectors. It is therefore very essential that all these sectors which contribute and play their role in maintaining/conserving the lake should be given equal importance for the ecological development of the lake and its surrounds. The various types of remedial measures suggested may be discussed as follows:

Catchment Treatment

The concentrated work of afforestation in the degraded catchment in the north, south and west of the lake should be taken on priority basis. Likewise, soil conservation and pasture development in parts of the exposed barren slopes may be undertaken on annual funding basis. The closure, afforestation, soil conservation and pasture works can produce excellent results in overcoming the sedimentation problem from the catchment. Besides this, fencing should be carried along with afforestation. The fencing in itself is a technical treatment in that it helps the refractive and non-responsive areas to become responsive to afforestation activities. The fencing invites grasses and new natural regeneration and goes a long way in stabilization and prevention of soil erosion from the sloping areas by checking over grazing and trampling activities of animals and such other biotic interferences.

Through pasture development, thousands of quintals of grasses and fodder will be made available from the afforestation units. This is meant to stop excessive pressure of grazing on the degraded hills besides contributing to economy of the people living in these catchment areas.

Proper Land Use System in Catchment

Due to availability of water for irrigation, the low lying plain areas all around Wullar with invariable contour are under agricultural cultivation with two major types of crops, i.e., rice in kharif and mustard in rabi season. The arable areas with gentle slope have also been brought under rice cultivation by terracing and contour bunding to a large extent.

However, the non-arable gentle to moderate slope areas are under dry land fanning, predominantly under horticulture for production of cash fruit crops like Apple, Pear, Peach, Plum, Grapes, Pomegranate and Almonds. These trees are grown on slopes, in more or less definite spacing. The non-arable land with moderate and sometimes steep slope, (with large chunks of encroached forest land), is again under dry land farming for production of maize, pulses and grasses/fodder etc. These areas of gentle slope and moderate to steep slope under dry land farming are strategic areas for soil erosion as the most unscientific agricultural practices like tilling across the contour are resorted to, in such areas by and large.

The forests occupy the upper areas with moderate to steep and very steep slope. Due to ruthless cutting of trees to satisfy human need and greed, for the last century, the frontline forest areas that form the direct catchment of Wullar Lake, have been rendered blank and treeless. The forests are limited mostly to some eastern and northern aspects with poor tree density.

Looking to the existing land-use system of the Wullar catchment, there is need to ameliorate it on scientific lines keeping in view the lake management. The farmers growing crops (maize, pulses etc) on moderate to steep barren slopes may be advised to go for plantation crops on such areas which can be made responsive by giving incentives in terms of free planting material and also technical know-how by the forest and social forestry department. Likewise, horticulture department can take a lead role in motivating farmers for cultivation of temperate fruits like apple, pear, peach, plum, apricot, almonds etc., which also give better financial returns. Such land-use activities can help a lot in overcoming the major catchment problems which affect the lake body.

Eco-friendly Farming

As already mentioned, the area adjoining the Wullar lake and its catchment is under farming with paddy and apple occupying the major portion of the agricultural land. A lot of chemicals are used in the crop production. The current rates of fertilizer use by farmers in the paddy production varies from 240-300 kgs per ha which is above than the recommended doses. Moreover, herbicides are also sprayed by farmers instead of manual weeding due to non-availability of labor in the peak season. Likewise, 7 – 8 sprays of fungicides / pesticides are sprayed on apple crop besides the fertilizer doses. These fungicides / pesticides are both systemic and non-systemic.

During rainy season, the leftover residues of fertilizers in top soil and pesticidal residue get leached down in the lake, thereby, causing excessive weed growth and eutrophication of lake body. The department of agriculture should make farmers aware of the implications of excess chemical use and such measures/practices should be adopted in farming which may help in the eco-restoration of lake.

MANAGEMENT OF THE LAKE

Keeping in view, the great magnitude of problems involved in the catchment and main lake body of Wullar Lake, there is an immediate need of a comprehensive multi-dimensional action plan, with both long term and short term objectives. The primary object of management here is to strike a healthy balance between the conservation of the lake at international standards and its economic use on sustainable basis. This can be achieved by the closely coordinated and integrated working of three principal actors:

(a) Applied ecologist and conservationist especially in the eco-management of the catchment, the governing factor in the conservation of lakes/wetlands.

(b) Scientist for the main lake body to guide the managers in the change detection and the balanced development of aquatic flora, fauna, fisheries, avifauna by ensuring water quality and a balance between inflow and outflow of the water.

(c) The administrative/engineering agencies for such basic activities like demarcation, delineation, removal of encroachments, checking of sewage and sewerage by engineering methods from point source, raising of peripheral bunding, fencing and protections, as also permitted eco-friendly development activities like eco-tourism, eco friendly farming and other economic aspects.

Lakes and wetlands representing dynamic and living aquatic eco-systems can in no case have isolated engineering or administrative solutions, nor can it be treated any other urban development activity. It is the primary and absolute mandate of applied ecologists/conservationists in co-ordination with aquatic biologists/limnologists.

The efforts at management/conservation of this important lake have been initiated as late as the mid nineties. Although, the grants on annual basis were made available by Government of India, though very meager, compared to the magnitude of the problem at hand; the approach has remained only an isolated one without an impressive impact.

It needs political will and coordinated legal, administrative and scientific measures coupled with earnest management and conservation strategies on a comprehensive scale to restore the lake to its pristine glory.

ACKNOWLEDGEMENT

The author is highly thankful to generous support of Sir rattan Tata Trust, Mumbai for carrying out the study under the IWMI-Tata Water Policy Program.

REFERENCES

Government of Jammu and Kashmir (2001), Wullar Integrated Conservation and Development Plan (2001), Directorate of Environment and Remote Sensing: Srinagar.

Lawrence, Walter R. (1996), Valley of Kashmir, Kashmir Kitab Ghar: Residency Road, Jammu Tawi.

Stein, M. A (1898), *Kalhana's Rajtarangni*, Oxford Publishing House: UK.

17

WHAT ROLE SHOULD THE EUROPEAN UNION PLAY IN THE DEVELOPMENT OF THE NIGER DELTA

Akongbowa Bramwell Amadasun

INTRODUCTION

Niger delta is located in Nigeria, Africa most populous nation, Nigeria is the 6th largest producer of crude in the world, with a production of 2.3 million barrels per day, it is the leading producer of crude oil in Africa, it has over 30 billion barrels of crude which is estimated to hit 40 billion barrels per day by 2010. It also has estimated 158 trillion cubic feet of gas. Representing 93% of West Africa 170 trillion cubic feet reserve, and ranking 7th in terms of total global reserve. Current production of gas stand at about 3.8 BCFD, which is estimated, will hit 7.1 BCFD by 2010. In addition crude oil accounts for over 90% of Nigeria export earning, and 80% of state revenue (Toyi, 1998). All these resources are gotten from the Niger delta which is currently at the tick of crude oil environmental and fiscal induced conflicts; occupies less then one quarter of Nigeria, and in a state economic quagmire, poverty and underdevelopment and lacking basic infrastructures.

Nigeria has four refineries, three petrochemical plants, two fertilizer plants and network of loading and off-loading terminals or depots and pipelines for petroleum product distribution and transportation. Crude oil extraction is principally located in the Niger Delta region. The extraction is done both onshore and offshore. Presently, effort is seriously stepped up to engage in gas marketing, though gas flaring remains a serious problem despite all effort to reduce the flaring and the consequent health, economic and ecological hazards (Akongbowa, 2005). The implication of this massive oil activities in the Niger Delta region culminates in suffering of the region from serious environmental degradation in terms of biodiversity loss, environmental neglect and pollution and radioactivity due to intensive crude oil exploration, loss of economic means of livelihood and other substantial negative externalities.

ENVIRONMENTAL HAZARD

Specifically, oil spills represent a major environmental problem in the Niger Delta. Some disastrous effects of it, which may not be on land or sea include discharges from oil tanks and pipes, which may lead into underground water, and spills and leaks in industrial facilities, refineries, gasoline service station and discharges from automatic crank case oil. Much of these may be washed into the water bodies containing drinking water supplies. In addition there is the loss of esthetic values of natural beaches due to unsightly leaks, damage to marine and wild life, modification to ecosystem through species elimination and decay in fauna and flora secession, and decrease in fishery resources reduced nutritional status. Also, gas-flaring emissions causes wide range of health hazards. Carbon monoxide has a high affinity for hemoglobin and is able to displace oxygen and this may have cardiovascular and/or neuro-behaviour and acute exposure which can cause inflammatory and permeability responses, lung function diseases in airways, radioactivity and general respiratory symptoms. Lastly, lead that is a heavy component of petroleum pose some localized health risk because of its prevalence in the Niger Delta. It affects health through pathways such as indigestion and inhalations. It causes enzymes changes, anemia, hyperactivity and other neuro-behavioural effects (Rilwanu, 2001).

REGIONAL UNDERDEVELOPMENT

The finding of crude oil seems to have undermined the growth in the economy of the region. This includes social and economic infrastructure, which are grossly deficient in all ramifications. Also, the region's education infrastructure is characterized by low literacy rate, low quality of students and graduates at all levels, and dilapidating physical infrastructures and/or ill equipped tertiary institutions. Also the out come of marked deficit in health delivery in the region have resulted in high infant mortality and low life expectancy. Similarly, physical infrastructure such as roads, pipe-borne water, electricity etc are either non existent in a large part of the region or are in a state of helpless decay that instead of assisting in the sustainable development of the region they have indeed assisted in compounding the state of helplessness of the region. Lastly, the region exhibits low per capital income and lacks viable manufacturing concerns. The oil multinationals are being the exception. The constraints to industrialization include infrastructural deficiencies, limited access to credit; militarisation of the region infidelity or government polices, high cost of funds and equipment and low purchasing power of the Deltians. Agriculture which use to be the main stay of the region before the discovery of crude oil is on a downward trend because of conflicting government initiatives; destruction of farm land and pollution of rivers and seas by oil exploration activities; marketing and pricing constraints; low purchasing power of small sale producers; farmers poor access to money and capital markets; low level of labour productivity; inappropriate technologies; a rigid land tenure system; inadequate agricultural extension services and lost of environment constraints (Akongbowa, 2005).

The regional State government, Local government and Federal interventionist agencies generally exhibit bad governance among others, such as pervasive corruption, lack of transparency public accountability and the capture of public services by the elites. This lack of transparency accountability in public transaction and the consequent corruption have had negative impact on the development of the region.

Niger Delta may be experiencing some gains but these are only grossly inadequate, particularly given the natural resources on the country's economic growth, low capacity utilization in the industrial sector, poorly performing utilities/infrastructures and the attendant increase in operating cost, among others. The regions economy is therefore embattled on all fronts and with crisis: the education crisis; the unemployment crisis; security crisis: extra judicial killing both from government, militias and robbers and politicians; food crisis, transportation crisis; debt crisis; energy crisis; and crisis of economic management. The overall consequence of these macroeconomic problems is the deplorable poverty profile of Niger Deltians.

To compound the regions poor micro-economic situation the region is at the fringe or margin of the Federal power configuration. Thus apart from its scant political representation undermined by national majority ethnic hegemony over the minority groups of the region, suffer from huge losses and disadvantages in the resources in 1953 and 50% in 1960, allocation progressively declined to 45% in 1970, 20% in 1982 1.5% in 1984 and 3 in 1992 and 13% in 1999 (Toyi, 1998). A poor fiscal relationship that has accentuated scarcity of funds needed for economic development of the region. To further compound the state of underdevelopment there was a state of growing economic crisis emanating from gross mismanagement of available resources (a pandemic corruption). Bad government accentuated by years of military dictatorship and domestic and international capital flight in the region as a result of insecurity occasioned by contestation not just for political space and resistance to authoritarian forms of state rule and accumulation in the region.

SUSTAINABLE DEVELOPMENT

To meet the challenges of the region social and physical infrastructural deficiencies, gross resources management, environmental degradation, macro-economic growth and prosperity within the framework of millennium development goals (MDGs) and New economic Partnership for Africa's development (NEPAD); there is the need for sustainable development to incorporate multi-dimensional approach which comprises the following (Oribu, 2002).

1. Concern for economic and social viability of future generations that is inter-generational equity.
2. Alleviation of poverty and concern for social life.
3. Local participation in planning of development projects.
4. Positive impact on environment and resource use.

The above demand that the quantitative dimension of sustainability should involve the assessment of the regions long-term ecological, social and cultural potentials for supporting economic activity and structural change. The formulation and implementation of policies that can contribute to new and awesome load of environmental problems, and the need for greater reality in economic decision making that ensure an integration of both economic development and environmental protection in an efficient allocation of scarce resources. This calls for adequate social and environmental cost benefit analysis, which requires the full environmental and social costs of resources used in addition to cost extraction and processing.

Sustaining the above point of view Egbon (2003) thinks that environmental policy planning and management in the region reflects:

1. Expectations to induce macroeconomic stability through demand and management in the short term as a basis of economic growth and balance of payment equilibrium in the medium term.
2. Enough evidence that macroeconomic and sectoral policies within the framework of IMF and World Bank supervised structural adjustment programme significantly impact on resources use. An approach to sustainable development will therefore make imperative the integration of growth and structural change with environmental protection and improvement.
3. Different policies effect (e.g. trade, economic, fiscal, monetary technology, national security, etc.) manifest in different forms of environmental resources degradation.

ROLE OF EUROPIAN UNION

The Niger Delta represent a paradox (rich but under developed and poverty stricken) present a unique challenge in terms of development partnering to the European union for three reasons: first it represents a region from which the crude oil that lubricates the industrial wheel of Europe and the United States comes from; secondly, the region represents the core from which the funding of Nigeria's peace or security initiative in Africa, West Africa in particular and the rest of the world. Thus a vital region such as the Europe union whom the responsibility for maintaining global, social political, ecological and economic stability cannot afford to toy with a critical region such as the Niger Delta. Finally, the Niger Delta has the potentials of exploding and becoming a security risk to the whole of the West African sub-region should urgent measures not be taken to tackle the underdevelopment of the region.

Consequently, this call for partnership in terms of development aids between Nigeria and the EU. The partnership present a serious challenges to the European Union in terms of invitation to deal with core questions of underdevelopment and poverty reduction and how to integrate the Niger delta into the global economy. Here the challenges present itself in the critical form of:

1. How can the EU partner with the region (or the Nigerian government) in terms of empowering the region with the desired level of physical and social infrastructure for transforming the regions economy from that of a low income to that of a middle income?
2. Given the obvious available financing gap between ascertainable needs and available resources, how can the EU in partnership with the Nigerian government and states government in the Niger delta region close the financing gap between needs and available resources.
3. How can the EU partner with Nigeria to build linkages that can drive the development of the Niger Delta region through the development of appropriate infrastructure. Specifically, how do we build linkages between the provision of infrastructure and the growth of the real sector, comprising agriculture, manufacturing and the small and medium scale industries?
4. How can the EU partner with Nigeria to optimize human capital input to support and optimize social capital development in the region's transformation and growth process, through the introduction of appropriate technology and adopting leveraging on skills, technology and entrepreneurship. Specifically, how to build linkage between the real sector and the social sector.

The appreciation of this magnitude of the challenges rides on a high priority on the need to define the strategic options for the development of the region. This is high on four critical pillars:

1. The real sector.
2. The active sector, comprising oil and gas as well as banking and finance.
3. Environmental protection.
4. Social and physical infrastructure (human capital and utilities inclusive).

The systematic dependence and interlocking dynamics creates a synergy to propel and drive the active and real sectors into accelerated growth. The constitutive growth drivers which constitute the in locking wheels for the strategic model for the Niger Delta development include:

1. Job creation.
2. Security to life and property.
3. Environmental protection.
4. Sound and reliable infrastructure.
5. Conducive investment climate.

Consequently, the EU should be partners in promoting poverty eradication and fighting under development in the region. Specifically this should include:

- Assisting in sustaining raise in manufacturing capacity enhancing labour productivity, sustaining political stability of life and property, and supporting the maintenance of a stable financial system in the region that is capable of supporting the private sector to realize its growth potential.
- Identifying and intensifying efforts aimed at rehabilitating and expanding economic and social infrastructures.
- Assisting the regions indigenous private sector to seek strategic alliances with foreign partners (EU partners in particular) in order to participate more effectively in the Niger delta economy.
- Given the previous dominant contribution of the region to agricultural output and to Nigeria's GDP, EU should partner with the region in infrastructural development which emphasize the improvement and upgrading competitiveness in the agricultural sector through the development of key process industries. In this regard four critical issues appear relevant (Anya, 2001).
  - Raising the level of productivity;
  - Reducing the population on the land ;
  - Raising the investment level in agriculture; and
  - Providing easier access to funds and capital.
- Identifying and assisting in the development of priorities in infrastructural development that are critical to fight underdevelopment in the region (Anya, 2001):
  - Vastly improved utilities (power, energy and communications)
  - Improved transportation (Rail, roads and air);
  - Better ports (The gate way to international trade)

   - Investment in human capital, with emphasis on entrepreneurial and technological skills, and information technology competencies.
   - Provision of quality water – for irrigation, with necessary reticulation portable water and water for industrial use.
- Vigorously implementing reform of federal and Niger Delta States budgetary process and intensifying measures to ensure value for money in public expenditures.
- Assisting and strengthening efforts geared towards quality education and health services, and community networks or social capital in support of all development initiatives.
- Pioneering the crusade of fiscal derivation equity and environmental regeneration and protection from massive crude oil exploration, which has been the cause of the insecurity in the region.

CONCLUSION

Niger Delta development indicators represent a paradox shift in the development of a complex and changing region. It will represent a unique development challenge to EU in her contribution to Nigeria's overall development in terms of transformation in the quality of life, environmental quality, and socio-political management. Particularly as Nigeria on its own, with low income economy, dysfunctional social, political and economic systems, and/or rent seeking elite class and a statist national security policy is incapable of generating the resources and the will to achieve poverty reduction and fight under development in the Niger delta. Therefore it is imperative that the EU partner with the region.

The immediate importance is:

- Unless the EU urgently restrategize basic parameters for her development role in the Niger Delta within the context of the MDGs in the direction of greater visibility and heightened will to act, the Niger delta will continuously be dogged by human disaster and misery, chronic underdevelopment and poverty.
- Unless the EU intervenes by assisting the Niger Delta to build development capacity and other policy capacity and capabilities, worsening economic hardship is likely to culminate in immigration flow from this region is likely to be directed mainly towards Europe.
- In crafting a new EU role in global development mindful of Nigeria's chronic underdevelopment and poverty, the EU should first accord high priority to Niger delta's development policy and

other policy fields, and second pay great need to the fundamental parameters of good governance, food security and environment, peace and security and socio-economic problems.

- The EU cannot shy away from Niger delta's poverty and underdevelopment problems, it need to devote more effort to come in terms with the quality of life in the region and its marked development deficits. On the whole, debates and policies on the role of the EU in the region in mitigating these deficits could be controversial in many respects but this should not be allowed to deteriorate into unfruitful "doctrinal dispute". What is needed are efforts directed primarily to specify at the regions level, the central issues of given problems which may develop key impacts or core problems rooted in funding or other fundamental bottlenecks.
- Finally, the importance of EU in lifting the Niger Delta out of its poverty Delta out of its poverty and underdevelopment should provide a ground for a rethink of EUs identification of further point for departure to assist Nigeria (particularly the Niger Delta) to build effective development expenditure with the means of development policy.

REFERENCES

Akongbowa Amadasun B. (2005), Crude Oil Induced Fiscal and Environmental Conflicts/Crisis in Nigeria: Issues in Strategizing Sustainable Development and Security Network for Nigeria and the West Africa Sub-region in the 21st Century, A Paper Presented at the European Association Development and Research Institute (EADI), 11th General Conference on Insecurity and Development, Bonn-Germany.

Anya O. (2001), Economic Infrastructure and National Development, in Conference Proceedings: Growing Nigerian Economy, Abuja: Central Bank of Nigeria.

Egbon. P. C. (2003), Unpublished Teaching Notes on Petroleum Economics, Delta State University, Abraka.

Oribu O. Christopher (2002), *Introduction,* In *The Petroleum Industry, The Economy and the Niger Delta Environment*, ed., Oribu Christopher O. Dicta Ogisi and Rosemary Okoh N., Abraka, Department of Economics, Delta State University.

Rilwanu Lukman (2001), Harnessing Nigeria Gas and Petrochemicals Potentials for Economic Development, In Conference Proceedings: Growing Nigerian Economy, Abuja: Central Bank of Nigeria.

Toyi Akinosho (1998), Petroleum in Nigeria's Energy Crisis, 2(4), October/ December, Abuja: Central Bank of Nigeria.

18

CHASING THE ELIXIR: CASE OF GROUNDWATER RESOURCE IN NORTH GUJARAT REGION

F. A. Shaheen and R. L. Shiyani

INTRODUCTION

India faces an overwhelmingly grave situation wherein it's planned economic growth and improved social and environmental security is threatened by a growing population and depleting water resources. The development of tube well irrigation has contributed significantly to the increase in food production and reduction in poverty; however, the over-exploitation of groundwater poses a major threat to environment, health, and food security. All over India, there is a severe crisis in groundwater management. In many areas water levels are going down rapidly, the quality of water is deteriorating. The North Gujarat region of the country has made a prominent place in groundwater over-exploited zones. The region is heavily dependant on the groundwater resources to meet the domestic, industrial and irrigation demand. The chronic overdraft of the resource with high rates of water table decline had many implications with regard to sustainability, equity, energy consumption and rising costs of groundwater extraction. The present study examines the investment costs incurred by the farmers in borewell as well as the pace with which the resource is chased at higher depths.

METHODOLOGY

Two highly over-exploited districts of North Gujarat viz., Mehsana and Banaskantha were chosen for the study purpose and a total of 160 borewell farmers were surveyed for the present problem. The necessary information was collected by a well designed and pre-tested questionnaire. The survey was done during March-April, 2003.

The life of well was estimated using the commonly used life tables in statistical theory. The general formula used was;

$$\text{Average life of well} = \sum_{i=1}^{n}(fi)(xi) \div \sum_{i=1}^{n}(fi) \quad \text{over} \quad i$$

Where, f = frequency of wells worked and

x = age of well (1, 2, 3, 4, ..., n, in years).

RESULTS AND DISCUSSION

Well Life

The average life of well was worked out by various permutations and combinations of working and failed wells with and without initial failures to reach at a common centric value, which is given in Table 1. The well life was found 7.22 and 6.77 years for Mehsana and Banaskantha districts, respectively by considering the working borewells with initial failures. The average life of well(s) increased by about half an year in Mehsana without considering the initial failures, while it increased by 1.94 years (almost 2 years) in Banaskantha district. The life of wells was also estimated by considering all working as well as past failed wells with and without initial failures. It was observed that the well life decreases in both the areas when past failed wells are also considered in the estimation procedure. By taking all the arrived values into consideration, 7 years were finally chosen as the co-centric value (average well life) around which all the estimated values fall. The average life of dug-cum-borewells was found 31.38 years, which were found operating only in Banaskantha district. Similar results were obtained by Sripadmini *et al* (2001) in Chitradurga district of Karnataka, where they found 8 years as average life for dug wells in an NGO based MYRAD watershed and 6 years for borewell in government operated (G.O) watershed.

Table 1
Estimated Well Life (years)

Particulars	*Mehsana*	*Banaskantha*
1. Average borewell life with initial failures	7.22	6.77
2. Average borewell life without initial failure	7.66	8.71
3. Average borewell life of all working an failed wells with initial failures	6.46	7.21
4. Average borewell life of all working and failed wells without initial failures	6.80	7.63
5. Average life of dugwells	–	21.13
6. Average life of dug-cum-borewells	–	31.38

Borewell Investment at Historical Prices

The general features and investment on well(s) at historical prices (Table 2) were worked out by considering the wells constructed upto the year

1999. The construction years ranged from 1985 to 1999 in Mehsana, whereas it varied from 1972 to 1999 in Banaskantha (Appendices 1 and 2).. The average depth of these bores was 436 feet in Mehsana, varying from 200 feet to 800 feet, whereas, the average depth was observed lower (310 feet) in Banaskantha which varied from 70 feet to 630 feet. On observing the data, the pump placement depth was found co-related with the depth of wells. Obviously, the pumps were placed at a much deeper depth in Mehsana (458 ft) in comparison to Banaskantha (258 ft).

Table 2
Investment on Borewell at Historical Prices

(Amount in Rs)

Particulars	*Mehsana*	*(%)*	*Banaskantha*	*(%)*
No. of wells	40		52	
Average depth (ft)	436		310	
Range (ft)	(200-800)		(70-630)	
Placement of pump from above (ft)	458		258	
Yield (GPH)	4098		1389	
Pump HP	36		15	
Modal year	1990		1997,98	
(A) Cost of drilling (Exploration cost)				
1. Drilling charge	87587	33.28	51443	34.52
2. Casing/lining	46967	17.84	25987	17.44
(B) Cost of pumpsets (Extraction cost)				
1. Pump cost	36278	13.78	18005	12.08
2. Electrical expenditure	10688	4.06	5879	3.95
(C) Other cost				
1. Pump house	1473	0.56	1469	0.99
2. Water storage structure	1117	0.42	1176	0.79
3. Fee to GEB	69765	26.52	42154	28.29
4. Conveyance cost	9325	3.54	2897	1.94
Total Fixed cost	263200	100	149010	100

The investment component on the wells (fixed cost) were categorized under three major heads: viz, (a) cost of drilling (b) cost of pumpsets, and (c) other items including pump house, water storage structure, conveyance pipes and fee to Gujarat Electricity Board (GEB). The investment per well at historical prices worked out to Rs. 2,63,200 in Mehsana and Rs. 1,49,010 in Banaskantha district. The proportion of drilling cost to the total fixed cost was 33.28 and 34.52 per cent in Mehsana and Banaskantha districts, respectively. Likewise, casing component also formed major portion of investment (>17%) in both the districts. The casing of bores is essential as the soil is sandy loam; there remains high chances of scouring. Besides this, perforated pipes are used to protect the bores, which are very costly, thereby increasing the exploration cost component. The share of pumpset in the total investment was 13.78 and 12.08 per cent in Mehsana and Banaskantha districts, respectively.

The electrical expenses, which include panel board, wiring and installation charges accounted for about 4 per cent of the total cost in both the districts. The cable wire cost works out to be highest for North-Gujarat region (particularly for Mehsana district) as the pumps were deeply placed. The cable wire cost at current prices (2003) ranged from Rs 20 to 30 per metre. Thus, the expenditure on wiring also forms a considerable portion to the total investment in absolute terms. From the table, it is observed that the exploration component is more expensive than the extraction component, which is in contradiction with the study done by Nagaraj and Chandrakanth (1995) in hardrock areas of Karnataka. In North-Gujarat region, the depth of borewells is very high as compared to hardrock areas of Karnataka, which makes the exploration cost component more expensive, besides high casing cost incurred on costly perforated pipes.

The cost on pump house and water storage structure formed a meagre portion of the total investment. The conveyance pipelines which were used by most of the farmers in Mehsana, whereas, very few in Banaskantha formed 3.54 per cent and 1.94 per cent of the total cost in the two districts, respectively. Fee to GEB at the time of getting new connection forms the major cost component after the drilling cost in both the areas.

Dug-cum-Borewell Investment

The dug-cum-borewells, very few in number, are still functioning in Banaskantha district with a very low yield (584 gph). The general features and investment structure are presented in Table 3. These dug-cum-borewells were initially open dug wells with 50 to 70 feet depth, constructed mostly in 60s to 70s. With the advancement in groundwater development and falling water table levels, these became obsolete and bores were done in these dug wells during the latter period of 70s and early periods of 80s. These bores are generally known as dug-cum-borewells. The average depth of these bores is 140 feet from the ground surface. These wells are operated with a lower pump HP, ranging from 5 to 10 HP, with an average HP of 7.

The total investment on dug-cum-borewells at historical prices was found Rs. 85,687. The drilling and casing cost accounted for 20.95 per cent and 8.44 per cent of the total cost, respectively. The average cost of deepening the wells was Rs 20,808 (24.28% of the total cost). This is the cost incurred on boring these open dug wells. It was observed here also that the total share of drilling cost is more than 50 per cent. Hence, in dug-cum-borewells too, the exploration cost was found more than the extraction cost.

Table 3
Investment on Dug-cum-Borewell at Historical Prices in Banaskantha

(Amount in Rs)

Particulars	*Banaskantha*	*(%)*
No. of wells	15	
Average depth (ft)	140	
Yield (GPH)	584	
Pump HP	7	
Modal Year	1970	
(A) Cost of drilling (exploration cost)		
1. Drilling charge	17948	20.95
2. Casing/lining	7235	8.44
3. Further deepening charges	20808	24.28
(B) Cost of pumpsets (extraction cost)		
1. Pump cost	15985	18.66
2. Electrical expenditure	5973	6.97
(C) Other cost		
1. Pump house	1378	1.61
2. Water storage structure	1032	1.20
4. Fee to GEB	14578	17.01
5. Conveyance cost	750	0.88
Total Fixed cost	85687	100

Well Investment at Current Prices

In order to examine the change in the features and rise in average cost/investment of borewells, the wells which were constructed in the year 2000 and thereafter were taken into consideration to reflect the current average investment. Here, the period after year 2000 is assumed to reflect the investment in present period. The number of successful borewells constructed from 2000 to 2003 by the sampled farmers were 47 and 21 in Mehsana and Banaskantha districts, respectively (Table 4). The modal year was 2002 in both districts. Even if all the sampled functioning borewells are considered (Appendices 1 and 2), the modal year will remain same for both the districts which shows that most of the wells were recently constructed.

The well investment at current price (2003) was Rs 3,68,329 for Mehsana and Rs 1,67,075 for Banaskantha. The difference of Rs. 1,05,129 per well investment was observed when the well investment at current price was compared with well investment at historical prices in Mehsana, whereas, the differential figure was relatively very less (Rs 18,064) in case of Banaskantha area. Very high increase in well investment in Mehsana is mainly attributed to deeper exploration levels with submersible pumpsets of high HP. The difference between the average depth of borewells at present and historical periods was 144 feet in Mehsana, whereas, the same figure for Banaskantha district was modest (15 feet). The share of each item in the total cost has remained more or less same, but in absolute terms, the cost on each component has increased sharply over the period. Fee to GEB forms a major portion of the investment. At present, to get a new connection, one has to deposit about one lakh rupee to GEB in advance, besides a lot of administrative and other problems in getting the connection.

Table 4
Investment on Borewell at Current Prices

(Amount in Rs)

Particulars	*Mehsana*	*(%)*	*Banaskantha*	*(%)*
No. of wells	47	21		
Average depth (ft)	580	325		
Yield (GPH)	4873	2014		
Pump HP	43	22		
Modal year	2002	2002		
(A) Cost of drilling (Exploration cost)				
1. Drilling charge	122894	33.37	57894	34.65
2. Casing/lining	65822	17.87	28843	17.26
(B) Cost of pumpsets (Extraction cost)				
1. Pump cost	48235	13.10	20623	12.34
2. Electrical expenditure	13832	3.76	6329	3.79
(C) Other cost				
1. Pump house	2034.319	0.55	2018	1.22
2. Water storage structure	1543.277	0.42	1639	0.98
3. Fee to GEB	96384.64	26.17	42892	25.67
4. Conveyance cost	17584	4.76	6837	4.09
Total Fixed cost	368329	100	167075	100

During the survey, it was found that there were a lot of administrative hurdles which take minimum one year to get new power connection sanctioned for pumping the groundwater. In some cases, unforeseen expenditures have also been incurred. If that amount is also taken into consideration, then the total amount for new connection will be worked out more than one lakh rupee in Mehsana district. The share on conveyance structure, which mainly comprise of PVC pipelines has increased in the present period in both the areas. Hence, the farmers are now placing more emphasis on minimizing water loss in field channels with the increase in scarcity of resource over a period and time. These results were supported by the findings of Nagaraj and Chandrakanth (1995) on the well investment in hard rock areas of Karnataka. They worked the investment per well to be Rs 60,155 at historical prices and Rs 75,756 at current prices with an increase of 10 per cent in the investment cost.

Chronological Change in Well Depth and Investment

The vintages of borewells at different points of time indicate the nature and degree of groundwater extraction. The sampled borewells were arranged in chronological order and the average depth and investment was worked out for the wells constructed in their respective years to reflect the increase in borewell depth and investment over the period, which is presented in Appendices 1 and 2. Trend lines based on the bore depth and investment data were fitted to examine changes in these two variables over time, the path they explain and the rate at which they have changed. Various trend lines such as linear, logarithmic, polynomial, exponential, power and moving averages were fitted and are depicted in Figure 1 to 9 (see Appendix 3). The periodical borewell depth graph of sampled wells for Mehsana was best explained by polynomial function with an R^2 value of 0.4067 (Figure 1). The power and exponential trend equations tried are shown in Figure 2 and 3, respectively. It can be concluded from these trend lines that the depth of sampled borewells has increased with varying rates over the time period. From the polynomial trend curve, which best explains the scattered points of the plane; it was observed that from 1985 to 1989, the curve increases at a decreasing rate and gets flattened thereafter, almost parallel to X-axis till year 1998. From 1998 year onwards, the curve shoots up at an increasing rate till 2003. Hence, during the last 5 years, the borewell depths have increased sharply in the area.

On the other hand, borewell investment was best explained by power function in Mehsana (Figure 4). Total well investment/cost equation was more explained by power function ($R^2 = 0.5097$) than the drilling cost trend equation ($R^2 = 0.4156$) (Figure 5). The polynomial and other functions were also tried but yielded low R^2 values.

Likewise, trend lines were also fitted on the well depth and investment data of sampled borewells in Banskantha district, which are presented in Figure 6 to 9. For the Banaskantha too, the well depth graph was best explained by polynomial trend equation (Figure 6). In this case, the polynomial curve assumes the same shape, showing an increase at decreasing rate in the initial stage of the curve (1972 to 1982) after which it slowly goes down upto year 1996, and then it shoots up with an increasing trend. Similarly, the trend equations were fitted again on the chronological borewell drilling cost. In this case, besides power function (Figure 7), exponential function (Figure 8) also explained the scattered points of the plane with same degree ($R^2 = 0.4387$). In case of polynomial function (Figure 9), value of R^2 was slightly lower (0.4131) than the power and exponential trend equations. These results on chronological well depth and investment cost were found in conformity with the findings of Nagaraj *et al* (2003) on average depth and well investments at different vintages in Tumkur district of Karnataka, where the average depth of wells has increased from 124 feet in 1985 to 250 feet in 1995, and 357 feet in 2001. Likewise, well investment was found to be Rs 12,289, Rs 14,871 and Rs 22,080 during 1985, 1995 and 2001, respectively at 2001 prices.

CONCLUSION

The study concludes that the investment in groundwater irrigation has increased sharply over the period with decline in water table in the region. If the pace of groundwater extraction chasing race moves in the same way, then the sustainability of resource will be at high stake. It will lead to major crisis in the region. The state government should think seriously over the present groundwater declining trend and make a concrete plan for sustainable groundwater management through various policies which may include incentives for water saving technologies and removal of subsidy on electricity supply to farm sector for groundwater pumping.

REFERENCES

Nagaraj, N. and Chandrakanth, M. G. (1995), Low Yielding Irrigation Wells in Peninsular India: An Economic Analysis, *Indian Journal of Agricultural Economics,* 50(1): 47-58.

Nagarj, N., Chandrashekar, H. and Yatheesh, H. S. (2003), Sustainability and Equity Implications of Groundwater Depletion in Hardrock Areas of Karnataka – An Economic Analysis, *Indian Journal of Agricultural Economics,* 58(3): 438-447.

Sripadmini, R., Chandrakanth, M. G. and Chandrashekar, H. (2001), Relative Economic Performance of Watershed Development Projects under Different Management Protocols in Karnataka, Department of Agricultural Economics, University of Agricultural Sciences: Bangalore: India.

Appendix 1
Construction Year, Average Depth and Investment on Wells in Mehsana

Construction Year	*No. of wells*	*Well Age (2003 - year of construction)*	*Average depth (ft)*	*Investment (Rs)*
1985	2	18	310	87500
1986	1	17	570	300000
1987	1	16	400	. 200000
1988	4	15	360	180000
1989	3	14	500	300000
1990	7	13	323	142857
1992	3	11	673	283333
1994	1	9	500	450000
1995	3	8	450	358333
1996	5	7	480	288000
1997	3	6	376	248333
1998	5	5	490	350000
1999	3	4	467	283333
2000	16	3	562	360667
2001	10	2	586	372500
2002	17	1	568	319412
2003	4	0	690	342500

Appendix 2
Construction Year, Average Depth and Investment on Wells in Banaskantha

Construction Year	*No. of wells*	*Well Age (2003 - year of construction)*	*Average depth (ft)*	*Investment (Rs)*
1972	2	31	90	70000
1980	2	23	87.5	82500
1982	2	21	350	60000
1987	2	16	485.5	130000
1988	2	15	405	162500
1990	4	13	287.5	126250
1991	1	12	362	175000
1992	5	11	312.8	180000
1994	1	9	240	70000
1995	2	8	337.5	162500
1996	4	7	247.5	128750
1997	10	6	388	172300
1998	10	5	315	165500
1999	5	4	264	153000
2000	6	3	195	110000
2001	4	2	300	128750
2002	10	1	379.8	184000
2003	1	0	650	350000

19

INSTITUTIONAL ASPECTS OF GROUNDWATER GOVERNANCE: EXPERIENCES FROM SOUTH AUSTRALIA AND LESSONS FOR INDIA[#]

Falendra K. Sudan and Jennifer McKay

INTRODUCTION

Water is vital for all life on earth – it is the most essential non-renewable resource that sustains all species in every ecosystem on the planet. Water is also the basic provision for our society's economic development and general wellbeing. The planet is rich in water. However, this is mainly seawater with only a small percentage of the total water available for human consumption. Of the total water resource of the planet: 2.5% of the total volume of water on Earth is fresh water; 69% of all freshwater is frozen (in glaciers and permanent snow cover); large quantities of water are held in underground aquifers; and 0.26% of fresh water is in rivers and lakes. Lakes and rivers are considered as renewable water resources because they have a natural circulation (recharge) time for water to go around the natural water cycle. The average circulation time of water is: 16 days in rivers; 17 years in lakes; and 1,400 years in groundwater reservoirs. Groundwater recharges so slowly that it is considered to be a non-renewable water resource. Since the annual global recharge rate of groundwater is so slow, sustainable consumption of groundwater can not be more than 2,000 km^3 per year for the world population (the estimated annual global recharge rate of groundwater).

This paper is based on research carried under 'International Training Program on Groundwater Governance in Asia (GGA)' during the field visit to South Australia. Thanks are due to Dr. Karen Villholth, Program Director, GGA, for her critical comments and suggestions while developing the research framework and writing the paper. The valuable discussions with Dr. D. Dutta and Prof. Bakhshal Lashari during the field visits are deeply acknowledged. The financial support provided by International Water Management Institute (IWMI) and CGIAR Challenge Program on Water and Food is thankfully acknowledged.

As the driest, inhabited continent on Earth, the challenges for water management in Australia are significant. Australia has extensively developed its groundwater resources. Australia has 25780 GL of groundwater that can be extracted sustainably each year and is suitable for potable, stock and domestic use, and irrigated agriculture. Groundwater's importance as a proportion of total use varies (e.g. in Western Australia groundwater use is twice surface water use; in New South Wales and Victoria, use is predominately from surface water). Australia has one of the world's larger aquifer systems: the Great Artesian Basin is an estimated 1.7 million km^2 in extent and stores 8700000 GL. Each year the Great Artesian Basin supplies 570 GL of water for a variety of uses-mainly grazing and mining. Australia's two largest rivers, the Murray and the Darling, have been dammed for power generation and irrigation. As a result, the flow to the sea has been reduced by three-quarters but provides three million people and 40% of Australia's farms with water. Around $90 billion has been invested in water infrastructure assets, which now underpins Australia's economy particularly in rural areas. Irrigation agriculture accounts for around 70% of the 21 GL of water per year used by Australians. These economic and social benefits have been achieved at the expense of some environmental degradation. Rising water tables as a result of over-irrigation and excessive land clearing has increased salinity in rivers and soil, resulting in 2.5 million hectares of severely salt-affected land with the potential to increase to over 15 million hectares in 50 years. Salt-affected land is now recognized as a problem of national significance. Over-allocation[1] of freshwater in rivers and aquifers has resulted in deterioration of water resources, both in terms of quantity and quality. All of these impacts are having a negative effect on environmental sustainability and hence Australia's economic development. Thus, in most areas of Australia, there are needs to improve water resource management to improve sustainability outcomes. Furthermore, the "global warming" is affecting the drought regime in Australia with drought periods in the future expected to be longer and more intense in many of the main agricultural areas. This includes the impact of reduced rainfall, higher temperatures and increased evaporation rates.

The ongoing institutional reform has had some success, but much remains to be done if the goal is to meet the reasonable expectations of future generations of Australians. A strong commitment and ongoing cooperation of governments and stakeholders will be a key factor in sustaining and driving the momentum of water reform. Continuing efforts on integrated water management, cross-border cooperation, improving irrigation practices and water use efficiency will be necessary to ensure improved productivity and environmental sustainability of water resources and secure access to water for all Australians. Better institutional arrangements leads to sustainable groundwater management and provide informed services to people without resource degradation. There is need to strengthen the social domain, building robust institutions and minimizing transaction costs in groundwater resource management. The lessons to be learnt from South Australian experience, which derive the 'best practice' and can be applicable in Indian context needs to be drawn and will be spelled out in the final outcome of the paper. The key issue is that of most effective groundwater governance, improved capacity and adequate financing to address the increasing challenge of satisfying human and environment requirements of water. In Indian context, there is a governance crisis, rather than a water crisis, groundwater governance is about putting integrated water resource management and public participation and community empowerment (i.e. better institutional arrangements) as critically important element of it and also to minimize the transaction costs of institutional change.

OBJECTIVES AND METHODOLOGY

The main objectives of the present paper are: (i) to study the groundwater economy and the institutional arrangements for its governance in South Australia; (ii) to present three case studies on the outcomes of visits to South-East Natural Resource Management (SENRM) Board and meetings with farmers, South Australia Natural Resource Management Council (SANRMC) meeting, and Virginia Pipeline Scheme (VPS) and the users, and (iii) to draw the lessons from South Australian experiences in groundwater governance in the context of the institutional perspectives which derive 'best practice' and can be applicable in Indian context.

The groundwater economy and the institutional arrangements for its governance in South Australia have been evaluated through collection of data and information from the published and unpublished sources including the various reports of the Government of South Australia and research studies conducted by individual researchers, consulting organizations and the Non-Government Organizations. The study has been supplemented by the primary data and information collected from the stakeholders and key informants during the field visits to SENRM Board and interactions with farmers, SANRMC meeting and VPS and users. The participatory appraisal techniques (PATs) and focal group discussions (FGDs) have been conducted to elicit the data and information regarding the institutional aspects of groundwater governance to prepare the case studies, for which a set of questions have been prepared and discussed with the participants and key informants.

GROUNDWATER ECONOMY OF SOUTH AUSTRALIA: AN OVERVIEW

Australia has 25,780 million m^3 (Mm^3)[2] of groundwater that can be extracted sustainably each year for domestic, stock, and irrigated agriculture uses.

Of this volume only 4,962 Mm^3 are currently being used. While this appears to be positive, there are regions of Australia with severe overallocation problems. Approximately 73% of the water used in Australia (~24 000 GL) is supplied by rivers, 21% by groundwater aquifers, and the remaining 9% by harvest of overland flows. Australian Water Resources Assessment (2000) defined 53 groundwater management units as a basis for reporting on water quantity, use and allocation in South Australia. There has been moderate development of water resources in South Australia with extraction/allocation[3] level of 30 – 70% and management and resource information requirement is moderate, compared with Queensland and Victoria, where the resource information and monitoring are vital and the development depends on installation of appropriate water markets to move water to higher value use and to provide surplus for development or the environment through efficiency gains (e.g. Pioneer River, Queensland with extraction/allocation level of 70 – 100%) and over-developed (more than 100% extraction/allocation level) in Wimmera-Avon Rivers, Victoria) where insufficient provision has been made for environmental and non-consumptive uses, and management intervention and information requirements are substantial.

Generally groundwater resources are not as well defined and their management is not as advanced or sophisticated as surface water resources. Implementation levels of best practice groundwater resource management vary. In Western Australia and Queensland, both high users of groundwater resources, knowledge and management of groundwater systems are well developed. In other States (e.g. New South Wales and Victoria), with the limit of surface water supply being reached, demands have substantially increased for access to groundwater resources.

Water requirements of groundwater-dependent environmental factors are significant in assessing sustainable yield[4]. Although a concept of sustainable yield has been agreed for groundwater, the basis of the calculation of sustainable yield varies greatly between the States and Territories, with varying degrees of consistency with the emerging approach to sustainable yield (incorporating consideration of all potential demands in the allocation of the resource). Approximately 96% of groundwater management units have use information and 57 of Australia's 538 groundwater management units are overused. On average across Australia, water allocation exceeds use by approximately 11%. However the variation across Australia is significant. The differences between allocation and use place particular challenges before water resource managers. Through the Audit, Australia has commenced identification and characterization of over-allocated groundwater systems. The allocation information for approximately 95% of groundwater management units is available. Eighty-three (15%) of Australia's 538 groundwater management units are judged to be over-allocated.

The Water Resources Act (1997) stipulates that water access and use be licensed in prescribed areas. In areas where water resources are not prescribed, there is no limit to the volume of surface water or groundwater that can be accessed. Water may be accessed from a watercourse, lake or well provided no detrimental affect results in the ability of other users to access the water or the enjoyment of the amenity of water in the water course or lake by another occupier of riparian land. Although not specifically defined, there is a provision in the Act 1997 (Part 4, c16) that reviews the inter-relationship between groundwater and surface water, in particular the impacts related to the accessing of water. The Act implies that surface water and groundwater are hydraulically connected and provisions are provided that may prohibit or restrict the accessing of water if the following occurs:

(a) the rate at which water is taken from a surface water resource (watercourse or lake) is having a serious affect on the level of water in a groundwater resource that depends on the surface water resource for its replenishment (recharge);

(b) the surface water is taken at a rate that seriously impacts on a groundwater resources (water levels) that depends on the surface water resource for its replenishment; and

(c) groundwater is taken at a rate that affects, or is likely to affect, the quality of surface water.

Trading markets for surface water and groundwater are independent. In particular, the scope for groundwater trade is limited to the resource defined within a Water Allocation Plan. The ephemeral systems in South Australia are relatively small and the relationships not sufficiently well understood to enable clear specification of water entitlements for transfers between surface water and groundwater. There are no specific policies in South Australia relating to groundwater – river interaction issues. There are no practical management guidelines in South Australia relating to the management of groundwater – river interaction issues. A water allocation plan for a catchment or region where the three types of water resources are prescribed and licensed are expected to take into account, and allow for, interaction under the allocation policy and resource management principles and criteria.

South Australia's topography and climate results in predominantly non-perennial watercourses. Whilst South Australia does have groundwater and surface water interaction it is not characterised by major groundwater extraction impacting significantly on river flows. There are no notable case examples of double allocation within South Australia. South Australian streams are charactered by relatively short duration, periodic, flow events. There aren't extensive water table, alluvial type aquifers which provide significant long term stream baseflow. It is characterised more by localised groundwater discharge areas from fractured rock aquifers or small sedimentary aquifers. Surface water capture in South Australia is via dams capturing overland flow reducing stream flow and off-stream dams into which water is diverted from watercourses opportunistically in "non-prescribed" (i.e. non-licensed) watercourses or in accordance with flow criteria under a water allocation plan. Bores may be located near streams to access better water quality which is recharged during winter stream flow for extraction during summer as summer stream flow cannot be relied upon. These sources adjacent to watercourses are generally low yielding.

In South Australia the main scenario is for surface water capture or diversion reducing stream flow, which would otherwise recharge groundwater. The Water Resources Act 1997 in South Australia is the current legislation framework under which Water Allocation Plans are written and licenses surface water, (over-land flow into dams) watercourses and groundwater from prescribed resources. There are no specific guidelines for groundwater-surface water interaction in South Australia. A water allocation plan for a catchment or region where the three types of water resources are prescribed and licensed are expected to take into account, and allow for, interaction under the allocation policy and resource management principles and criteria.

In South Australia, the potential for groundwater development remains in the south-west of the State, the alluvial systems in the north-east and the South Australian/Victoria border zones in the north and south. While many of the major groundwater resources are already fully allocated there remains significant unused allocation for development in the State. The majority of the groundwater development potential lies in the south-east of the State and the Murray Basin. However, in these areas soil conditions, salinity and depth to water table may ultimately constrain some developments. As the large resources are becoming increasingly regulated, other resources (e.g. the Mt Lofty Ranges) are coming under increasing pressure to meet demands for irrigation water. Development pressures are increasingly leading to regulation of the groundwater resources.

In order to achieve sustainable water resources management, work is under way towards sustainable water resources management. Examples of initiatives include: Environmental, economic, social and engineering assessment criteria for assessing development proposals. Further development through efficiency gains include the South Australian Highland Irrigation districts with 15% efficiency gains through infrastructure improvements; piping of part of the Wimmera Mallee stock and domestic system; and the pipelining of the Woorinen supply system in Victoria.

Improved and coordinated management of groundwater is also being attempted. Management of the groundwater resource of the Great Artesian

Basin is shared between four jurisdictions and has suffered from lack of coordination in the past-it has tended to be managed as four separate resources. Key issues for management are the continued decline in artesian pressure, the deterioration in the water infrastructure to the point where extractions are substantially in excess of requirements, the difficulty in making water available for new and high-value users, and loss of key groundwater-dependent ecosystems.

The water management agencies are working towards meeting requirements for environmental allocations through legislation, planning and assessment processes (e.g. stream flow management plans, stressed rivers programmes, riparian vegetation management and nutrient management strategies). The water use in South Australia has increased significantly during the last more than three decades. This was mostly due to increases in irrigated agriculture. Urban centres have shown either low increases or net decreases in water consumption per person (AATSE, 1999). Urban water use per person has declined over the 1990s mainly due to an increased awareness of the need to reduce water wastage and changes in water pricing. Much remains to be achieved, with changes to urban gardening practices and water efficiency in toilets being obvious opportunities for improvement. Industrial use is not large and is falling as industries become more water efficient, often coupled with increased emphasis on recycling or with efficiency gains in energy use as part of overall industry environmental responsibility.

South Australia has put water allocation systems in place. Entitlements ensure security and reliability of supply. Trading maximizes the value of water that is provided and water can be moved to high value uses. Catchments across South Australia are at differing levels of water resource development and use. In parallel with the need to move towards full volumetric allocations is the need to track allocations by use type. As a minimum, for those water resource systems where total extractive use is approaching sustainable limits and water quality thresholds, metering and water quantity management need to be implemented across the state.

To enable and facilitate water trading, changes in water allocation and definition of rights to water, use monitoring is important. Water use monitoring will also provide information for managers to track and to evaluate the effectiveness of allocation policies particularly in relation to highly and over-committed resources. Targeted monitoring of water use is an important component of water resource management as Australia seeks to maximise economic and ecological benefits from its water resources.

GROUNDWATER INSTITUTIONS AND POLICIES

In the recent past, there is growing concern in all aspects of groundwater resource problems, especially the role of groundwater institutions[5] and their impact on economic and environmental outcomes. Institutional arrangements are important because they define the rules for determining who has access to groundwater resources, who can claim income occurring from their use, and who must bear the costs incurred. Traditionally, neoclassical economists have treated institutions as fixed and determined economic efficiency in that context. Under these conditions efficiency rests on the notion of "Pareto optimality," wherein at least one party benefits and no one is worse off and no rational individual would engage in market activity that worsen their position. However, this attractive normative criterion cannot be used to analyze institutional change (Bromley, 1982). Institutional changes, by their nature, redistribute the outcomes, because institutions establish initial endowments, any change results in losses as well as gains. Economists frustrated by the limitations of Pareto optimality normally use the far more controversial criterion of 'potential Pareto optimality' to evaluate changes that involve losses, that is, a change is efficient when those who gain could compensate those who lose and still be better off.

On the other hand, institutional economics relies on "methodological collectivism", which is based on the argument that the fate of the individual and the whole are inextricably linked. Aggregate economic outcomes depend heavily on the distribution of resources among individuals (Peach, 1987), and good policies structure distribution of opportunity such that individuals can flourish and contribute to the whole, otherwise, good policies lead to aggregate outcomes which promote individual welfare. The ultimate function of economic systems and the institutional arrangements upon which they rest is the survival and well being of both individuals and the system over time. Therefore, desirable resource allocation policies are defined as those that foster the long-term availability of both economic and environmental systems. For example, groundwater resource allocation that maximizes net present value in the short run (say 50 years), may be destructive for the long run productivity and environmental sustainability and therefore, seriously jeopardize the prospects of future generations. Despite the lack of precision and direct measurability, the idea of sustainability has begun to be used to evaluate natural resource allocations (Dixon and Fallon, 1989).

The policy prescriptions that derive from these alternative views will obviously differ. Neo-classical economists evaluate institutions based on their market like properties. A water policy that imitates markets, or enables an increase in net present value will be preferred. Any attempt to temper with the market is likely to result in 'government failure,' wherein net benefits are decreased. On the other hand, for institutionalists, there is no first principle for the form of institutions. The public/private mix resulting from institutional change is regarded as incidental (Klein, 1985), that is, no a priori deference to public or private solutions. The relevant issue is whether or not a particular institutional arrangement serves as a vehicle or an impediment in reaching instrumentally grounded outcomes. Institutionalists are pragmatic in the sense that "The policies offered ... are not held as universal truths or as a flowing from immutable laws, but are seen instead as humanity's best stab as societal improvement given the factual base of a particular situation and the possibilities, and limitations, of the human intellect" (Petr, 1984, p. 9). The normative views held by neoclassical and institutional economists concerning what institutions should be like does not necessarily square with their positive descriptions of how institutions are actually determined.

Water is an input into human use including economic production and recreational activities that generate many externalities, and also provides many valuable ecosystem services only partially captured by the market. The challenge, across space and through time is to find the most efficient and equitable way to deploy the nation's water resources in economic and human use and in the maintenance of environmental processes. Robust institutional arrangements[6] are required that are dynamically efficient and low cost, and robust in an institutional sense. These arrangements must be able to optimally manage and deliver benefits arising from consumptive uses, in providing ecosystem services from non-consumptive uses, and in addressing the full suite of externalities that result from water use. When they are compromised, investment is less efficient than it otherwise would be and community welfare is less than it could be.

Water Institutions and Policy: Early Initiatives

The implications of stream-aquifer connectivity and the need for a conjunctive management approach[7] have been known in Australia for at least forty years. The first national review of water resources (AWRC, 1965) stressed that groundwater should not be regarded as a resource separate from surface water, noting that: "In some localities base flows in streams are maintained by drainage into them from underground sources while in other places streams are a source of replenishment for underground supply. Consequently, exploitation of the one can affect the availability of supply from the other. It is desirable, therefore, and likely to become increasingly important, that the conservation and utilizations of both resources should be planned jointly." A similar review of water resources in Australia undertaken a decade later (AWRC, 1976) came up with the same conclusions: "In the past, groundwater and surface water have tended to be viewed as separate resources, as a result, no doubt, of inherent differences in their modes of occurrence, assessment and development. Yet they are often hydraulically connected, and in any event, are complementary components of a larger single system. Thus, in assessing the water resources of a region, independent measurements of groundwater and surface water yield are not necessarily additive." This was again reiterated in the 'Water 2000' report of water resource priorities undertaken in 1983, which specified that conjunctive use of surface water and groundwater supplies should be encouraged in present and future water resources planning (Department of Resources and Energy, 1983). The report also advocated that analyses of surface water-groundwater relationships are required to provide an appropriate systems approach basis for conjunctive use.

Water Institutions and Policy: Recent Initiatives

Over the last decade, the significance of the connectivity between surface water and groundwater systems has occupied more significance in Australia. The following paragraphs are devoted to highlight the recent water institutions and policy initiatives and how these relate to conjunctive water management.

In February 1994, all governments across Australia, recognizing inefficient and inappropriate use of water, agreed that the management and regulation of water resources required significant policy and institutional change. This agreement resulted in the endorsement of a national policy by the Council of Australian Governments known as the 'COAG Water Reform Framework', to achieve an efficient, economically viable and environmentally sustainable urban and rural water industry. The key policy elements included: (i) water pricing based on full cost recovery and the amount of water used; (ii) establishment of clearly specified water entitlements and the arrangements to enable trade in those entitlements; (iii) allocation to the environment as a legitimate user of water; (iv) adoption of an integrated catchment management approach to water resource management; (v) establishment of regulatory and water service institutions that have clear roles and responsibilities; and (vi) public education and consultation.

Although the management of groundwater was included in the provisions of the 1994 reforms, the framework had a focus on surface water resources and was not explicit about which aspects applied to surface water and to groundwater. It was not until 1996 that the issues surrounding groundwater management such as trading, pricing, drilling, well construction and institutional arrangements were formally accepted into the reform framework by COAG and thus introduced into national water policy. Also recognized at that time was the notion of considering the linkages between surface water and groundwater. The problems of managing connected water resources through different programmes were acknowledged, noting that: "groundwater and surface water resource management should be better integrated, including approaches to pricing (especially adjacent to public surface water regulated schemes), water allocations and trading to ensure consistency"(ARMCANZ, 1997: Recommendation 3).

According to a policy paper by the Framework Taskforce: "In many situations, groundwater and surface water are interconnected and interchangeable resources where decisions made in one area affect the other. Consider, for example, the situation which exists in parts of Australia where no groundwater license is required for a well adjacent to a stream, whereas the stream is fully allocated. Single resource policies embracing conjunctive use and integrated management of groundwater and surface water through allocation and pricing mechanisms need to be more actively pursued by the States" (ARMCANZ 1997:3). In addition, Resolution 6 (ARMCANZ, 1998) states that: "Progress has been made in some limited areas in meeting environmental needs of groundwater, but further progress has been constrained by a poor understanding of the location, extent and processes associated with groundwater/surface water interactions and associated ecosystems."

In August 2003, the COAG agreed that there was a pressing need to consolidate and refresh the 1994 water reform agenda to increase the productivity and efficiency of water use, sustain rural and urban communities and to ensure the health of river and groundwater systems. COAG agreed to develop a National Water Initiative (NWI)[8] to: (a) improve the security of water access entitlements; (b) ensure ecosystem health by implementing regimes to protect environmental assets at a whole-of-basin, aquifer or catchment scale; (c) ensure water is put to best use by encouraging the expansion of water markets and trading; and (d) encourage water conservation in cities. At its meeting in June 2004, COAG agreed to establish a National Water Commission (NWC) to assess progress in implementing the NWI and advise on actions required to better realize the objectives of the NWI Agreement. There are a number of references to a conjunctive management approach within the intergovernmental agreements relating to the NWI, namely: (i) One of the primary objectives is to recognize the connectivity between surface and groundwater resources and connected systems managed as a single resource (NWI, 2004, clause 23x); (ii) With reference to water resource accounting, the key outcome is to ensure that adequate measurement, monitoring and reporting systems are in place to support public and investor confidence in the amount of water being extracted for consumptive use (NWI, 2004, clause 80).This outcome is to be achieved, in part, by consolidated water accounts, that can be reconciled annually and aggregated to produce a national water balance (NWI, 2004, clause 82 iii); (iii) These accounts are to be inclusive of systems to integrate the accounting of groundwater and surface water use where close interaction between aquifers and streamflow exist (NWI, 2004, clause 82 iiib). Also States and Territories to immediately establish common arrangements in the case of significantly inter-connected groundwater and surface water systems (NWI 2004, clause 79 i(c)); and (iv) The NWI also stipulates that all States and Territories agree to identify by the end of 2005 situations where close interaction between groundwater aquifers and stream flow exist and implement by 2008 systems to integrate the accounting of groundwater and surface water use (NWI, 2004, clause 83).

Fullagar (2004) has recognized the importance of institutions in terms of water laws, policies and organisational arrangements for water management in Australia. The institutional arrangements for managing water resources differ among the States in Australia (NGC, 2004). Discontinuity in government institutional arrangements between (and within) jurisdictions should not be an excuse for failing to identify and address issues associated with connected systems, or for not progressing opportunities associated with conjunctive water management (Fullagar, 2004). Institutional factors such as the rules governing water use and the organisational arrangements for water management are likely to play important roles in determining whether, when, and how conjunctive management programs develop and perform. Institutions facilitate the ease with which multiple stakeholders interact in complex situations, prescribing what actions are allowed, required, or forbidden in given situations (Crawford and Ostrom, 1995). Thus, institutions are especially significant when a task requires coordination. Given the organisational and physical complexity involved with conjunctive management, it is likely to require considerable amounts of coordinated behaviour. In brief, there are institutional issues, which affects at every step in even the simplest of conjunctive management projects, however, these institutional arrangements can be conducive to success, or they can present substantial barriers.

Water Institutions and Policy: Some Other Initiatives

This recognition of connectivity has also been supported by other recent water policy discussions. The study, Water and the Australian Economy (ATSE, 1999) concluded that: "Policies are needed to ensure the resource is managed in its totality, including groundwater, unregulated rivers and water quality. Water markets must be regulated to ensure individual trades do not impose external effects on third parties. Current regulatory systems do not offer sufficient protection to third party interests or the proper regulation of whole water systems". The National Land and Water Resources Audit (NLWRA) stressed the importance of developing and applying methods to understand and then manage at a basin scale conjunctive use, and to integrate surface water and groundwater management within an overall context of sustainability (NLWRA, 2001). The Audit recommended an Australia-wide initiative in partnership with State and Territory water management authorities that included understanding and managing interactions between surface water and groundwater quality and quantity.

In 2002, a series of groundwater policy papers was developed by the High Level Steering Group on Water in conjunction with the National

Groundwater Committee (NGC). In particular, one of these papers on groundwater quality protection noted the importance of connectivity. In July 2003, a Blueprint for a National Water Plan was released. In addition to other aspects of water resource management, the blueprint acknowledged the importance of considering groundwater-surface water interactions: "... water accounts also need to take into account... the interconnected nature of groundwater and surface water systems... Entitlements to access water must therefore address run-off, land use, river water and groundwater... In the 21st century we will need systems that recognize hydrological connections".

An important component of water reform is the management of water quality. In 1992 a National Water Quality Management Strategy (NWQMS) was introduced by the Commonwealth, State and Territory governments. In 1994, the strategy was incorporated into the COAG Water Reform Framework. The main policy objective of the NWQMS is based on the philosophy of ecologically sustainable development to 'achieve sustainable use of the nation's water resources by protecting and enhancing their quality while maintaining economic and social development'. The NWQMS comprises policies, a process and 21 national guidelines for water quality management. The Natural Heritage Trust (NHT) and National Action Plan for Salinity and Water Quality (NAP) are two other major programmes that aim to address environment and natural resource issues in Australia. The NHT was set up by the Australian Government in 1997 and funding has since been extended to 2007 to support the sustainable management of Australia's natural resources, including measures to improve water quality. The NAP was announced in October 2000 to address salinity and water quality in some of Australia's worst affected areas. The NAP is designed to improve land and water management in 21 priority regions. In relation to surface and groundwater management, the Action Plan promotes: (i) caps to be set for all surface and groundwater systems identified as over-allocated or approaching full allocation; and (ii) introduction of a new approach to groundwater and surface water administration that recognises their interdependency and the need for their joint management for salinity and water quality outcomes.

Thus, significant water reforms began in Australia in 1994 when the Council of Australian Governments (COAG) agreed to implement a strategic framework to achieve an efficient and effective water industry. The next major development was the National Water Initiative (NWI) in 2003 designed to establish a nationally compatible system of water access entitlements, efficient water markets, institutional arrangements for the recovery and management of water for the environment, improved accounting and best practices for water pricing, and urban water issues. All these policy initiatives and institutions have recognized the conjunctive use and management of groundwater and surface water resources.

CONJUCTIVE WATER MANAGEMENT FRAMEWORK: ROLES AND RESPONSIBILITIES

A primary function of the conjunctive water management framework is to identify who does what, where and how. It seeks to clarify and recognise the scales of involvement and complementary roles for participants at catchment and regional levels, and within Commonwealth, State and Territory, and local government in Australia. These roles are not new or constructed, but they are a reflection of existing and emerging roles. The framework also promotes strong and effective partnerships between the various stakeholder groups in the catchment. Such partnerships are needed for conjunctive water management to become an effective tool for achieving integrated water management goals and bringing practical benefits to water users.

Role of Water Users

The framework enables water users to view their activities in a catchment sense. Users can support the progressing of conjunctive water management through: (a) representation of local (community and environmental) needs and objectives; (b) development of locally applicable conjunctive water management options (e.g water trade); (c) identifying potential trade-offs (e.g less water if there are more licence flexibilities) at the regional level; (d) management options that reflect an understanding of how communities will be affected by different options (local environment and socio-economics); and (e) local critique and quality assurance to technical outputs.

Role of Catchment and Regional Groups

Catchments and regions are increasingly recognised as the most appropriate operational scale for managing natural resource issues. Governments have focused on this scale to achieve tangible environmental improvements over the medium to longer term. As a result, most efforts at natural resource management priority setting and monitoring are being undertaken at the catchment level. Catchment planning should be based on the best available science and be inclusive of the needs of the community that lives within the region or catchment. It should also be responsive to the expectations of the communities that live outside. Catchment groups can: (i) participate in catchment conjunctive water planning that is cooperative, makes effective use of best science information, is inclusive of all stakeholders, and fully integrates social, economic and environmental objectives; (ii) support the delivery of community-agreed objectives for integrated groundwater-surface water management outcomes; (iii) support capacity-building activities and ensure the necessary resources are in place to enable goals to be met; (iv) form strategic partnerships and encourage industry to participate in and deliver on planning processes and outcomes for conjunctive water management; (v) actively seek information and support from reliable sources (including relevant biophysical and economic expertise) and apply it to the catchment water resources planning and management process; (vi) prepare catchment integrated water management plans in sufficient detail to enable private landholders to link to catchment-scale outcomes; and (vii) regularly monitor and publish the progress towards meeting catchment goals and objectives. The framework enables landholders to view their farm-based activities in a catchment sense. It enables communities to be confident in regional water planning and its ability to address broad national issues of integrated water management and natural resource management outcomes.

Role of Governments

As discussed earlier, the conjunctive water management framework is consistent with the objectives of the NWI. The NWI was brought about by governments recognising that water availability has always been important in Australian life and management of water resources is extremely challenging. The initiative foreshadows expanded water trading across State borders, increased security of water entitlements for land owners, introduction of new arrangements for environmental flows and greater recognition of the connectivity between surface water and groundwater resources. The framework seeks to build on these existing water resources management initiatives and to facilitate the development and implementation of conjunctive water management.

Australian, State and Territory governments share a range of roles that support conjunctive water management. The role of governments should be consistent with existing legislation and with intergovernmental initiatives, such as the NWI and other water reforms. This is the case even for activities that do not rely on legislation. Assessment against National Competition Policy principles is useful in considering whether government intervention is applied in a consistent manner that does not impinge on private enterprise. The proposed roles of governments are: (a) providing and facilitating water user access to education and information on conjunctive water management; (b) coordination; (c) developing integrated water management policy; (d) research and development; and (e) monitoring and evaluation.

There are also some roles that are unique to a particular sphere of government depending on the responsibilities and scale at which the action is needed. For example, The *Australian Government* has a strong role in: (i) providing leadership, coordination and support through a mixture of policies and incentives to drive the widespread adoption of conjunctive water management; (ii) supporting regional bodies to engage water users to contribute to integrated water management that supports natural resource and environmental outcomes; (iii) promoting a consistent intergovernmental approach to conjunctive water management in agriculture that is consistent with major natural resource policy direction in other sectors such as mining, fisheries, forestry and urban; (iv) assessing and supporting national natural resource management priorities; (v) promoting a coordinated approach to conjunctive water management across industry sectors; (vi) assessing the impact on water resources and management of decisions in other policy sectors (such as immigration, defence, transport, regional development); (vii) actively participating in water resource issues that cross jurisdictional boundaries; and (viii) promoting Australia's approach within the international water arena.

State and Territory governments, with constitutional responsibility for water resource management, are in a primary position to influence natural resource management at the catchment level and to support integrated water management by: (a) providing leadership and support through a mixture of State water policies and incentives to drive the widespread adoption of conjunctive water management; (b) provide research and development support to primary producers for designing more sustainable production systems and practices; (c) supporting regional bodies to engage water users to contribute to landscape natural resource and environmental outcomes through integrated water management linked to sufficiently detailed natural resource management plans; (d) assessment of, and support for, state and regional natural resource management priorities; and (e) working across state boundaries where catchments dictate.

Local governments are the sphere of government 'closest to the people' and can facilitate the voluntary adoption of conjunctive water management by: (a) acting as local information brokers for community and landholders on conjunctive water management; (b) being a local facilitator and advocate by providing meeting rooms, secretarial support, administrative support and facilitating negotiations; (c) assessing and supporting local natural resource management priorities and working with regional and catchment bodies to enable them to link conjunctive water management to landscape natural resource and environmental outcomes; and (d) where agreed and resourced, providing infrastructure or undertaking capital works to support local actions and changed management practices arising from integrated water management plans.

The conjunctive water management framework recognizes a need for a simple check list of activities set out in a logical order. This gives the opportunity for those involved in water resource management to assess and revise their position within the overall scheme of sustainable water management. The philosophy of *adaptive management*[9] is followed where policies and practices are continually improved by learning from the outcomes of previous work. The process is iterative and aspects of the management process are revisited and reviewed. A feedback loop is incorporated into the management process. This recognizes that there are inherent uncertainties in our understanding of catchment processes, the priorities and perspectives of water users, the impact of water management options and future changes and threats. There is constant change in our knowledge about complex natural systems which themselves are dynamic and community priorities, perceptions and expectations also change (Pagan and Crase, 2004). This means that conjunctive water management needs to be flexible and be able to evolve.

The ideals of adaptive management have recently become popularised in natural resource management. Ludwig *et al* (1993) provide some practical advice when working towards effective management: (i) Include human motivation and responses as part of the system to be studied and managed; (ii) Act before scientific consensus is achieved. Calls for additional research may be delay tactics; (iii) Rely on scientists to recognize problems but not to remedy them. The judgement of scientists is often heavily influenced by their training in their respective disciplines, but the most important issues involving resources and the environment involve interactions whose understanding must involve many disciplines; (iv) Claims that basic research will lead to sustainable resource use may lead to false complacency; (v) Effective policies are possible under conditions of uncertainty, but they must take uncertainty into account. Consider a variety of plausible hypotheses about the world; consider a variety of possible strategies; favour actions that are robust to uncertainties; hedge, favour actions that are informative; probe and experiment; update assessments and modify policy accordingly; and favour actions that are reversible.

CASE STUDIES ON GROUNDWATER INSTITUTIONS

The field visits have been made to three different locations dealing with groundwater use and governance in South Australia during November and December 2006. The main objective of the field visits was to understand the institutional mechanisms in place for groundwater governance in South Australia and to draw some lessons to be applicable in Indian context. The organisations visited were SENRM Board and farmers, SANRMC meeting, and VPS. As discussed earlier, the PATs and FGDs were organised to elicit the data and information regarding the phenomenon under observation to report the case studies. The case studies were also supplemented by the use of secondary data and information provided by the participating members and the publications of the Government of South Australia. The exciting and enriching experiences on groundwater institutional arrangements were gained during the field visits to the aforementioned organisations and institutions, which are given in the following paragraphs.

Case Study-I: Visit to SENRM Board and Farmers

In 2004, the South Australian Government established the *Natural Resources Management Act 2004* (the NRM Act), introducing a new framework for the integrated use and sustainable management of the State's natural resources. The NRM Act provided for the creation of eight NRM regions within South Australia and the establishment of a regionally based, community driven NRM board for each region. SENRM Board was established in May 2005. The members of the SENRM Board are both community and government members with knowledge, skills and experience across many areas of NRM, including water, soil, biodiversity or pest plant/animal management, primary production, business, local government, community and Indigenous interests. SENRM Board plays an active role in the management of the region's soil, water, ecosystems, coast and marine environments.

The South East is one of the major groundwater provinces in South Australia, with availability of large supplies of groundwater. Groundwater resources in the region are contained within two regionally extensive aquifer systems – an upper, unconfined aquifer and a deeper confined aquifer. The unconfined aquifer is the principal source of groundwater for the South East, and is used for municipal supplies for towns, stock and domestic water, industrial use, and widespread irrigation. Groundwater from the deeper confined aquifer is utilised mainly by the agricultural industry and for municipal water supplies. The surface water resources of the region are characterised by four ephemeral streams with their catchments originating in Victoria. To control widespread annual inundation, an extensive drainage network has been developed. Management considerations for the water resources of the South East include water quality (including diffuse and point source pollution, and salinity), protecting water dependent ecosystems, and the sustainable use of the resource.

With a favourable climate, suitable soils and underground water, the South East has a strong history as a highly productive area that supports a diverse and profitable industry base. The key economic activities in the South East supported by natural resources include plantation forestry, wine, agriculture, dairy, potatoes, fishing/aquaculture and their associated industries. Natural resources-based tourism is also a significant activity in the region. There is a strong interdependence between the social, economic and environmental elements of natural resources management. The involvement of and connections between all natural resources managers is central to effective NRM in the South East.

When adopted, the SENRM Plan will be a statutory document under South Australia's *Natural Resources Management Act 2004*. The Regional NRM Plan should also be consistent with the State NRM Plan 2006. SENRM Plan will guide and lead strategic NRM activity in the region, as well as the operation of the NRM Board. Importantly, the SENRM Plan will be a plan for all involved in NRM in the region, including: Landowners and managers, Indigenous communities, Community groups, State Government agencies, Australian Government, Industry, State Government, Non-government organizations, Local Government, Urban communities, Research institutions, Natural resources users, and Rural communities.

Box 1

NRM Group's Formations, Proposed Functions and Expected Collaborations

Process for formation of NRM Groups	*Proposed functions of NRM Groups*	*NRM Group's functions and expected collaborations*	*Requirements of NRM Group under Act*	*Demonstration by member of NRM group*	*Qualities of an effective NRM group member*
Advertising in local media to invite applications for group membership	Be actively involved in development and implementation of any relevant regional NRM plan at local level	Any other neighbouring NRM group	Keep proper accounts of its financial affairs and must prepare financial statements each financial year	Clear understanding of NRM issues of area of NRM group	Necessary skills and experience to contribute to issues raised at group meetings, and interested in NRM and community issues
Clear criteria for member selection that relate to functions of group	To develop implement or participate in programmes associated with NRM at the local level	Other State agencies, and if relevant, agencies of Commonwealth, and agencies of other States and Territories	Accounts and financial statement audited by Auditor-General each financial year	Knowledge, skills and experience needed to address functions assigned to NRM group	Thinks strategically and understands the implications of decisions made, able to devote time require to meet commitment
Consistency and objectivity in evaluating candidates for NRM Group membership against criteria	To promote public awareness of importance of integrated and sustainable NRM within its area and to support educational initiatives with respect to NRM	Constituent councils for areas of NRM Group and other councils as may be relevant	Provide to regional board a report on its activities for financial year ending preceding 30 June, including a copy of audited accounts and financial statements	At least one member of NRM group must be a woman and one member a man	Makes decisions in best interests of community within its group area, communicates and connects effectively with community, capacity to learn, and support innovation
	To provide advice to SENRM Board, and other bodies or agencies, with respect to assessment of various activities, proposals, situations or circumstances within its area	Relevant industry, environmental and community groups and organizations, and if relevant, persons who own or occupy land within area of NRM Group		Approximately equal representation of men and women on NRM groups across the region	Knowledge of financial provisions and funding arrangements, and familiar with legislation under which Board and Groups operate

SENRM Board is committed to effectively engaging the South East community in managing the precious natural resources. As an essential step in involving the community the Board is setting up NRM Groups within the South East region. These groups will be an important part of the Board's organisation and will play a major role in communication with local communities and implementing natural resource programs in their local area. As part of a major initiative in South Australia, the State Government is strongly promoting a more integrated approach to natural resource management with a strong focus upon community input and coordination at the regional level through the introduction of the NRM Act in late 2004.

During the PATs and FGDs with the members of the SENRM Board, they have revealed that the SENRM Board is enthusiastic about setting up NRM Groups in the South East, for which an information and promotion campaign is being initiated to encourage people to participate. The results of the PATs and FGDs regarding the process for formation, proposed functions, expected collaborations, requirements of NRM group under Act, etc. are presented in Box 1.

Box 2

Interactions with Farmers and Members of SENRM Board

During field visit and interactions with farmers and members of SENRM Board, it was revealed that South East Australia is agriculturally well developed and farmers are very rich. The size of farms was very large ranges from 1000 acres to 3000 acres. The farmers were using highly mechanized agricultural machineries for sowing, irrigating, harvesting and packaging of final produce and relying on groundwater irrigation for crop and no surface water was available to them in Coonwara. However, along the Murray River, small patches of agricultural lands were put under flood irrigation, which were highly fertile flood plains. The groundwater irrigation was also highly mechanised and farmers were using sprinklers (Centre Pivot Techniques) for irrigating potatoes and drip irrigation for large size vineyards and cherry fields. On the cherry fields, the farmers were also found using automatic bird hooters to save crop from the birds.

Three big farmers were main focus of the Case Study, who have given a lot of information on the use groundwater irrigation in the South East. They were: (i) Mr. Terence Buckley (with a potato farm size of 600 Acres, 17 Centre Pivots, and 8000 lambs); (ii) Mr. Gary Widdison (with a potato farm size of 460 Acres, 15 irrigation points of various category, and 4600 lambs); and (iii) Mr. Dale Lewis (owning a big size vineyard farm and also a member of SENRM Board). During the PATs and FGDs, it was revealed that farmers have knowledge of institutional arrangements that facilitate integrated water resource management (IWRM) techniques and were active stakeholders of SENRM Board. They were in regular interactions with each other and knowing the activities and agenda for formation of NRM Groups. The importance of participatory resource management (PRM) was highly recognised by them and as such they were supporters of ongoing institutional reforms in groundwater governance. The sources of funding for groundwater use were known to them. However, due to availability of own funds, they were very less frequently relying on external funding and meeting all expenses from their own funds.

They were strongly of the opinion that local people should be given due consultation in evolving institutional arrangements and changes and there is need for groundwater advocacy to transcend local interests to overcome government inertia. At the same time, farmers as well as Mr. Kent Martin (the President of South Australia Farmers Association -SAFF) have highlighted the problem of severe conflicts between farming and forestry. In their opinion, development of forestry plantations by Forestry Department in South East Australia is jeopardising the future availability of groundwater for irrigation and therefore they were highlighting their opinion and voice on different forums. SAFF was lobbying for putting check on forestry plantations to save the farming community from the clutches of frequent droughts.

The agricultural economy of South East Australia is tied with farming and animal husbandry, and for sustainability of both, the sustainable groundwater use practices are required to put in place, which were visible and practiced by farmers. At the same time, they approve the need for IWRM practices and the need for leadership skills in groundwater management organizations. The farmers were highly enlightened of innovations in agricultural sciences and technology (S&T) and keen to adopt as soon as they noticed them. It was also observed that water of the River Murray is being use wherever possible in conjunction with groundwater for irrigating crops. Thus, the farmers were of the opinion that the river basin organizations should be used when water use is dominated by groundwater extraction. However, they were equally favouring the existing institutional arrangements for development and sustainable use of groundwater and advocating and promoting participatory groundwater governance approach (PGGA).

The existing groundwater institutional arrangements were based on the principle of cost-sharing mechanism, for which the government is charging a suitable levy on farmers for the use of groundwater resource and a system of metering has been put in practice. The recent changes in institutional arrangements has advanced the notion of volumetric allocation and pricing as well as trading of the groundwater resources to overcome the problem of over allocation, over use, and declining and/ or rising water tables. All this has been attempted with community consultations and to achieve the goal of the sustainability and keeping in view the need for water allocation of the environment also.

The sub-basin level management solutions are put in use and the Water Allocation Plans (WAPs) have been framed for small areas falling in South East Australia. Thus, the decentralised participatory groundwater management (PGM) generates more administrative costs, however, at the same time; it has the attributes of good groundwater governance practices. At the same time, there is no denying the fact that financial and human resource availability is vital for good groundwater governance. The farmers were equivocally supporting the idea of private sector participation in groundwater management as being practiced in some parts of South Australia. In areas, where size of farm was comparatively very low, the notion of social equity was emphasised in participatory groundwater management (PGM). At the same time, the farmers were supporting the attribute of accountability for good groundwater governance and the existing legal and jurisdictional setting affecting groundwater governance. However, the SAFF was trying to put pressure on the Forest Department to check the expansion of its plantation derive at different forums and was a source of growing conflicts affecting institutional reforms in groundwater governance, which need to be attended by the relevant authorities on priority basis.

The members of SENRM Board were of the opinion that the ongoing financial and infrastructure support provided by the government for groundwater governance through community participation is very significant component of institutional reform. There was a system of regular meetings of various NRM Boards and through these boards the government is keen to operationalise various NRM Groups for sustainable use of natural resources including groundwater. The process for formation of the NRM Groups has already been initiated in South East Australia. Besides, the WAPs have been prepared and operationalised in various sub-regions for effective interactions between groundwater users and government functionaries to reap the advantages of decentralised resource governance. The farmers' representatives were participating very effectively in whole process and promoting the interests of the community. The farmers have knowledge of the notion of groundwater trading and markets functioning in institutional perspective, however, during the field visits they reported that because of ongoing drought and high water demand, they have little surpluses to trade and markets.

The whole processes of institutional change in groundwater governance in South East Australia was participatory and community consultations have been held at frequent intervals to discuss the modalities of proposed reforms and therefore, the transactions costs involved in putting new institutional regime in practice was reportedly by high. However, once the new institutions starts functioning smoothly and economic and environmental outcomes started accruing to the farmers and the society, the lower administrative costs of institutional change would neutralised the heavy expenses incurred during initial phases of institutional change. Besides, high initial transaction costs of putting new institutional arrangements in place, the other problematic issue is the conflicts between the farmers and the forestry, which needs to be overcome on priority basis so that the farming and forestry should go along and sustainable NRM should be the outcomes.

Case Study-II: Visit to South Australia Natural Resource Management Council

In 2004, the South Australian government introduced a new legislation – The NRM Act 2004. The NRM Act provides a more integrated and sustainable structure for managing the State's natural assets and establishes the NRMC as the statewide peak body for natural resources management. The first skills based Council was appointed in April 2005 based on the range of knowledge, skills and practical experience identified in the Act. The State NRM Plan (the Plan) identifies a 50-year vision for NRM in South Australia, and sets out policies, milestones and strategies to achieve that vision. The Plan recognises the importance of a staged approach, as some actions will take time to show benefits. Many milestones are to be achieved within the next five years, whereas others will take longer. Current trends in the condition of the natural resources show that, despite considerable and laudable efforts by many sections of the community, the condition of the natural resources is declining in many areas. Increasing efforts from all sectors to step up innovation, science-based decision-making, development and exchange of knowledge and technology, and leadership and commitment are critical to safeguarding the condition of South Australia's natural resources and the future wellbeing.

The Plan is based on a set of Guiding Principles (page 5) that aim to clarify the thinking and intent behind successful and sustainable NRM. Three themes are prominent because they inform the philosophy of integrated NRM: (i) managing all resources within relevant natural resource system boundaries, (ii) seeking to involve all parties with interests in a natural system in open planning and decision-making to build shared awareness, understanding and knowledge and develop ownership of, and commitment to, better NRM, and (iii) involving an ongoing cycle of "planning, doing, checking" to ensure continuous learning from actions and improve management practices. It is also essential that this Plan engage a wider audience. Achieving ecologically sustainable development will depend on all users and managers of natural resources working to realise a shared vision. Hence, this Plan is the product of wide-ranging input from community, government and industry bodies and as such represents broad ownership of and commitment to NRM in South Australia. Such ownership is needed to establish common ground and foster innovative partnerships for delivering the Plan's vision and goals. The NRM Council, in conjunction with the community, identified a range of key priorities that have been included in the Plan. Implementing these priorities will occur through agency programmes, regional NRM plans and other NRM activities.

Box 3

Experiences from Meeting of NRMC, South Australia

On Friday, 1 December 2006, a visit has been made to attend the 21st Meeting of NRMC, South Australia, which was held at Plant Research Centre, Hartley Grove, Urrbrae. The aim was to ascertain the institutional arrangements issues related to NRM including groundwater governance especially in the context of prevailing drought in South Australia. To begin with, some of the strategic issues like drought and community engagement and volunteerism to mitigate the problems were discussed, after which the standing agenda items such as Progress Report on NRM Programme, which includes State NRM Plan Implementation update, Review of the NRM Act, and its Monitoring, Evaluation and Reporting were discussed. In between a presentation on Strategic Issues of Drought in South Australia highlighted the emerging scenario due to prevailing droughts in Murray River areas and other areas of the State. The issue of serious deficiency of rainfall was also debated. The crop yields in bad season have declined from 60% of 10 year average in 1994 to less than 45% of 10 year average in 2006. The years between 2002 and 2005 were 'exceptional' years. The issues concerning the National Drought Policy (NDP) were also debated, which includes objectives and key consideration in drought response. It was highlighted that impact of drought is not known in advance and its impact is extended over the period and it affect the whole community. The need to manage the drought risk was also discussed and the role of Drought Response Management Committee (DRMC) was also emphasised.

Two phases of actions initiated by the Government of South Australia were deliberated upon. In the first phase the Government has established a Drought Help Line (DHL) and developed a website to provide the drought related information to the stakeholders. It was apprised to the members that an integrated drought information workshop was also organised and steps were initiated to print and distribute the related information to the farming families. In the second phase, a $4 million package programme has been announced, which includes additional rural financial counselling support to the farming families, stamp duty relief on mortgage extensions due to drought, community support grant, additional counselling support, integrated production system and land management technical support and development of web-based application for managing rural industries. It was also highlighted that NRM Leaders' Forum, Business Leaders, Industry Leaders, and Social Leaders Forum were held to discuss the drought issues and ideas collected through these processes will be analysed for consideration by High Level Task Force.

The decline in the flow of River Murray successively in the last five months was highlighted and informed that out of 60% allocations made, nothing has been allocated to the cotton growers of South Australia, which is a cause of concern and necessary steps are required to attend to the problems confronting the cotton growers. The issue of water marketing for irrigation purposes and need to curtail the irrigation of the households' lawns in the prevailing drought situation has been emphasised. The environmental issues related with River Murray and the opportunities to look at wetland management alongside the River Murray have also been highlighted. The need to pursue the short-term arrangements for storm water management (SWM) and NRM were focused evolving multi-stakeholders' partnerships of the industry, community, agricultural forums, etc. The farmers' representative raised the issue of water levy in South East, as this region has been worse affected by the drought and informed that the impact of the drought on the community is being examined and emphasized the issue of Kangaroo vis-à-vis drought and people's perceptions needs to be examined. The problem of climate change and its impact on biodiversity and the Government's strategies concerning irrigation has also been discussed and members advocated the provisioning of subsidy and incentives to the farmers. The policing of use of groundwater by urban households as a part of drought mitigation strategy was also highlighted. It was emphasised that if management practices are good, there will be no trouble, and if management practices are bad, there will be big troubles.

The Meeting also discussed NRMC Priorities for 2007, which includes strategies to tackle the problem of ongoing drought, support for regional NRM boards, water allocation as part of land-use planning, further integration of NRM, monitoring and information (M & E) and role of science and technology information system (STIS), biodiversity decline, climate change, sustainable production/land use, and State-wide pest/weeds strategy beyond bio-security. The priority areas for NRM in next twelve years, which includes water allocation in wider context, natural resource management, monitoring and evaluation, role of science and technology, biodiversity decline, climate change, pest plan and animals, land use plans including water mapping and planning have also been discussed. The community consultations in NRM were very boldly emphasised. It was suggested that community awareness of the water scarcity needs to promoted by the Government agencies to mitigate the problems of drought and more emphasis should be laid on community participation and volunteerism under Drought Management Strategy (DMS) and its strict implementation.

Case Study III: Visit to Virginia Pipeline Scheme

The Virginia Pipeline Scheme (VPS) is the largest scheme of its type in Australia and was established in 1999 to deliver unrestricted Class A reclaimed water to irrigators on the Northern Adelaide Plains (NAP) in South Australia. Virginia - about half an hour drive to the north of Adelaide - is home to highly productive market gardens and Australia's largest concentration of greenhouse production. The scheme is a co-operative undertaking of the Virginia Irrigation Association (representing market gardeners and other irrigators), SA Water and Water Reticulation Systems Virginia (a private sector subsidiary of Tyco International). As part of its *Environment Improvement Programme*, SA Water constructed a $30 million filtration/disinfection plant (DAFF) to treat lagoon effluent from the Bolivar wastewater treatment plant, producing Class A reclaimed water which can be used for irrigation of Virginia's crops. The quality of the water is closely monitored in accordance with procedures set down by the Department of Human Services to ensure public health standards are maintained.

Box 4

Farmers' Perceptions on Use of Treated Waste Water for Irrigation

During the PATs and FGDs, it was revealed that growers became aware of the scheme from other people, attending meetings, field excursions, the media, and from the Virginia Irrigation Association (VIA). The Department of Health communicated the safety aspects of produce grown with reclaimed water to retailers and wholesalers. A crisis management team met as a matter of urgency to address rumours or negative feedback. The farmers generally thought that awareness meetings relating to scheme were informative. The lack of trust in government, inadequate knowledge of farmers and language barrier were main impediments faced by farmers.

The farmers have given the following reasons for connecting to the VPS: to access a greater volume of water; additional source to replace the existing 'channel water' or to make use of land with no water source; security of supply that allowed expansion; and a cheaper option to main water. The reasons given by the farmers who did not want to connect to the scheme include: they did not need the water, bore water is cheaper, salt levels are too high, or it was difficult to estimate future requirements.

The farmers were initially concerned how irrigating with reclaimed sewage effluent would affect their markets. They compared produce grown with reclaimed water to other sources and confirmed that there is no difference. However, others were concerned about the effects on produce. There were some concerns relating to 'chemical' elements in the water, nutrient levels, and some references to people drinking the water. The quality of Class A reclaimed water in relation to public health is generally accepted as meeting the standards set by the regulators. The farmers were guaranteed water quality negotiated with them during the development of the scheme.

The five-year commitment to purchase of the water was a concern for some and reassuring for others. All farm types use dams and tanks to store reclaimed water.

At least 24 hours storage is required to provide balancing storage, however, there is some confusion over the period of storage required and the purpose of holding the water. There is uncertainty in relation to the rules governing the transfer of water. Some farmers believe the transfer of water is not allowed and one would like an amalgamated system so that the water can be transferred between properties, yet, others confirm that they are already doing this.

It is claimed by some farmers that the BOOT scheme was originally proposed for transferring the scheme back to the growers. The fact that it was established to transfer back to the SA government (which the farmers understand as being 'SA Water') has implications for farmers' concern for future pricing of the reclaimed water. The majority of farmers are concerned about paying for an allocation whether the water is used or not, and some of them specifically advocating a 'user pays' system. Some suggest that a review of the payment system should be undertaken. Few changes were noted that specifically relate to reclaimed water. In some cases the introduction of reclaimed water provided an opportunity for farmers to improve their general farm management skills and knowledge.

On the whole, the farmers have changed their irrigation practices since connecting to reclaimed water such as irrigating for longer periods to leach salts from the soil. Management strategies need to be implemented to counteract the effects of salt on crops; although some bore water also has high salt levels. United Water has a continuous monitoring system to ensure that the water delivered to WRSV meets contractual requirements. General comments relating to the scheme are: its bringing people back into the district; it has made farmers drought proof and sustainable; the farmers were providing a public benefit but their understanding of the scheme was not great; good idea to get rid of the effluent water; and some concerns relating to unknown effects of reclaimed water on the land. However, there has been an overall reduction in groundwater use as the reclaimed water has been substituted for it, but there has also been a clear increase in production and increase in total water demand.

The farmers still hold a general concern regarding salinity issues, although these may not necessarily relate to reclaimed water. Salinity is an ongoing problem in the region and is a difficult issue to address, as there are several potential contributing factors. These include the rising water table, existing saline bore water, questionable fertiliser and soil ameliorant use, and efficiencies relating to the volume of water applied. Stakeholders agree that there needs to be a more holistic view and coordinated approach to monitoring and reporting of data in the region; and strengthening of relationships amongst and between supporting agencies and the farmers in order to achieve sustainability.

Several factors lead to the establishment of the VPS including: (i) The depleting groundwater resources on the NAP causing a 'cone of depression', (ii) The Environment Protection Authority (EPA) ruled that SA Water needed to reduce nitrogen discharged to Gulf St Vincent from the Bolivar plant, (iii) The current irrigators were looking for an alternative water source to the then used Class D effluent and to supplement bore water, (iv) It was also seen as an opportunity to expand horticulture production within the region. The funding was obtained from a Federal grant ($10.8 million) through the Better Cities program and Multi Function Polis (MFP); the State Government through SA Water ($6.7 million); and the private sector via Water Reticulation Systems Virginia (WRSV). The project involves over 100 km of pipeline at a cost of $22 million and a $30 million Dissolved Air Flotation and Filtration (DAFF) plant to achieve Class A quality water (SA Water 2005). The scheme was implemented under a Build Own Operate and Transfer (BOOT) system where ownership transfers back to the SA government from WRSV after 20 years, the term of the BOOT contract.

COMPARISON OF GROUNDWATER INSTITUTIONS IN SOUTH AUSTRALIA AND INDIA

The layers of groundwater institutions in South Australia and India are presented in figures 1 and 2. A perusal of figures 1 and 2 reveals that the Parliament and State legislatures are the higher level institutions of groundwater governance in South Australia as well as India, which empowers the Prime Minister and the Cabinet (the minister of water resources and other concerned ministers) to legislate, enact, and govern the groundwater resources. After the CoAG reforms in 1994, the Government of South Australia has restructured and introduced new levels of decentralized institutions, for example, the creation of 8 NRM Boards has been entrusted with the tasks of implementation of NRM plans, water resource plans, water allocation plans, etc. Keeping in view the acute problem of salinity in South Australia, the National Action Plan for Salinity and Water Quality has been initiated with an allocation of $18.6 million. The recent issue of groundwater requirements for environment has motivated the government to form community groups with grants from National Heritage Trust. The groundwater institutions vary across the States of Australia. In South-East South Australia, the individual farmers have water rights through a system of volumetric allocation and water entitlement, which differs seasonally (more draught – more entitlement – more allocation). Thus, with the increase in environmental awareness of the community, there have been more proposals for a shift from a system of area allocation to volumetric allocation of groundwater.

In India, the Ministry of Water Resources is the national organization responsible for overall planning and management of groundwater resources, for which the technical support, and research and training support is provided respectively by Central Water Commission, Central Groundwater Board, and the National Water Development Agency, and Water and Land Management Institutes (name differs in different States), agricultural universities and other research institutions. At the national level, the Planning Commission provides the project clearance and approves financial allocation to various water projects in different States. Besides, different ministries of the Government of India have their role in influencing the groundwater governance. However, the actual implementation of policies and programmes of groundwater management at the State level are done by the various departments (such as public works, water resources, etc.). At the local level, the pollution control boards, municipalities, and panchayats play a vital role in decentralized groundwater governance. Overall, compared with South Australian groundwater institutions, the institutions in India lacks vigor and calls for major reforms, which will also be evident from Box: 5.

Figure 1
Layers of Groundwater Institutions in South Australia

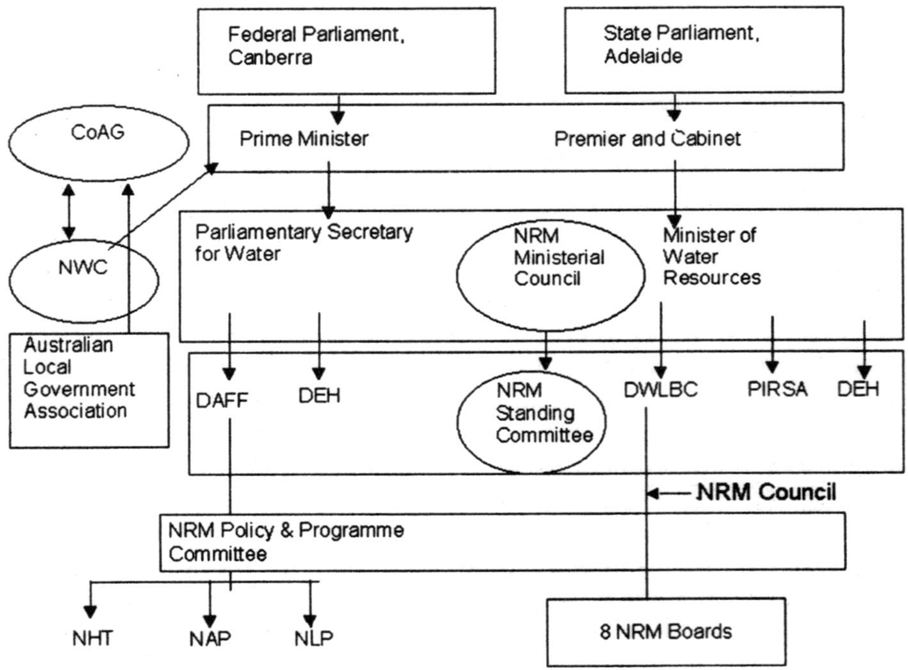

Note:

CoAG: Council of Australian Governments
NWC: National Water Committee
DAFF: Department of Agriculture, Fisheries and Forestry
DEH: Department of Environment and Heritage
NRM: Natural Resource Management
DWLBC: Department of Water, Land and Conservation
PIRSA: Primary Industries and Research in South Australia
NHT: National Heritage trust
NAP: National Action Plan
NLP: National Land Policy

Figure 2
Layers of Groundwater Institutions in India

Note:

CWC: Central Water Commission
CGWB: Central Ground Water Board
NWDA: National Water Development Agency
WLMI: Water and Land Management Institute
AUs: Agricultural Universities
ORIs: Other Research Institutes
NWRC: National Water Resources Council
NWB: National Water Board
MoEAs: Ministry of Economic Affairs
MoA: Ministry of Agriculture
MoRD: Ministry of Rural Development
MoEF: Ministry of Environment and Forests
MoHUD: Ministry of Housing and Urban Development
MoH: Ministry of Health

Box 5
Groundwater Institutional Arrangements in Comparative Perspectives

Institutional arrangements	*India*	*South Australia*
Ownership rights	Does not have any explicit institutional and legal framework specifying water rights, individual rights to GW recognized only indirectly through land rights, allocation rights of percolating GW rest with ownership of land overlying it, allows private rights on GW by viewing it as easement connected to land, allowed	When flowing down a river or sitting in an aquifer, water rights are vested in the State, if GW is subject to restrictions, irrigators and bulk suppliers of irrigation water require an entitlement or an access right to extract CoAG (1994) recommended separation of water

Table Contd...

Institutional arrangements	*India*	*South Australia*
	unlimited withdrawals of GW beneath land by owners, followed by landowners as a customary right which has created inequitable distribution of GW resources among various stakeholders, and ownership rights can be transferred only if dominant heritage i.e. land is also transferred	licenses from land title, allowing water access entitlements and allocations to be deployed to uses generating greater economic returns, and in South Australia 'use-it or lose-it' policy is in place
Trading and Markets	Highly dynamic and complex informal GW markets, very crucial in alleviating rural poverty and livelihood amelioration	Development of GW markets is in its infancy, however, there is considerable potential for GW trading to expand as it offers the potential to solve difficult management issues as demand for use increases. Markets offer an alternative to 'command and control' approaches to moving water use to areas of higher economic value and higher efficiency
Pricing	Subsidized/free electricity due to which GW economy has boomed, but at the cost of energy economy and price of water is decided by free-market forces	Fees and charges on GW management are either zero or at a very nominal level and a broad range and types of fees and charges exist, for example each and every irrigator pays for electricity
Regulation	Inadequate user participation, mainly top down approach expect some parts of country	More institutionalized and decentralized, introduced system of user consultation in GW governance
Implementation of Laws	GW bills have been making rounds for several decades, there is no bill to make them into laws, perhaps precisely because these will be difficult to implement. Some states such	Does not specifically address water except as a limited reference and as a result legislative responsibility for water use and reuse is largely

Table Contd...

Institutional arrangements	*India*	*South Australia*
	as West Bengal, Maharashtra and Andhra Pradesh have enacted GW laws, but these are yet to be implemented in any significant way	left to the State and Territory governments, and each State has developed legislation to manage water resources
People's initiatives	Quite significant popular people's movement for GW recharge, especially in water scarce states like Rajasthan and Gujarat and might indeed offer a window of opportunity for better governance of GW resource not aimed at demand management, rather focused on supply augmentation	Farmer's lobby is much stronger, which makes it possible to apply a wider repertoire of instruments to manage GW, face stiff opposition from farmer's lobby
Monitoring	Practical difficulties with some 25 million well owners scattered over the length and breadth of the country	Switching to volumetric allocation to overcome the problem of over-allocation/under-allocation and use

CONCLUSION AND LESSONS LEARNT

A national agreement on water reform was reached in 1994 and the COAG agreed to a comprehensive water reform agenda that explicitly linked, for the first time, economic and environmental issues with a package of reform measures. The agreement focused on establishing water allocations and entitlements separate from land tenure, backed by secure access to water. It also provided for trading in water entitlements, making water available for ecosystems, as well as institutional reform, public consultation and education, and research. Based on the 1994 COAG agenda, there has been some progress across all jurisdictions. For example, the pricing reforms and separation of government regulatory and management functions and commercialization of government water businesses have begun. The water trading, particularly temporary water trading has expanded and the water management arrangements have been developed to take account of the range of water uses including: extractive uses, environmental needs and the needs of stressed and over-allocated river systems. The better arrangements for examining proposals for new rural water infrastructure (against the tests of economic viability and ecological sustainability) are now applied and water legislation to underpin these reforms has been enacted. The greater levels of accountability, transparency and reporting have been instituted and stakeholder consultation and community engagement have been improved; and water access entitlements are being separated from land titles – an almost revolutionary achievement in the context of South Australia's historical treatment of water.

The groundwater governance reform programme in South Australia has had some success, but much remains to be done if the goal is to meet the reasonable expectations of future generations. A strong commitment and ongoing cooperation of governments and stakeholders will be a key factor in sustaining and driving the momentum of groundwater governance reform. The continuing efforts on integrated water management, cross-border cooperation, improving irrigation practices and water use efficiency will be necessary to ensure improved productivity and environmental sustainability of water resources and secure access to water for all.

The above discussion draws attention to a number of significant policy issues, which broadly includes the search for a robust set of institutional arrangements; and an efficient and equitable transition pathways to such a set of robust institutional arrangements. Thus, if arrangements for managing groundwater allocation and use are to be robust, the components of existing systems must be separable from one another. In unseparated systems, whenever any problem emerges, the entire system comes under review, negotiations are complex, and an opportunity is provided to reopen old agendas. The clue to the robust resolution of many of South Australia's water resource problems lies more with separation than in integration. The robust system would need to facilitate: (a) Resolution of resources allocation between consumptive use and the environment, among consumptive users, and of issues related to distribution and use; (b) Secure, economically efficient and low cost trading and administration; (c) Assignment of risks making it clear where responsibility lies, under what circumstances compensation is due, and specifying the processes for obtaining redress; and (d) Management of externalities associated with use – the interests of third parties, future generations and the environment – with minimum controversy. A robust system must pass the conventional tests of efficiency and fairness. For this to occur in a changing world, the system must not only be build on a solid conceptual foundations, but also be flexible and adaptive, transparent and equitable.

Robust design principles also require that careful attention be given to the allocation hierarchies. In many parts of South Australia, entitlements and licences have been distributed both to water supply companies and trusts in the form of a bulk licence and to irrigators as an individual use licence. Robust systems recognize that only one interest can be primary. All other interests must be defined via an encumbrance on the water access entitlement or use licence. From an efficiency perspective, it does not matter whether the primary interest is held by a water supply company periodically selling water to irrigators, or by individual irrigators. Either way, each party can be expected to contract with the other party to gain access to the services they require. If irrigators are the primary holders, bulk access entitlements are unnecessary for all but supply system losses.

Other design detail includes options to enable trade between surface and groundwater and for the development of markets for channel capacity (with entitlement shares) and salinity (with use licences). Where annual and perennial land-use systems coincide, there is an option for high security and general security entitlements shares to be issued. There is also need to carefully consider whether or not entitlements should be defined as share of the entire system or part of the system. If entitlements are not defined as shares of the entire system then trading environment arbitrate opportunities emerge as people trade to take advantage of flawed exchange rates and system reviews. One pragmatic way of dealing with this problem is to introduce tagged entitlement systems so that the exchange rate risks and reform risks cannot be passed via a trading system to others. The tagged entitlement system had been used in South Australia for many years. The administrative costs of managing such a system could become considerable. Clearly, the fewer the tags the cheaper the system will be to administer and the easier it will be to communicate implications of any changes in exchange rates and system charges to entitlement holders. Deciding upon the optimal number of types of entitlement is a non-trivial issue that requires very careful analysis.

The sequencing of reform process is equally important. South Australian water policy makers are currently showing a preference for the removal of impediments to permanent water access entitlements trading and securing additional environmental flows for river system and groundwater systems. The presence of market impediments, even given the comparative absence of externality cost obligations, means that the value of water access entitlements, allocations and use licences are less than they would otherwise be. In over-allocated systems, the result is sequencing opportunity and a sequencing trap. Remove market impediments to permanent trading before a robust entitlement and allocation system is put in place and the cost of reform increases – especially if all water is to be purchased using a voluntary offer or similar market-like schemes. However, if water is secured for the environment at the same time as market impediments are removed and entitlements are respecified, then the net market gain to entitlement holder from this process can be used to lessen the need for transitional assistance payments. Failure to pursue these reform sequencing opportunities and, in particular, to implement the politically easy options without addressing the underlying flaws is likely to result in the emergence of problems that will hinder opportunities to make further progress. As a general rule, it will be less costly for governments to acquire additional water to enhance the environment before impediments to permanent water trading are removed and before a new robust water access entitlements system is put in place.

In most parts of South Australia, market and market-like arrangements for water access entitlements and use licences are in their infancy and characterized by arrangements that are designed to prevent trade from resulting in the movement of water into highly saline areas and to ensure that existing delivery expectations can be met. The private markets for access to channel capacity during peak periods and salinity impacts are virtually non-existent. The market arrangements that allow counter-cyclic trading of water environmental allocations do not exist. As a result, the removal of so-called market impediments to regional and interstate trading without the development of new market arrangements could result in emergence of a new set of water quality and adjustment problems. In short, the simple removal of trading restrictions without the introduction of new control mechanisms could result in ineffective trading.

Within regions, the main reason for current restrictions on temporary trading with consequent high transaction costs relates to the need to check that a temporary trade will not generate increased environmental impacts, and to ensure that the traded water can be delivered. Between regions and types of licence, the necessary checks remain but require more time consuming coordination among agencies, regional offices and officers. In all fully allocated systems, the further development of seasonal allocations (temporary trading) coupled with a freeze on the expansion of access entitlement (permanent) trading could be used to limit arbitrage opportunities until robust accounting systems are put in place. Otherwise, administrators must expect the cost of structural adjustment and transitional payments to be much higher than otherwise could be the case. There are economies of scale in reform processes. As a general principle, the more reforms undertaken at once mean the lower the total transaction costs and opportunities for arbitrage.

An interesting way of facilitating permanent trade without the reassignment of risk is to tag all water entitlement by the region they are sourced from. If this is done, then volume of allocations that attach to that entitlement at any location would always be calculated by reference to the place from which the entitlement came from and the exchange rates that operate at each allocation period. This mechanism makes it clear that the risk associated with any trade would always be assigned to the person who purchases the entitlement and not the system as a whole. The process of dealing with each licence is expensive as most involve third parties and complex registration procedures. As robust systems guarantee the nature of all dealings associated with an entitlement and/or a use licence, these transaction costs are significant. In many cases, it will also be advantageous to reduce the number of types of entitlement – a high security entitlement and a general security entitlement. With two types of entitlement, water users can mix the two options to manage allocation risk. Some regional tagging may also be justified but as more and more regional tags are introduced the cost and complexity of the resultant system must be expected to increase exponentially.

Lastly, the extent of the net financial impact of entitlement reduction and upgrade to a robust entitlement and allocation system will depend both on the volume secured for the environment and the extent and nature of the existing imperfections. If market and market-like mechanisms are used, then the financial resources necessary will need to be sourced. Determining how much of the cost of securing environmental water should be borne by water licence holders and how much by the rest of the community is a distributional policy issue. Significant benefits from the suggested reforms will flow to irrigators in the form of valuable, fully specified and fully tradable water access entitlements and allocations.

A number of important lessons have been learnt by analysing the institutional reform process in South Australia, which can be applicable in Indian context too and these includes:

1. integrated water resource management must account for the total water cycle, recognising the connectivity between surface and ground water systems, if it is to ensure security for consumptive users and the environment;
2. integration must address a range of cross-cutting issues such as: large scale impacts of land use on the water-balance of catchments, increased water efficiency, and the impact of upstream activities on downstream water quality and availability;
3. effectively balancing management of the environment and the need for certainty of access for water users - water users need sufficient certainty and confidence about on-going access to the resource to enable investment, particularly in water efficient technologies and practices;
4. water markets and water trading provide mechanisms for the more efficient use of water. However, the impediments to water trading includes different types of water access rights regimes that exist between states and regions, and the restrictions placed by water authorities on trading upstream and downstream;
5. continual improvement of data and scientific information is needed to support decisions and actions, although it is likely we will never have all the information we would prefer;
6. extensive public consultation engaging all stakeholders is essential for effective decision making; and
7. the costs of remediation far outweigh the costs of a precautionary approach to development or protection.

There remains a critical need for better knowledge and understanding of water resources including basic data, particularly in the areas of groundwater systems and interconnected surface/groundwater systems. The better information on groundwater resources, based on agreed scientific assessment and targeted research, and the development of nationally compatible water resource accounts will help to underpin better water resource management. The development of water balances at a variety of geographic scales and monitoring the factors influence water balances over time will be important for managing the risks to water resources and changes in supply and demand. The changes to water balances over time will result from changes in land use, climate, demography, industry and water policies.

With the right investment climate, there is scope for developing and implementing a wide range of technologies for improving water use, including water metering and water measurement systems, better piping and channel systems and more water-efficient irrigation systems. There is also scope for developing technologies that provide real-time monitoring and application of water to suit the requirements of individual crops, with possible extension to the paddock and catchment scale. There is greater scope for a much improved investment climate in groundwater sector, particularly with provisions for more secure water access entitlements and for a substantially improved framework for managing the risks of any future reduction in water availability. The potential therefore exists for the groundwater governance reform agenda to drive greater investment in water smart technologies and infrastructure. However, issues will continue to emerge around the possible negative impacts of greater water use efficiency on groundwater accessions and return flows to rivers and streams. The clarity about ownership of groundwater saved by the new technologies and the availability of the groundwater for consumptive or environmental uses will also be important.

NOTE

1. Historically, water has been allocated on an ad hoc basis with little regard to its scarcity. As a result, many water systems are over-allocated, in that the total volume of water that can be extracted by entitlement holders exceeds the sustainable level of extraction for that system. For example, it is estimated that in New South Wales, licences and water allocations equal 120 per cent of total available water resources (Melville and Broughton, 2004). Many water users have historically held licences providing for higher water extraction amounts than they have actually used. For as long as the actual amount extracted was below the sustainable level of extraction, this was not a problem. However, as irrigators have increased production and used more of their licences, over-allocation has led to overuse, in that the total volume of water physically extracted from the system exceeds the sustainable level of extraction. Also contributing to the emergence of overuse is the fact that as individual irrigators adopt

practices such as drip irrigation to increase the efficiency of their water use, less excess water re-enters the water system. The over-allocation and subsequent overuse of river systems has left many of Australia's water resources under significant pressure. Of Australia's 325 surface water basins, 84 are currently overused or close to it. These systems account for about 55 per cent of total water use in Australia. The areas under the most pressure from overuse are predominantly located in the eastern States, with the Murray Darling Basin showing the greatest signs of pressure (NLWRA,2002).

2. Australia uses Megalitres (ML) as the primary measure of water volume. 1 megalitre = 1,000,000 litres or 1,000 cubic meters, 1 Gigalitre (GL) has 1000 megalitres.

3. Water extraction (diversion for surface water or abstraction for groundwater) and/or allocation as a percentage of the sum of sustainable flow regime (surface water) and sustainable yield (groundwater).

4. The sustainable yield for groundwater systems is defined as the level of extraction measured over a specified planning timeframe that should not be exceeded to protect the higher value social, environmental and economic uses associated with the aquifer. Operationally this definition would be interpreted as groundwater use being sustainable where groundwater level and pressure are maintained. In May 2000, the National Groundwater Committee defined the 'sustainable yield' as the groundwater extraction regime, measured over a specified planning timeframe, that allows acceptable levels of stress and protects the higher value uses that have a dependency on the water.

5. Schmid (1972, p.893) defines institutions as "set of ordered relationships among people which define their rights, exposure to rights of others, privileges and responsibilities."

6. Institutional robustness requires a focus on some new allocation mechanisms and assemblages of these mechanisms that differ from those considered by most policy makers at the time a system is under review and from those previously encountered. Robust systems have an architecture that can be expected to produce efficient and politically acceptable outcomes in an ever-changing world (Jen, 2003). Robust systems persist, are adaptable, and can withstand the test of time. Robust design principles also require that careful attention be given to the allocation hierarchies. In many parts of Australia, entitlements and licences have been distributed both to water supply companies and trusts in the form of a bulk licence and to irrigators as an individual use licence. Robust systems recognize that only one interest can be primary. All other interests must be defined via an encumbrance on the water access entitlement or use licence. From an efficiency perspective, it does not matter whether the primary interest is held by a water supply company periodically selling water to irrigators, or by individual irrigators. Either way, each party can

be expected to contract with the other party to gain access to the services they require. If irrigators are the primary holders, bulk access entitlements are unnecessary for all but supply system losses.

7. Conjunctive water management requires a large amount of joint and coordinated effort among catchment communities and organisations; and institutions are the key to such coordination. Even where conjunctive water management is physically possible and economically feasible, whether and how it is pursued depends on the institutional setting. By better understanding, identifying and clarifying the relationship between institutional arrangements and water management outcomes, institutional researchers can help advance the larger goal of improving the sustainable use of water resources. Recently, Blomquist *et al* (2004) investigated the feasibility of conjunctive water management as a means of improving the water management, as well as how institutions affect conjunctive water management.

8. The NWI agreement includes objectives, outcomes and agreed actions to be undertaken by governments across inter-related elements of water management. A key aim of the Initiative is to restore surface and groundwater systems to environmentally sustainable levels. Water sharing plans will help to bring certainty for consumers, and allow them greater scope to plan agricultural and other activities. The NWI will work towards the removal of institutional barriers to trade in water. Water trading systems will have the widest possible geographic scope, and will not be restricted to within catchment areas. Water pricing and institutional arrangements under the NWI will promote economically efficient and sustainable use of water resources, water infrastructure assets, and government resources; ensure sufficient revenue streams to allow efficient delivery of services; facilitate the efficient functioning of water markets; give effect to the principles of consumption-based pricing and full cost recovery; and provide appropriate mechanisms for the release of unallocated water. There is need to identify within water resource planning frameworks the environmental and other public benefit outcomes sought for water systems and to develop and implement management practices and institutional arrangements that will achieve those outcomes. The outcome of water resource accounting is to ensure that adequate measurement, monitoring and reporting systems are in place in all jurisdictions, to support public and investor confidence in the amount of water being traded, extracted for consumptive use, and recovered and managed for environmental and other public benefit outcomes. The NWI will ensure healthy, safe and reliable water supplies; increase water use efficiency in domestic and commercial settings; encourage the re-use and recycling of wastewater; facilitate water trading between and within the urban and rural sectors; encourage innovation in water supply sourcing, treatment, storage and discharge; and achieve improved pricing for metropolitan water. The NWI identifies areas where there is significant

knowledge and capacity building needs for its ongoing implementation. Signatories to the Initiative have agreed to identify the key knowledge and capacity building priorities needed to support ongoing implementation of the Agreement, and identify and implement proposals to more effectively coordinate the national water knowledge effort. Government are to engage water users and other stakeholders in achieving the objectives of the Initiative by improving certainty and building confidence in the reform processes; transparency in decision making; and ensuring sound information is available to all sectors at key decision points. New and improved measuring, monitoring, reporting and accounting procedures will be introduced, and improved public access to information will increase public acceptance of the Initiative.

9. Adaptive Management is an approach that involves learning from management actions, and using that learning to improve the next stage of management (Holling, 1978). It is "learning to manage by managing to learn" (Bormann *et al*, 1993).

RFERENCES

Australian Academy of Technological Sciences and Engineering (1999), Water in the Australian Economy, Australian Academy of Technological Sciences and Engineering, Parkville, Victoria.

ARMCANZ (1997), Allocation and use of groundwater: a national framework for improved groundwater management in Australia, Occasional paper number 2, Policy position paper for advice to States and Territories, Commonwealth of Australia, Agriculture and Resource Management Council of Australia and New Zealand.

ARMCANZ (1998), Special Water Meeting: Resolution No. 6, Groundwater Management, Adelaide, Agriculture and Resource Management Council of Australia and New Zealand.

AWRC (1965), Review of Australia's water resources 1963 - Stream flow and underground resources, Australian Water Resources Council, Department of National development, Canberra.

AWRC (1976), Review of Australia's water resources 1975, Australian Water Resources Council, Canberra.

Blomquist W., Edella S. and Tanya H. (2004), Common Waters, Diverging Streams: Linking Institutions and Water Management in Arizona, California, and Colorado, Resources for the Future, Washington, D.C.

Bormann, B. T., Cunningham, P. G., Brookes, M. H., Manning, V. W., Collopy, M. W. (1993), Adaptive ecosystem management in the Pacific Northwest, USDA Forestry Services General Technical Report PNW-GTR-341.

Bromley, D. W. (1982), Land and water problems: An institutional perspective, *American Journal of Agricultural Economics*, Vol. 64: 843-44.

Crawford S, Elinor O. (1995), A grammar of institutions, *American Political Science Review*, Vol. 89(3): 582-598.

Dixon, J. A. and L. A. Fallon (1989), The concept of sustainability: Origins, extensions, and usefulness for policy, *Natural Resources*, Vol. 2: 73-84.

DRE (1983), Water 2000, A perspective on Australia's water resources to the year 2000, Department of Resources and Energy, Australia.

Fullagar I. (2004), Rivers and Aquifers: Towards conjunctive water management, Workshop Proceedings, 6-7 May, Bureau of Rural Sciences, Adelaide.

High Level Steering Group on Water (1999), *Implementing the COAG Framework for Water Reform*, Natural Resource Management Policy Division, Agriculture, Fisheries and Forestry, Australia.

Holling, C.S. (1978), *Adaptive Environmental Management and Assessment*, Wiley: Chichester.

Jen, Erica (2003), Stable or robust? What's the difference, Santa Fe Institute, Santa Fe.

Klein, P. A. (1985), Economic activity and the public sector, *Journal of Economic Issues*, Vol. 19: 419-23.

Ludwig, D., Hilborn, R., Walters, C., (1993), Uncertainty, resource exploitation and conservation: Lessons from History, *Science*, Vol. 260: 17-18.

Melville, F. and Broughton, P. (2004), Trading in water rights, water and the Australian economy, Committee for Economic Development of Australia, Melbourne.

NGC (2004), Knowledge gaps for groundwater reform, A strategic directions paper for water researchers, National Groundwater Committee, Australia.

NLWRA (2001), Australian Water Resources Assessment 2000, Surface water and groundwater – availability and quality, National Land and Water Resources Audit, Canberra.

NLWRA (2002), *Australians and Natural Resource Management*, Natural Heritage Trust, Canberra.

NWI (2004), Intergovernmental Agreement on a National Water Initiative, Council of Australian Governments, Australia.

Pagan, P., Crase, L. (2004), Does adaptive management deliver in the Australian water sector, 48th Annual Conference of Australian Agricultural and Resource Economics Society, 11-13 February, Melbourne.

Peach, J. T. (1987), Distribution and economic growth, *Journal of Economic Issues*, Vol. 21: 1495-1529.

Petr, J. L. (1984), Fundamentals of an institutional perspective on economic policy, *Journal of Economic Issues*, Vol. 18: 1-17.

Schmid, A. A. (1972), Analytical institutional economics, *American Journal of Agricultural Economics*, Vol. 54: 893-901.

20

INSTITUTIONAL AND LEGAL GROUNDWATER MANAGEMENT FRAMEWORK: LESSONS LEARNT FROM SOUTH AUSTRALIA FOR PAKISTAN

Bakhshal Lashari, Jennifer McKay and Karen Villholth

INTRODUCTION

Significant populations of South and Southeast Asia have come to increasingly depend on groundwater for use in agriculture as well as other economic sectors, including domestic supplies. Small holders in developing agrarian economies of India, China, Nepal, Bangladesh and Pakistan have huge stakes in groundwater reduction programs in recent decades. However, this sudden boom in groundwater development has also triggered the secular and seasonal groundwater level declines, wells running dry and well failures, rising energy use and pumping costs, weakened drought protection, salinity ingress in coastal area and health hazards due to arsenic, fluoride or other naturally inherent, -toxins or waste-or agriculturally derived chemicals and these impacts are seriously threatening the long-term sustainability of the use of the resource. Groundwater has been developed in the South and South East Asia primarily during the last 40 years. The rate and scale at which this has, and still is, occurring is so intense that it is causing concerns, not only within the countries themselves, but also at an international level. This is because unsustainable groundwater use potentially influences livelihoods and food security for huge numbers of people dependent on groundwater for subsistence or commercial farming in these regions (many millions of people) as well as potentially influencing international food trade and associated policies(Villholth and Sharma, 2006).

Groundwater problems emerge slowly and incrementally. The impacts are lowering of groundwater tables and pollution spreading to surrounding and downstream areas are long. Conversely, the timescales for remediation are also long, and impacts noticed today will persist for some time, even after the reversal of the original stresses. Hence, emerging problems, which are indeed evident in many parts of these countries today, need to be taken seriously and confronted with a degree of priority (Burke and Moench, 2000). Global groundwater use is about 1000 km3/year, which is around 8.2 per cent of annually renewable groundwater resources (Shah, 2000), but its contribution to human welfare is huge. India, China and Pakistan alone account for one-third of global groundwater use. Globally, growth in groundwater irrigation has had little to do with the occurrence of the resource, as its intensive development has tended to occur in arid and semi-arid regions with relatively poor groundwater endowments. There appears to be a good correlation between high population densities and high tubewell densities in India, Pakistan, Bangladesh and China. India alone is adding about 1 million new tubewells every year since last 15 years and there is no sign of deceleration (Shah et. al., 2003).

The five countries Pakistan, Bangladesh, India, China and Nepal are to a large extent similar and parallel in groundwater use and management. The groundwater development in these countries started in the 1970's, with the introduction of tube well and pumping technology, rural electrification and demand for increasing crop production due to population increases. Groundwater development has occupied an important place in poverty alleviation policies because of its role in stabilizing the agriculture and ensuring food supplies and livelihoods for farmers of which many are in the lower brackets of household income. The major use of groundwater is considered for agriculture, though it is being used for other purposes like livestock, municipality and industries, but here is focused on irrigated agriculture use because this sector is utilizing more share of water. The use of groundwater of some countries of South Asia and Australia is shown in table 1. The extraction of groundwater is mostly being done by huge number of tube wells which are shallow and deep. In many countries more than 50 per cent population is dependent on the groundwater (Table 2). Australia alone has groundwater availability more than 25 million cubic meter, from which only 10 per cent is currently being utilized. Most of the population of Australia is directly or indirectly dependent on the groundwater. It is very interesting to know that groundwater of Australia is delineated/divided in to units. These units are basically explained on the basis of water availability. The socio ecology of groundwater use in Australia is explained in table 3.

Table 1
Use of Groundwater for Agriculture in Various Countries

Country	*Groundwater (%)*
Pakistan	45
India	53
Bangladesh	69
Nepal	50
China	25
Australia	34

Table 2
Socio Ecology of Groundwater Use

Country	*Annual use (BCM)*	*No. of Tubewells (M)*	*% of Polpulation dependent on GW*
India	150	20	55-60
Pakistan	68	>1.0	60-70
Australia	Na	0.5 (about)	Na
USA	100	0.2	<1-2
China	112	3.5	20-25
Bangladesh	Na	1.23	64
Nepal	1.1	0.86	na

Table 3
Socio Ecology of Groundwater Use in Australia.

Groundwater available (MCM)	25.78
Current use (MCM)	0.25 (10%) Agri, livestock and domestic use
Population dependent	1.0 million directly and remaining indirectly
Groundwater management units	538, Federal Govt criteria set 2000
Management units close to over allocated	168, allocation is based on CWR and other usages
Management units over used	161

OBJECTIVES AND METHODOLOGY

The main objectives of the paper are:

1. To review institutional and regulatory groundwater framework in South Australia and Pakistan;
2. To compare approaches of groundwater governance in two countries and find any coherence between them; and
3. To recommend, from the experience of South Australia, any possibility of replication of any approach in the context of Pakistan for sustainable groundwater management.

The available reports on groundwater governance including institutions, legal and socio economic aspects and some articles on water allocation procedures were reviewed and internet material were collected on the relevant aspects of groundwater governance. Besides, numerous institutes were visited and meetings with groundwater experts and farmers were organized to collect relevant information.

GROUNDGROUNDWATER GOVERNANCE MECHANISM IN AUSTRALIA

To understand practical aspects of management of groundwater resources an exposure visit to South Australia was made which was organized by the International Water management Institute (IWMI) with the help of Department of Business, University of South Australia (UniSA). During visit literature survey, visiting various institutes, meeting with groundwater resource managers and users were made. The outcome of the visit and literature survey has been incorporated in the following sections. These sections cover the legal and institutional framework to manage the groundwater in South Australia. Also some aspects on socio economic have been highlight.

Groundwater is the main resource of water that can be extracted as per need for all kind of usages at any time like agriculture, livestock, municipality and industries. Groundwater management considers the aspects of socio-economic dimensions (demand- side management), hydro-geological dimensions (supply-side management), social development goals and policies and regulatory framework that look into water rights, permits, allocation, water balance and others.

The management of groundwater in Australia is primarily the State Governments who are the legislative bodies. But, since 1995 Federal Government has been influential for setting up the state legislative agenda-called Council of Australian Govt. Reforms 1995. Since 1896 several legislative and regulatory frame work have been adopted in Australia to manage the groundwater. These are summarized in Table 1.

Table 4
Regulation of Groundwater up to 1995 in Australia

Regulations	*Main features/contents of regulations*
Common law up to 1896	• Individual growers is the owner of water below his/her feet • Neighbor has no right to go against him even if he runs well affecting his aquifer
Vesting of power to Crown (Govt.) 1896-1997	• Issuing license and allocation system for groundwater use and manage • Certainly created the power to require licenses for wells and powers to manage
Council of Australian Governments reforms 1995-ESD	• Enhance individual and community economic development • Provide equity among generation and • to protect biological diversity and maintain ecological process

Source: McKay JM 2006

Reforms 1995 and Onward

In 1995 and onward the reforms have been introduced in Australia to make sustainable use of natural resources. These reforms are highlighted as under.

- The reforms were initiated in 1995 but Act was passed in 1997 as Water Resources Act 1997.

- In this Act water entitlements initially tied to land.
- These entitlements were subsequently separated operationally from land to allocation in River Murray from 1983 and in the Northern Adelaide Plains groundwater area from 1984, and legislatively with the enactment of the Water Resources Act 1997.
- This Act laid out a comprehensive assessment, allocation, management and planning framework for environmentally sustainable water resource management.
- It is in consistent with 1995 Council of Australian Government (COAG) water reforms.
- Over a period of a decade South Australia has come up with Natural Resources Management Act 2004 which came into effect on 1 July 2005, and abolished Water Resources Act 1997.

Main Features of Natural Resources Management Act 2004 for South Australia.

- The main objective of the Act *2004*, is to achieve ecologically sustainable development through an integrated approach to promote the use and management of natural resources.
- NRM Act reforming current institutional arrangements and decision-making processes.
- It also provides integrated and sustainable structure for managing State's natural assets:
- soils, water, plants and animals and the diversity of landscapes and ecosystems.
- NRM runs by a council as apex body-called NRM Council that advising the Minister on issues and policies. NRM has eight *regional NRM Boards.*

Natural Resources Management

The Water Resources Act 1997 formalised a number of previous ad hoc operational and legislative arrangements for the management of watercourse and groundwater resources in South Australia. This Act established water access entitlements that were initially tied to land. These entitlements were subsequently separated operationally through the transfer of water allocations separate from land along the River Murray from 1983 and in the Northern Adelaide Plains groundwater area from 1984, and legislatively with the enactment of the *Water Resources Act 1997.*

Although not specifically provided for in the 1976 legislation, the management and allocation regime was underpinned by local community developed management policies and guidelines for each of the controlled water resources, which were subject to periodic reviews and became the first statutory water allocation plans under the 1997 Act.

The *Water Resources Act 1997*, laid out a comprehensive allocation and planning framework for water resource management in South Australia, consistent with 1994 Council of Australian Government (COAG) water reforms. The cornerstone of the legislation was a hierarchy of plans for the assessment, allocation, management and monitoring of water resources in the State, including environmental water.

Over the past decade South Australia has undergone an extensive consultation process seeking input on what's important and how to deliver natural resource management in South Australia. The outcome is the *Natural Resources Management Act 2004* which came into effect on 1 July 2005.

This Act simplifies natural resource management in SA by reforming current institutional arrangements and decision-making processes with the development of an integrated partnership model. The NRM Act brings together three Acts or legislations, the *Animal and Plant Control (Agricultural Protection and Other Purposes) Act 1986, Soil Conservation and Land Care Act 1989* and *Water Resources Act 1997*.

The NRM Act incorporates the powers and functions in these Acts as well as enabling the establishment of a new framework for NRM institutions and decision-making and the administered process that delivers the Commonwealth/State programs such as the Natural Heritage Trust, the National Landcare Program and the National Action Plan for Salinity and Water Quality.

The objective of the *Natural Resources Management Act 2004*, is to achieve ecologically sustainable development in the State by establishing an integrated scheme to promote the use and management of a range of natural resources.

The State Natural Resources Management Plan, developed in consultation with the community, sets the overall strategic direction for the management of all water resources in the State and addresses the holistic management and integration with other natural resources.

The responsibility for developing Regional Natural Resources Management Plans, including detailed water allocation plans that define the consumptive / non consumptive balance for each catchment, is devolved to local Natural Resources Management Boards. These plans are developed using hydrologic and hydro geological models based on historical, current and predicted data with input from relevant experts and extensive community consultation.

Water allocation plans for different water resources are designed to address all aspects of resource management at the catchment level and set out the criteria for determining water access entitlements together with appropriate site use and trade conditions with a focus on sustainability.

Groundwater Allocation Principles

- To make sustainable allocation and use of groundwater, the following recommendations are made.

- Groundwater should only be allocated within the limits of the defined sustainable yield.
- All extraction bores in Groundwater Management Units other than Domestic and livestock bores should be licensed.
- In systems which are under-allocated, if need has not been demonstrated over a specific time frame then the unused portion of allocations should be forfeited. Criteria need to be set for recovering the unused portions of allocations and a set of rules developed, including consideration of the threat of degradation of the resource.
- Regulation and accounting of Domestic and Stock use should be introduced in fully allocated and over-allocated systems where use is significant and needs to be managed.
- In over-allocated systems States should be proactive in reducing allocations to within the sustainable yield as early as possible.
- The time frame for reductions of allocations to the sustainable yield should be assessed based on consideration of risks of the degradation of the resource (including the impact on groundwater dependent ecosystems) as compared with the economic and social impacts of the reduction.
- States should ensure that allocations are reduced to the sustainable yield as specified in the management plan. The management plan should ensure use is managed to the sustainable yield.
- A community-based process should be undertaken which identifies methods to reduce over-allocation. However, States should establish a default mechanism which is publicly understood.
- States are encouraged to develop metering programs for all high yield bores where over allocation or over-use is believed to be occurring.

Natural Resources Management Council, 2004

To manage all the natural resources in integrated approach in South Australia, and to implement this mechanism, the South Australian government has introduced new legislation – *The Natural Resources Management Act 2004*. The NRM Act provides a more integrated and sustainable structure for managing State's natural assets which includes but not limited to soils, water, plants and animals and the diversity of landscapes and ecosystems.

The new Act has established a State NRM Council as the apex body that advising the Minister on NRM issues and policy. It comprises nine members, four are drawn from the SA Farmers' Federation, the Conservation Council of SA, the Local Government Association and Aboriginal interests. The remaining five members are community nominees appointed by the Governor on the recommendation of the Minister.

Regional Natural Resources Management in South Australia

To manage the natural resource in sustainable development, various Board/units have been developed based natural resources. Each Board or unit has different characteristics and ecological regions. These boards have their own administrative body which looks after the natural resources but will have to work under the NRM Council. There are eight regional Natural Resource Management Boards in SA under the umbrella of Natural Resource Management Council. These boards are appointed by the Governor under the provisions of the *Natural Resources Management Act 2004*. The set up of the board is almost same for all eight boards. However the formation of all boards has taken place at different times.

SOCIO-ECONOMIC DIMENSION OF GROUNDWATER USE IN SOUTH AUSTRALIA: EXAMPLES OF FIELD VISITS TO SOUTH AUSTRALIA.

South Australia is one of the states of Australia which is dependent more on groundwater that is being used for agriculture, livestock and domestic. There are several departments working on groundwater management that includes groundwater hydrology, groundwater governance, groundwater allocation, rights and policies.

During three weeks visit to South Australia where several departments, organizations and institutions were visited. The outcome of the visit and meetings are narrated in the following paragraphs.

In South Australia groundwater is the only and main source of water resources which is being used for all type of irrigated agriculture, livestock and domestic use. In irrigated agriculture and main livelihood is the agriculture and livestock. As it is main source of water and survival of human beings this source of water is used very efficiently.

The farmers or land owners of South Australia are well educated and well prepared for managing the natural resources in sustainable way, particularly the groundwater. They are economically self sufficient and well equipped with technical information.

The farmers and civil society people are well aware of regulatory framework for groundwater management. During the development process of regulatory framework most of the stakeholders are being consulted and therefore the community own the rules and regulation being implemented.

Water users understand the importance of groundwater use and therefore most of the farmers have opted high irrigation efficiency systems-such as sprinkler, and drip systems. No doubt, the economy and agriculture return play a major role to use high irrigation efficiency system.

Marketing of the products are very helpful to the community. For example: if they produce potatoes it is being sold at the price fixed before sowing, because they make an agreement with the buyer before cultivating crop specially cash crops. However, they use fodder for the livestock which is the main source of income from the agriculture farming.

Sustainable management of groundwater is the main concentration at both the levels of management groups –Government and the users. Therefore every pump has been put the metering system which is regularly being monitored. Besides, water table is also regularly being recorded by both parties.

Protecting and preserving the groundwater resources, South Australia have started recycling of wastewater- as the case of the North Adelaide plain (Virginia piping project).

GROUNDWATER GOVERNANCE IN PAKISTAN

Groundwater in Pakistan

The massive development of groundwater from the Indus Basin aquifer started about 30 years ago. This source is tapped by about 20,000 public and 550,000 private tube wells. The total abstraction of groundwater in Pakistan is estimated at 50 BCM out of potential of 63 BCM (Kahloon and Majeed 2004). Most of this is used for agriculture. Also more than 70% of the population of Pakistan depends on groundwater for drinking water and domestic use.

Much of the economy in Pakistan depends on groundwater. Agriculture employs 47% of the workforce and contributes 25% of GDP and 60% of export. With the restraints on large-scale surface water development since 1976, agriculture expansion in the country has been driven to a very large extent by the development of more than 600,000 private tube wells (PTWs) in the country. It is estimated that 75% of the increase in water supplies in the last twenty-five years is to be attributed to public and private groundwater exploitation. The investment on these private tube wells is of the order of Rs. 30-40 billion whereas the annual benefits in the form of agriculture production are estimated at Rs. 200 billion, roughly equivalent to 5% of GDP. In addition, many industries rely on relatively clean groundwater for their supply.

In drinking water the major concern is groundwater quality. There are several problem spots- mainly in urban areas and around rural industries-where groundwater quality is compromised with sewerage effluents, oil residuals, chromium or other contaminants. This holds true for major cities, such as Karachi and Lahore, where a large number of households still depends on individual wells for water supply (Raman 1996. Mashhadi and Anwar 2000). Recent surveys have recorded the extent of arsenic in groundwater, with results from Sindh being particularly worrisome. Finally, for some urban areas the absolute shortage of drinking water is posing questions on the future. The prime example is Quetta (Brich et al 1998).

In 1960, groundwater accounted for only 8% of the irrigation water supplies in Punjab. Twenty-five years later, in 1985, this amount had gone up to 40%. Fresh groundwater was exploited to be used conjunctively with surface water. This way, intense shallow tubewell development in the Indus Plain made up for many of the deficiencies of unreliable and insufficient surface water supplies in the Indus irrigation system. Outside the large- scale irrigation in the Indus Plain, groundwater development changed the face of many arid areas, particularly, in Balochistan and North West Frontier Province, and in the unirrigated tracts of Sindh and the Punjab. According to official estimates, the development of shallow tube wells and deep tubewells has played the main role in the expansion of the area under cultivation in Pakistan, with 75% of the increase in water supplies from 1960 to 1985 being attributable to public and private groundwater exploitation. During the same period, the cropping intensity increased by 39% (MINFA 1988).

Private tube wells densities in Sindh are far below those in Punjab. The estimated number of tube wells in Sindh is slightly over 28,000 (Qureshi et al 2003). Tube well densities in different canal commands are in the order of 2 to 3 private tube wells per 100 ha (Lashari and Memon 2000, van Steenbergen 2002).

In the Punjab Province of Pakistan, most groundwater supplies are fresh. The main exceptions are the areas of saline groundwater in the centre of the interfluviums ('doab' in Pakistani terminology), particularly those between Multan and Faisalabad, around Sargodha, and in the south-eastern part of the province.

In the southern part of the Indus Plain, in Sindh, groundwater supplies are more problematic. With the exception of a small strip along the Indus River, groundwater supplies are highly saline. The discharge from the aquifers in Sindh is generally less than that in Punjab (Survey of Pakistan 1989). Supplies are even less in the eastern deserts and the western limestone ridges bordering the Indus Plain.

Balochistan and North West Frontier Province are geologically more complex. In Balochistan, land formations consist of limestone, sandstone, and shale formations alternating with sand, silt, and gravel deposits. The aquifers are confined and generally yield only limited quantities of water. The quality of groundwater is unsuitable for agriculture in parts of Balochistan, particularly in some of the coastal areas, in patches in the large, alluvial Kacchi plain in the east, and in several valleys in the uplands.

Although the picture for North West Frontier Province resembles that for Balochistan, irrigation with groundwater in North West Frontier Province is more recent and, until recently, less extensive. This is owing to the less arid conditions and ample surface water resources. Another difference is in the areas around Bannu and Mardan, where there are thick and extensive aquifers. In Bannu, however, the fresh water is overlain by saline water.

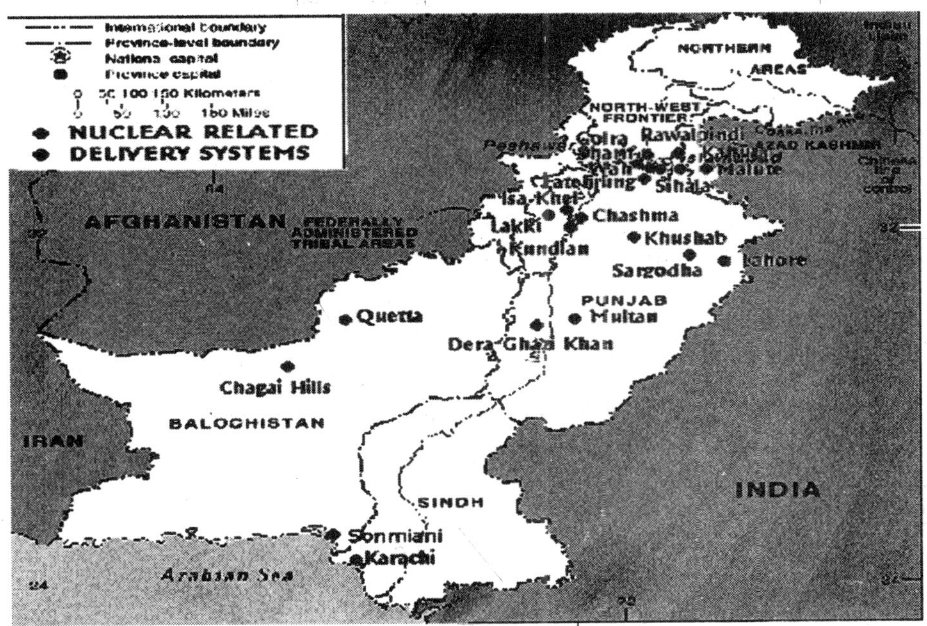

Table 5
Density of Private tube Wells in Punjab

Institution	*TWs/1000 ha*
WAPDA (1980)	10
NESPAK (1990)	14
Ag Statistics of Pak (1999)	27
PPSGDP (2000-01)	29
IWMI (2002)	32

Table 6
Province Wise Tube Wells and Type of Prime Mover in Pakistan

Province	*Total no. of wells*	*Diesel*	*Electric*	*Electric %*
Punjab	566,446	501,092	65,354	11.5
Sindh	28,079	25,086	2,993	10.6
NWFP	11,077	5,539	5,538	50.0
Balochistan	24,000	13,871	10,129	42.2
Total	629,602	545,588	84,014	13.3

Source: Qureshi, Shah and Akhtar (2003)

After promulgation of the Acts, many SCARP Projects were undertaken. The SCARP projects typically involved the installation of deep tube wells (Public tube wells) that pump water from depths of 40 to 120 m '. By 1995, there were more than 15,000 public deep tube wells in the Punjab and Sindh and some in North West Frontier Province.

Table 7
Salinity Control and Reclamation Projects (SCARP).in Pakistan

S#	Project	*Period of construction*	*Cultivable area(mi acres)*	*Tube wells Saline groundwater*	*Fresh groundwater*
PUNJAB					
1.	SCARP-I	1960-63	1.141	–	2,069
2.	SCARP-II (FGW)	1961-83	1.490	–	2,205
3.	SCARP-II (SGW)	1961-83	0.550	821	–
4.	SCARP-III (FGW)	1969-81	0.952	–	1,635
5.	SCARP-III (SGW)	1969-81	0.096	61	–
6.	SCARP-I Muridki	1969-73	0.544	–	935
7.	SATINA	1975-77	0.096	51	20
8.	Shorekot Kamalia	1975-77	0.154	–	101
9.	Allabad	1975-79	0.202	–	623
10.	Minchinabad	1976-80	0.073	23	203
11.	Shahpur	1975-79	0.112	–	258
	Total	5.388	956	8,049	–
N.W.F.P					
1.	Peshawar	1972-82	0.118	–	218
2.	Bannu	1976-82	–	176	
3.	Y.W. Sharif	1979-80	–	97	
4.	Khanwand	–	–	–	
	Total	491			
SINDH					
1.	Khanpur	1963-70	0.380	365	175
2.	N. Rohri	1969-79	0.686	–	1,192
3.	Shikarpur	1973-74	0.014	–	50
4.	Larkana	1974-75	0.005	–	35
5.	Sukkur	1977-78	0.004	–	18
6.	Kandhkot	1977-78	0.012	–	26
7.	Sukkur R.B.	1975-79	0.130	–	400
8.	Ghotki	1976-90	0.400	–	1,050
9.	South Rohri	1976-89	0.375	–	1,215
	Total		2.006	365	4,161

Source: Ahmed, N. (1998).

Community based management of tube wells

Transfer of SCARP tube wells in fresh groundwater (FGW) zones

Transfer of SCARP tube wells to Community Tube Well Groups in Pakistan are as under

Table 8
Transfer of SCARP tube wells

	Number and transfer modus	*FGW or SGW zone*
SCARP Transition Pilot Project (1986-1992)	315 SCARPs into private tube wells	SCARP-IFGW
Second SCARP Transition Project (1991-1997)	1,346 SCARPs into private and community tube wells	SCARP-IFGW
PPSGD Project [1997-2002]	4,233 SCARPs into community tube wells	SCARP-II,-III,-IV,-V FGW + SGW

Groundwater Regulation and Institutional arrangements in Pakistan

Over the years a number of laws and acts have been issued to support the management of groundwater, but in reality they have not been implemented. The following are the series of regulations promulgated time to time.

Table 9
Groundwater regulation in Pakistan

Regulations	*Main features/contents*
Punjab Soil Reclamation Act 1952	• To control waterlogging and salinity through tube wells. • To instigate licensing procedure to permit land owners to install private tube wells
Water and Power Development Authority Act 1958	• WAPDA shall control Pakistan's groundwater resources, and • Issue official area-specific rules
Groundwater Rights Administration Ordinance 1978	• District water committee should issue the permits for the development of Karezes, dug wells and tubewells. • Estimate safe yield, than issue new permits, but • This regulation did not made noticeable impact on GW rush.
Environmental Protection Agency Act 1996.	• This act strongly focuses on giving permits based on quality setting standards, but • it has had difficulty in getting implemented
Provincial Irrigation and Drainage Authority (PIDA) Act 1997	• PIDA act says all Fresh groundwater tube wells will be transferred to farmer organizations but never talks about sustainability of GW

Economics of Groundwater of Pakistan

Groundwater is main and driving force of rural economy where poor people have also excess to get it. The latest estimate have indicated that more than 68 BCM of groundwater is currently being used for irrigation and other purposes. This all extraction is being done through the use of shellow and deep tubewells which are more than 1 million in the country. Major number of tubewells are diesel operated engine and also electric.

Presently the investment on this private tubewell and groundwater extraction is estimated more than Rs 25 billion. However, the contribution to the economy of Pakistan is estimated to about Rs 80 billion which is three time greater than the investment.

Impact of Groundwater Use on Agriculture

In Pakistan, the use of groundwater is about 40 years old, but more intensive is the recent phenomena because of severe drought and population pressure and also fragmentation or distribution of land in families which wants to irrigate all lands in every season to feed the family.

The use of groundwater has increased the cropping intensity from 70 – 150 per cent, though the groundwater is expensive as compared to surface water. But the timely availability of this water has motivated the people to go with this option. However the use of this water has tremendously increased the crop yield by 150 – 200 per cent that compensate the expenditure incurred to get the groundwater.

Rural population are dependent on agriculture therefore the extraction and use of groundwater has strengthened the livelihood the only option.. This has also significantly impacted on the rural poverty and sustainable irrigated agriculture in Pakistan.

Implementation of the Policy in Pakistan

Policies and laws are the main vehicles for sustainable management of natural resources including groundwater and surface water. Government time to time introduce new policies and laws to manage the resources in a sustainable manner but some times these policies are not effective due to some reasons. In Pakistan policies were given for groundwater management and controlling the water logging problems but these were not being implemented with time. The reasons are:

- Groundwater produces reliable supply and increases yield 100 per cent more (Food security)
- Smallholders have easy excess to this resource that helps them for sustainable livelihood (Reduce poverty)
- Farmers use their own funds-no pressure on Government.

Issues of Groundwater use in Pakistan

When laws and policies are not effective and the only source available for the servival of the people then it influences on the management of the sources. This has happened in Pakistan that the extrection of groundwater has been made very extensively and is still continued that has created following issues.

- Mining of the resource by about 1.5 m per year

- Development of secondary salinization (sodicity and salinity)
- Influencing on or no implementation of regulations
- Pollution due to agricultural, industrial and human activities
- Increase pumping cost because of groundwater table decline

Table 10
Management of Groundwater in Various Countries

Australia	*India*	*Pakistan*
Ownership Crown in right of State	State for surface water Private for groundwater	State for surface water Private for groundwater
Land nexusnot linked	Land nexuslinked	Land nexuslinked strongly
AllocationVolumetric now % share of consumptive pool	Groundwater bill adopted in 7 States but difficult in implementation	Warabandi for surface water only
Review policy after every 5 year	No license for GW	No License for GW
PriorityEqual losses to satisfy ESD	None	None
Australia Rights issued byCrown	India GW Act for drillers but difficulty in implementing	Pakistan no License for GW but surface system managed by Provincial Irrigation Department
Conflict resolution by Court of each State	Conflict resolution by water users Associations at watercourse level, then members of Panchyats then civil court	Conflict resolution at Farmers organisations level then to irrigation department and lastly civil court

Source: Jennifer McKay: Summary workshop held in New Delhi 2007

CONCLUSIONS

Groundwater Governance in SA

Groundwater is the main source of Australia that is used for agriculture and other all purposes. The management of groundwater in Australia and South Australia has been improved and changed with time through government laws and policies.

Until 1995 the Federal Government was the authority to managing the water resources, but In 1997 reforms were introduced and all the states including South Australia were given authority to manage the water resources. Progressively the regulatory framework for surface and groundwater were improved and effectively implemented.

State has power and replaced the common law riparian rule and groundwater ownership rule(land owner owns water and can use all he likes) with licensing system soon as agriculture provoked conflict.

Australia moved from riparian system to licensing system area based to volumetric to now % of consumptive pool.

Land and water not tied together anymore since 1994 which has created flexibility and retirement of some land from irrigation.

Facilitated water markets and revisions of water allocation amounts, this will reduce water use as growers will only get a % of the permissible annual volume PAV allowed from a GW or SW source. The amount of water is not fixed as before but will vary annually.

Socio-economic dimension SA

In South Australia, the formation of any policy and laws, the consultative process is more common and all stakeholders are involved in the process of formation of policies that has impacted on sustaining the policy and proper implementation of it.

People are well aware of the sustainable level of use of groundwater. They have good coordination system between farmers, agriculturists, environmentalists and irrigated agricultural engineers.

The population in the country is in fact in a acceptable limit that can be very helpful to manage the resources in sustainable limits. As the population increases the pressure continuously goes up and sources becomes in the stress.

South Australia has established a very strong data base that helps to the managers and others to make sources more profitable through sustainable use of groundwater. And the sprinkler and drip systems of irrigation are the best examples of managing the resources in sustainable and in safe limits.

Groundwater Governance in Pakistan

Since 1952 various acts have been enacted in Pakistan but not all are focusing to sustainable groundwater management. They are for controlling the waterlogging and salinity. Groundwater regulation is linked to land rights and therefore more extraction is being taking place in the country.

The implementation of groundwater regulation is not effective, because it also helps in food security through reliable supply of water and more yield from the crops grown, Also it provides sustainable livelihood of smallholders. Moreover, there is no any pressure on Government for financing.

Socio-economic Dimension

The political set up in Pakistan has always been hindrance in implementing the regulation. For this there are several reasons like political agenda, social set up, influence of big land holding, and most important is the poverty.

People are less familiar with hydro-geology of aquifer which plays major role in the management of groundwater resource. It is well established fact that the reliable data have importance in managing the groundwater for which this country is lacking.

The use of groundwater in Pakistan has shown tremendous impact on increase of yield which most of the time goes up from 150 to 200 per cent and the cropping intensity so rapidly has gone from 70 to 150 per cent. This clearly indicates that groundwater use has strengthened the livelihood of the poor and marginalized rural populace and consequently the rural poverty has reduced.

RECOMMENDATIONS

The paper finally recommends that some policies and legal framework of Australia can be replicated in Pakistan for sustainable management of groundwater. These are:

The allocation for groundwater be developed as per directions set out in this paper.

Proper groundwater assessment be made and data base be developed so that every user will be aware of the availability of groundwater.

Monitoring mechanism be developed and strong coordination be made with various stakeholders.

The policy may be developed through the process of consultation with all users and a consensus be developed once it is finalized this will help in implementation.

ACKNOWLEDGEMENT

Authors are highly thankful to International Water Management Institute for financial support under the CGIAR challenge program for Food and Water.

REFERENCES

Burke, Jacob J., Moench, Marcus H. (2000), Groundwater and Society: Resources, Tensions and Opportunities., United Nations publication No. E.99.11.A.1, New York.

Shah, T., Deb Roy, A., Qureshi, A. S. and Wang, J. (2003), Sustaining Asia's groundwater boom: An overview of issues and evidence. Natural Resources Forum, 27 (2003): 130-141.

Karen G. Villholth., Sharma, B. R. (2005), Groundwater Research and management: Integrating Science into Management Decision: Proceedings of IWMI-ITP-NIH International Workshop on "Creating Synergy between Groundwater Research and management in South and Southeast Asia" 8-9 February 2005, Roorkee, India.

ACE, HALCROW (2002a), Exploitation and Regulation of Fresh Groundwater. Lahore: NDP.

Ahmed Nazir (1993), Water Resources of Pakistan and Their Utilization. Lahore: Shazad Nazir.

Birch, David, Yaver Abdi and Martin Perkins (1998), Quetta-Growing Water Demand from Declining Resources. Paper at 24th WEDC Conference Islamabad.

Johnson Robert (1989), Private Tube Well Development in Pakistan's Punjab: Review of Past Public Programmes/Policies and Relevant Research, Pakistan Country Paper1. Lahore: IWMI.

Kahlown M. A. and Abdul Majeed (2004), Pakistan Water Resources Development and Management. Islamabad: Pakistan Council for Research in Water Resources.

Kruseman G. P. and Naqvi S. A. H. (1988), Hydrogeology and Groundwater Resources of the NWFP, Peshawar and Delft: WAPDA and TNO.

Lashari B. and Memon Y. (2000), Use of shallow groundwater in water short period: the case study on Dhoro Naro Minor Nawabshah, Sindh, Pakistan. Proceedings of Regional Groundwater Management Seminar, October 9-11 2000, Islamabad. GWP Pakistan Water Partnership. 241-249.

Mashhadi S. N. H. and Mohammad Anwar (2000), Recharging the depleting aquifer of Lahore Metropolis. Proceedings of Regional Groundwater Management Seminar, October 9-11 2000, Islamabad. GWP Pakistan Water Partnership. 209-219.

Qureshi A. S., Shah T. and Akhtar M. (2003), The Groundwater Economy of Pakistan. Working paper 64. Lahore: IWMI.

Van Steenbergen F. and W. Oliemans (2002), A Review of Policies in Groundwater Management in Pakistan: 1950-2000. Water Policy 4: 323-344.

INDEX